Fuel Cell Systems Explained

Second Edition

Fuel Cell Systems Explained

Second Edition

James Larminie
Oxford Brookes University, UK

Andrew Dicks
University of Queensland, Australia
(Former Principal Scientist, BG Technology, UK)

WILEY

Other Wiley Editorial Offices

John Wiley & Sons Inc., 111 River Street, Hoboken, NJ 07030, USA

Jossey-Bass, 989 Market Street, San Francisco, CA 94103-1741, USA

Wiley-VCH Verlag GmbH, Boschstr. 12, D-69469 Weinheim, Germany

John Wiley & Sons Australia Ltd, 33 Park Road, Milton, Queensland 4064, Australia

John Wiley & Sons (Asia) Pte Ltd, 2 Clementi Loop #02-01, Jin Xing Distripark, Singapore 129809

John Wiley & Sons Canada Ltd, 22 Worcester Road, Etobicoke, Ontario, Canada M9W 1L1

Wiley also publishes its books in a variety of electronic formats. Some content that appears
in print may not be available in electronic books.

Library of Congress Cataloging-in-Publication Data

Larminie, James.
    Fuel cell systems explained / James Larminie, Andrew Dicks. – 2nd ed.
        p. cm.
    Includes bibliographical references and index.
    ISBN 0-470-84857-X (alk. paper)
    1. Fuel cells. I. Dicks, Andrew. II. Title.

TK2931.L37 2003
621.31′2429 – dc21

                                                      2002192419

British Library Cataloguing in Publication Data

A catalogue record for this book is available from the British Library

ISBN-13: 978-0-470-84857-9 (H/B)

Typeset in 10/12pt Times by Laserwords Private Limited, Chennai, India

Contents

Preface

The success of the first edition of *Fuel Cell Systems Explained* has been shown both by very pleasing sales and the positive comments of reviewers and critics. Nevertheless, with fuel cell technology, no one can rest on their laurels, and the last few years have seen rapid developments in the field, making a second edition necessary.

Although the basic principles of fuel cells remain the same, this new edition provides the opportunity to describe some of the significant developments that have emerged over the past three years. Examples are to be found in both low- and high-temperature fuel cell types. In particular, we have expanded our descriptions of the *Direct Methanol Fuel Cell*, which has an entirely new chapter of its own (Chapter 6), and the *Intermediate Temperature Solid Oxide Fuel Cell*, which forms part of an expanded description of high-temperature fuel cells (Chapter 7). We have substantially updated other chapters including those concerning the proton exchange membrane fuel cell (Chapter 4) and fuel processing (Chapter 8). In the latter case, new sections concerning hydrogen storage (including the possible use of nanofibres), biological hydrogen generation, and electrolysis may help to lay the foundation for an understanding of fuel cells in sustainable energy systems.

One of the most frequent comments by reviewers of the first edition has been to commend its clarity. To make the many topics encompassed by fuel cell technology clear to newcomers was one of our most important aims. Nevertheless, there is always room for improvement, and many changes have been made to make the text and diagrams even clearer. Many new diagrams have been added, and others improved. We have added an entirely new chapter that brings together the issues involved in system design and analysis. This introduces concepts such as well-to-wheel analysis, which is now a vital element of systems studies in transportation applications. In this new chapter, practical examples of both stationary and mobile applications completes our analysis of fuel cell systems.

It is clear that the first edition has been widely appreciated and used by engineers in the fuel cell industry, and among those who supply components to that industry. In addition, many of our readers are also students and teachers at universities and in technical schools, and some of them have requested a resource of questions to test understanding. We will supply such material, but because there are many advantages in being able to renew and rapidly update it, we have decided to provide this through the good offices of the publisher's website. A steadily expanding resource of questions, both numerical and more discursive, to test students' understanding will therefore be made available at the website linked to this book, which can be found at *www.wiley.co.uk/fuelcellsystems*.

We wish all readers well, and hope that our efforts here meet with success in helping you understand better this most interesting and potentially helpful technology.

James Larminie, Oxford, England
Andrew Dicks, Brisbane, Australia
January 2003

Foreword to the first edition

By Dr Gary Acres OBE, formerly Director of Research, Johnson Matthey plc

A significant time generally elapses before any new technological development is fully exploited. The fuel cell, first demonstrated by Sir William Grove in 1839, has taken longer than most, despite the promise of clean and efficient power generation.

Following Bacon's pioneering work in the 1950s, fuel cells were successfully developed for the American manned space programme. This success, together with a policy to commercialise space technology, led to substantial development programmes in America and Japan in the 1970s and the 1980s, and more recently in Europe. Despite these efforts that resulted in considerable technical progress, fuel cell systems were seen to be 'always five years away from commercial exploitation'.

During the last few years of the twentieth century, much changed to stimulate new and expanding interest in fuel cell technology. Environmental concerns about global warming and the need to reduce CO_2 emissions provided the stimulus to seek ways of improving energy conversion efficiency. The motor vehicle industry, apart from seeking higher fuel efficiencies, is also required to pursue technologies capable of eliminating emissions, the ultimate goal being the zero emission car. The utility industries, following the impact of privatisation and deregulation, are seeking ways to increase their competitive position while at the same time contributing to reduced environmental emissions.

As these developments have occurred, interest in fuel cell technology has expanded. Increasing numbers of people from disciplines ranging from chemistry through engineering to strategic analysis, not familiar with fuel cell technology, have felt the need to become involved. The need by such people for a single, comprehensive and up-to-date exposition of the technology and its applications has become apparent, and is amply provided for by this book.

While the fuel cell itself is the key component and an understanding of its features is essential, a practical fuel cell system requires the integration of the stack with fuel processing, heat exchange, power conditioning, and control systems. The importance of each of these components and their integration is rightly emphasised in sufficient detail for the chemical and engineering disciplines to understand the system requirements of this novel technology.

Fuel cell technology has largely been the preserve of a limited group consisting primarily of electro and catalyst chemists and chemical engineers. There is a need to develop more people with a knowledge of fuel cell technology. The lack of a comprehensive review of fuel cells and their applications has been a limiting factor in the inclusion of this subject in academic undergraduate and graduate student science and engineering courses. This book, providing as it does a review of the fundamental aspects of the technology, as well as its applications, forms an ideal basis for bringing fuel cells into appropriate courses and postgraduate activities.

The first three chapters describe the operating features of a fuel cell and the underlying thermodynamics and physical factors that determine their performance. A good understanding of these factors is essential to an appreciation of the benefits of fuel cell systems and their operating characteristics compared with conventional combustion-based technology. A feature of fuel cell technology is that it gives rise to a range of five main types of systems, each with its own operating parameters and applications. These are described in Chapters 4 to 7.

The preferred fuel for a fuel cell is hydrogen. While there are applications in which hydrogen can be used directly, such as in space vehicles and local transport, in the foreseeable future, for other stationary and mobile applications, the choice of fuel and its conversion into hydrogen-rich gas are essential features of practical systems. The range of fuels and their processing for use in fuel cell systems are described in Chapter 8. Chapters 9 and 10 describe the mechanical and electrical components that make up the complete fuel cell plant for both stationary and mobile applications.

This book offers those new to fuel cells a comprehensive, clear exposition and a review to further their understanding; it also provides those familiar with the subject a convenient reference. I hope it will also contribute to a wider knowledge about, and a critical appreciation of, fuel cell systems, and thus to the widest possible application of an exciting twenty-first century technology that could do much to move our use of energy onto a more sustainable basis.

Gary Acres
February 2000

Acknowledgements

The point will frequently be made in this book that fuels cells are highly interdisciplinary, involving many aspects of science and engineering. This is reflected in the number and diversity of companies that have helped with advice, information, and pictures in connection with this project. The authors would like to put on record their thanks to the following companies or organisations that have made this book possible:

Advanced Power Sources Ltd, UK
Advantica plc (formerly BG Technology Ltd), UK
Alstom Ballard GmbH,
Armstrong International Inc, USA
Ballard Power Systems Inc, Canada
DaimlerChrysler Corporation
DCH Technology Inc, USA
Eaton Corporation, USA
Epyx, USA
GfE Metalle und Materialien GmbH, Germany
International Fuel Cells, USA
IdaTech Inc., USA
Johnson Matthey plc, UK
Hamburgische Electricitäts-Werke AG, Germany
Lion Laboratories Ltd, UK
MTU Friedrichshafen GmbH, Germany
ONSI Corporation, USA
Paul Scherrer Institute, Switzerland
Proton Energy Systems, USA
Siemens Westinghouse Power Corporation, USA
Sulzer Hexis AG, Switzerland
SR Drives Ltd, UK
Svenska Rotor Maskiner AB, Sweden
W.L. Gore and Associates Inc, USA
Zytek Group Ltd, UK

In addition, a number of people have helped with advice and comments to the text. In particular, we would like to thank Felix Büchi of the Paul Scherrer Institute; Richard Stone and Colin Snowdon, both from the University of Oxford; Ramesh Shah of the

Rochester Institute; and Tony Hern and Jonathan Bromley of Oxford Brookes University, who have all provided valuable comments and suggestions for different parts of this work. Finally, we are also indebted to family, friends, and colleagues who have helped us in many ways and put up with us while we devoted time and energy to this project.

James Larminie, Oxford Brookes University, Oxford, UK
Andrew Dicks, University of Queensland, Australia

Abbreviations

AC	Alternating current
AES	Air electrode supported
AFC	Alkaline (electrolyte) fuel cell
ASR	Area specific resistance, the resistance of $1\,cm^2$ of fuel cell. (N.B. total resistance is ASR *divided* by area.)
BLDC	Brushless DC (motor)
BOP	Balance of plant
CFM	Cubic feet per minute
CHP	Combined heat and power
CPO	Catalytic partial oxidation
DC	Direct current
DIR	Direct internal reforming
DMFC	Direct methanol fuel cell
EC	European Community
EMF	Electromotive force
EVD	Electrochemical vapour deposition
FCV	Fuel cell vehicle
FT	Fischer–Tropsch
GHG	Greenhouse gas
GNF	Graphitic nanofibre
GT	Gas turbine
GTO	Gate turn-off
HDS	Hydrodesulphurisation
HEV	Hybrid electric vehicle
HHV	Higher heating value
IEC	International Electrotechnical Commission
IGBT	Insulated gate bipolar transistor
IIR	Indirect internal reforming
IT	Intermediate temperature
LHV	Lower heating value
LH_2	Liquid (cryogenic) hydrogen
LPG	Liquid petroleum gas
LSGM	Lanthanum, strontium, gallium, and magnesium oxide mixture
MCFC	Molten carbonate (electrolyte) fuel cell

MEA	Membrane electrode assembly
MOSFET	Metal oxide semiconductor field-effect transistor
MWNT	Multi-walled nanotube
NASA	National Aeronautics and Space Administration
NL	Normal litre, 1 L at NTP
NTP	Normal temperature and pressure (20°C and 1 atm or 1.01325 bar)
OCV	Open circuit voltage
PAFC	Phosphoric acid (electrolyte) fuel cell
PDA	Personal digital assistant
PEM	Proton exchange membrane or polymer electrolyte membrane – different names for the same thing which fortunately have the same abbreviation.
PEMFC	Proton exchange membrane fuel cell or polymer electrolyte membrane fuel cell
PFD	Process flow diagram
PM	Permanent Magnet
ppb	Parts per billion
ppm	Parts per million
PROX	Preferential oxidation
PURPA	Public Utilities Regulatory Policies Act
PTFE	Polytetrafluoroethylene
PSI	Pounds per square inch
PWM	Pulse width modulation
SCG	Simulated coal gas
SL	Standard litre, 1 L at STP
SOFC	Solid oxide fuel cell
SPFC	Solid polymer fuel cell (= PEMFC)
SPP	Small power producer
SRM	Switched reluctance motor
SRS	Standard reference state (25°C and 1 bar)
STP	Standard temperature and pressure (= SRS)
SWNT	Single-walled nano tube
TEM	Transmission electron microscope
t/ha	Tonnes per hectare annual yield
THT	Tetrahydrothiophene ($C_4H_8O_2S$)
TLV	Threshold limit value
TOU	Time of use
UL	Underwriters' Laboratory
WTT	Well to tank
WTW	Well to wheel
YSZ	Yttria-stabilised zirconia

Symbols

a	Coefficient in base 10 logarithm form of Tafel equation, also Chemical activity
a_x	Chemical activity of substance x
A	Coefficient in natural logarithm form of Tafel equation, also Area
B	Coefficient in equation for mass transport voltage loss
C	Constant in various equations, also Capacitance
c_p	Specific heat capacity at constant pressure, in $J\,K^{-1}\,kg^{-1}$
$\bar{c}_p$	Molar specific heat capacity at constant pressure, in $J\,K^{-1}\,mol^{-1}$
d	separation of charge layers in a capacitor
e	Magnitude of the charge on one electron, 1.602×10^{-19} Coulombs
E	EMF or open circuit voltage
E^0	EMF at standard temperature and pressure, and with pure reactants
F	Faraday constant, the charge on one mole of electrons, 96,485 Coulombs
G	Gibbs free energy (or negative thermodynamic potential)
ΔG^0	Change in Gibbs free energy at standard temperature and pressure, and with pure reactants
ΔG_{T_A}	Change in Gibbs free energy at ambient temperature
$\bar{g}$	Gibbs free energy per mole
$\bar{g}_f$	Gibbs free energy of formation per mole
$(\bar{g}_f)_X$	Gibbs free energy of formation per mole of substance X
H	Enthalpy
$\bar{h}$	Enthalpy per mole
$\bar{h}_f$	Enthalpy of formation per mole
$(\bar{h}_f)_X$	Enthalpy of formation per mole of substance X
I	Current
i	Current density, current per unit area
i_l	Limiting current density
i_n	Crossover current within a cell
i_o	Exchange current density at an electrode/electrolyte interface
i_{oc}	Exchange current density at the cathode
i_{oa}	Exchange current density at the anode
m	Mass
$\dot{m}$	Mass flow rate
m_x	Mass of substance x

N	Avogadro's number, 6.022×10^{23}, also revolutions per second
n	Number of cells in a fuel cell stack or number of moles
$\dot{n}$	Number of moles per second
P	Pressure
P_1, P_2	The pressure at different stages in a process
P_X	Partial pressure of gas X
P^0	Standard pressure, $100\,\text{kPa}$
P_{SAT}	Saturated vapour pressure
P_e	Electrical power, only used when context is clear that pressure is not meant
R	Molar or 'universal' gas constant, $8.314\,\text{J}\,\text{K}^{-1}\,\text{mol}^{-1}$, also electrical resistance
r	Area specific resistance, resistance of unit area
S	Entropy
$\bar{s}$	Entropy per mole
$(\bar{s})_X$	Entropy per mole of substance X
T	Temperature
T_1, T_2	Temperatures at different stages in a process
T_A	Ambient temperature
T_c	Combustion temperature
t	Time
V	Voltage
V_c	Average voltage of one cell in a stack
V_a	Activation overvoltage
V_r	Ohmic voltage loss
W	Work done
W'	Work done under isentropic conditions
$\dot{W}$	Power
z	Number of electrons transferred in a reaction
α	Charge transfer coefficient
Δ	Change in ...
ε	Electrical permittivity
γ	Ratio of the specific heat capacities of a gas
η	Efficiency
η_c	Isentropic efficiency (of compressor or turbine)
ϕ	Relative humidity
λ	Stoichiometric ratio
ω	Humidity ratio
μ_f	Fuel utilisation

1

Introduction

1.1 Hydrogen Fuel Cells – Basic Principles

The basic operation of the hydrogen fuel cell is extremely simple. The first demonstration of a fuel cell was by lawyer and scientist William Grove in 1839, using an experiment along the lines of that shown in Figures 1.1a and 1.1b. In Figure 1.1a, water is being electrolysed into hydrogen and oxygen by passing an electric current through it. In Figure 1.1b, the power supply has been replaced with an ammeter, and a small current is

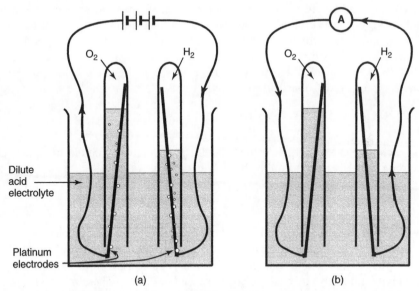

Note that the arrows represent the flow of negative electrons from – to +.

Figure 1.1   (a) The electrolysis of water. The water is separated into hydrogen and oxygen by the passage of an electric current. (b) A small current flows. The oxygen and hydrogen are recombining.

Fuel Cell Systems Explained, Second Edition James Larminie and Andrew Dicks
© 2003 John Wiley & Sons, Ltd   ISBN: 0-470-84857-X

flowing. The electrolysis is being reversed – the hydrogen and oxygen are recombining, and an electric current is being produced.

Another way of looking at the fuel cell is to say that the hydrogen fuel is being 'burnt' or combusted in the simple reaction

$$2H_2 + O_2 \rightarrow 2H_2O \qquad\qquad [1.1]$$

However, instead of heat energy being liberated, electrical energy is produced.

The experiment shown in Figures 1.1a and 1.1b makes a reasonable demonstration of the basic principle of the fuel cell, but the currents produced are very small. The main reasons for the small current are

- the low 'contact area' between the gas, the electrode, and the electrolyte – basically just a small ring where the electrode emerges from the electrolyte.
- the large distance between the electrodes – the electrolyte resists the flow of electric current.

To overcome these problems, the electrodes are usually made flat, with a thin layer of electrolyte as in Figure 1.2. The structure of the electrode is porous so that both the electrolyte from one side and the gas from the other can penetrate it. This is to give the maximum possible contact between the electrode, the electrolyte, and the gas.

However, to understand how the reaction between hydrogen and oxygen produces an electric current, and where the electrons come from, we need to consider the separate reactions taking place at each electrode. These important details vary for different types of fuel cells, but if we start with a cell based around an acid electrolyte, as used by Grove, we shall start with the simplest and still the most common type.

At the anode of an **acid electrolyte** fuel cell, the hydrogen gas ionises, releasing electrons and creating H^+ ions (or protons).

$$2H_2 \rightarrow 4H^+ + 4e^- \qquad\qquad [1.2]$$

This reaction releases energy. At the cathode, oxygen reacts with electrons taken from the electrode, and H^+ ions from the electrolyte, to form water.

$$O_2 + 4e^- + 4H^+ \rightarrow 2H_2O \qquad\qquad [1.3]$$

Clearly, for both these reactions to proceed continuously, electrons produced at the anode must pass through an electrical circuit to the cathode. Also, H^+ ions must pass through the electrolyte. An acid is a fluid with free H^+ ions, and so serves this purpose very well. Certain polymers can also be made to contain mobile H^+ ions. These materials are called *proton exchange membranes*, as an H^+ ion is also a proton.

Comparing equations 1.2 and 1.3 we can see that two hydrogen molecules will be needed for each oxygen molecule if the system is to be kept in balance. This is shown in Figure 1.3. It should be noted that the electrolyte must only allow H^+ ions to pass through it, and not electrons. Otherwise, the electrons would go through the electrolyte, not a round the external circuit, and all would be lost.

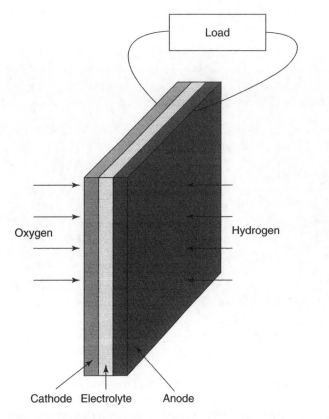

Figure 1.2 Basic cathode–electrolyte–anode construction of a fuel cell.

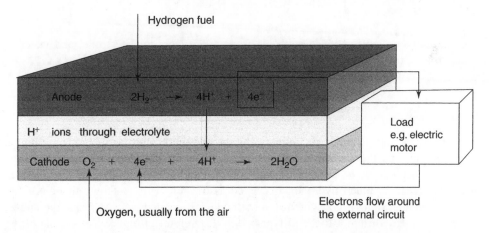

Figure 1.3 Electrode reactions and charge flow for an acid electrolyte fuel cell. Note that although the negative electrons flow from anode to cathode, the 'conventional current' flows from cathode to anode.

Positive Cathodes and Negative Anodes

Looking at Figures 1.3 and 1.4, the reader will see that the electrons are flowing from the anode to the cathode. The **cathode** is thus the electrically **positive** terminal, since electrons flow from − to +. Many newcomers to fuel cells find this confusing. This is hardly surprising. The Concise Oxford English Dictionary defines cathode as

"**1.** the negative electrode in an electrolyte cell or electron valve or tube, **2.** the positive terminal of a primary cell such as a battery."

Having two such opposite definitions is bound to cause confusion, but we note that the cathode is the correct name for the positive terminal of **all** primary batteries. It also helps to remember that cations are positive ions, for example, H^+ is a cation. Anions are negative ions, for example, OH^- is an anion. It is also true that the **cathode is always the electrode into which electrons flow**, and similarly the anode is always the electrode from which electrons flow. This holds true for electrolysis, cells, valves, forward biased diodes, and fuel cells.

A further possible confusion is that while negative electrons flow from minus to plus, the 'conventional positive current' flows the other way, from the positive to the negative terminal.

In an **alkaline electrolyte fuel cell** the overall reaction is the same, but the reactions at each electrode are different. In an alkali, hydroxyl (OH^-) ions are available and mobile. At the anode, these react with hydrogen, releasing energy and electrons, and producing water.

$$2H_2 + 4OH^- \rightarrow 4H_2O + 4e^- \qquad [1.4]$$

At the cathode, oxygen reacts with electrons taken from the electrode, and water in the electrolyte, forming new OH^- ions.

$$O_2 + 4e^- + 2H_2O \rightarrow 4OH^- \qquad [1.5]$$

For these reactions to proceed continuously, the OH^- ions must be able to pass through the electrolyte, and there must be an electrical circuit for the electrons to go from the anode to the cathode. Also, comparing equations 1.4 and 1.5 we see that, as with the acid electrolyte, twice as much hydrogen is needed as oxygen. This is shown in Figure 1.4. Note that although water is consumed at the cathode, it is created twice as fast at the anode.

There are many different fuel cell types, with different electrolytes. The details of the anode and cathode reactions are different in each case. However, it is not appropriate to go over every example here. The most important other fuel cell chemistries are covered in Chapter 7 when we consider the solid oxide and molten carbonate fuel cells.

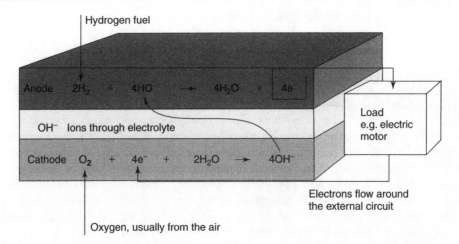

Hydrogen fuel

Anode $2H_2$ + $4HO^-$ → $4H_2O$ + $4e^-$

OH^- Ions through electrolyte

Cathode O_2 + $4e^-$ + $2H_2O$ → $4OH^-$

Load
e.g. electric
motor

Electrons flow around
the external circuit

Oxygen, usually from the air

Figure 1.4 Electrode reactions and charge flow for an alkaline electrolyte fuel cell. Electrons flow from anode to cathode, but conventional positive current flows from cathode to anode.

1.2 What Limits the Current?

At the anode, hydrogen reacts, releasing energy. However, just because energy is released, it does not mean that the reaction proceeds at an unlimited rate. The reaction has the 'classical' energy form shown in Figure 1.5.

Although energy is released, the 'activation energy' must be supplied to get over the 'energy hill'. If the probability of a molecule having enough energy is low, then the

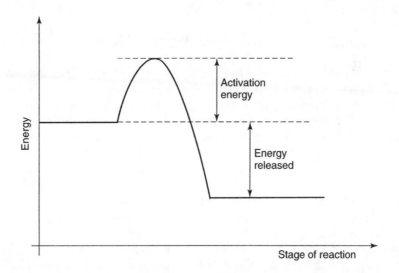

Activation
energy

Energy
released

Energy

Stage of reaction

Figure 1.5 Classical energy diagram for a simple exothermic chemical reaction.

reaction will only proceed slowly. Except at very high temperatures, this is indeed the case for fuel cell reactions.

The three main ways of dealing with the slow reaction rates are

- the use of catalysts,
- raising the temperature,
- increasing the electrode area.

The first two can be applied to any chemical reaction. However, the third is special to fuel cells and is very important. If we take a reaction such as that of equation 1.4, we see that fuel gas and OH^- ions from the electrolyte are needed, as well as the necessary activation energy. Furthermore, this 'coming together' of H_2 fuel and OH^- ions must take place **on the surface of the electrode**, as the electrons produced must be removed.

This reaction, involving fuel or oxygen (usually a gas), with the electrolyte (solid or liquid) and the electrode, is sometimes called the *three phase contact*. The bringing together of these three things is a very important issue in fuel cell design.

Clearly, the rate at which the reaction happens will be proportional to the area of the electrode. This is very important. Indeed, electrode area is such a vital issue that the performance of a fuel cell design is often quoted in terms of the current *per cm²*.

However, the straightforward area (length × width) is not the only issue. As has already been mentioned, the electrode is made highly porous. This has the effect of greatly increasing the effective surface area. Modern fuel cell electrodes have a microstructure that gives them surface areas that can be hundreds or even thousands of times their straightforward 'length × width' (See Figure 1.6.) The microstructural design and manufacture of a fuel cell electrode is thus a very important issue for practical fuel cells. In addition to these surface area considerations, the electrodes may have to incorporate a catalyst and endure high temperatures in a corrosive environment. The problems of reaction rates are dealt with in a more quantitative way in Chapter 3.

1.3 Connecting Cells in Series – the Bipolar Plate

For reasons explained in Chapters 2 and 3, the voltage of a fuel cell is quite small, about 0.7 V when drawing a useful current. This means that to produce a useful voltage many cells have to be connected in series. Such a collection of fuel cells in series is known as a 'stack'. The most obvious way to do this is by simply connecting the edge of each anode to the cathode of the next cell, all along the line, as in Figure 1.7. (For simplicity, this diagram ignores the problem of supplying gas to the electrodes.)

The problem with this method is that the electrons have to flow across the face of the electrode to the current collection point at the edge. The electrodes might be quite good conductors, but if each cell is only operating at about 0.7 V, even a small voltage drop is important. Unless the current flows are very low, and the electrode is a particularly good conductor, or very small, this method is not used.

A much better method of cell interconnection is to use a 'bipolar plate'. This makes connections all over the surface of one cathode and the anode of the next cell (hence 'bipolar'); at the same time, the bipolar plate serves as a means of feeding oxygen to the

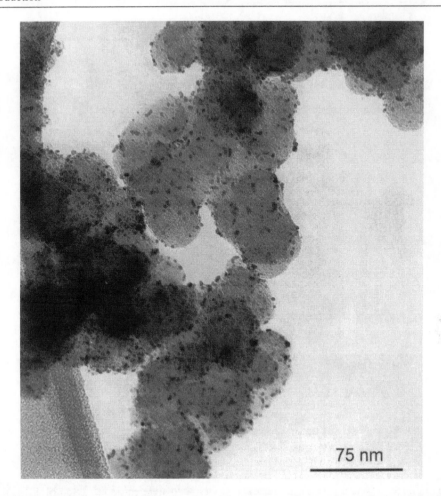

Figure 1.6 TEM image of fuel cell catalyst. The black specks are the catalyst particles finely divided over a carbon support. The structure clearly has a large surface area. (Reproduced by kind permission of Johnson Matthey Plc.)

cathode and fuel gas to the anode. Although a good electrical connection must be made between the two electrodes, the two gas supplies must be strictly separated.

The method of connecting to a single cell, all over the electrode surfaces, while at the same time feeding hydrogen to the anode and oxygen to the cathode, is shown in Figure 1.8. The grooved plates are made of a good conductor such as graphite, or stainless steel.

To connect several cells in series, 'bipolar plates' are made. These plates – or cell interconnects – have channels cut in them so that the gases can flow over the face of the electrodes. At the same time, they are made in such a way that they make a good electrical contact with the surface of each alternate electrode. A simple design of a bipolar plate is shown in Figure 1.9.

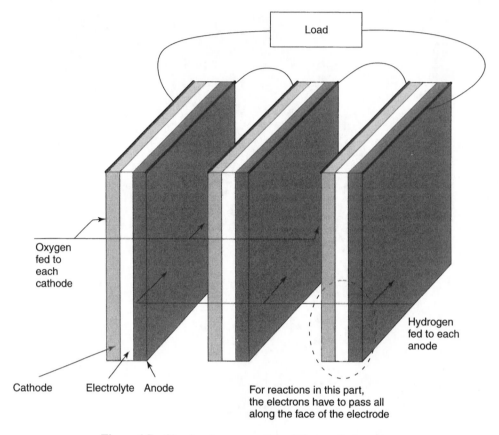

Oxygen
fed to
each
cathode

Hydrogen
fed to each
anode

Cathode   Electrolyte   Anode

For reactions in this part,
the electrons have to pass all
along the face of the electrode

Figure 1.7   Simple edge connection of three cells in series.

To connect several cells in series, anode/electrolyte/cathode assemblies (as in Figure 1.2) need to be prepared. These are then 'stacked' together as shown in Figure 1.10. This 'stack' has vertical channels for feeding hydrogen over the anodes and horizontal channels for feeding oxygen (or air) over the cathodes. The result is a solid block in which the electric current passes efficiently, more or less straight through the cells rather than over the surface of each electrode one after the other. The electrodes are also well supported, and the whole structure is strong and robust. However, the design of the bipolar plate is not simple. If the electrical contact is to be optimised, the contact points should be as large as possible, but this would mitigate the good gas flow over the electrodes. If the contact points have to be small, at least they should be frequent. However, this makes the plate more complex, difficult, and expensive to manufacture, as well as fragile. Ideally the bipolar plate should be as thin as possible, to minimise electrical resistance and to make the fuel cells stack small. However, this makes the channels for the gas flow narrow, which means it is more difficult to pump the gas round the cell. This sometimes has to be done at a high rate, especially when using air instead of pure oxygen on the cathode. In the case of low-temperature fuel cells, the circulating air has to evaporate and carry away the product water. In addition, there usually have to be further channels through the bipolar plate to carry a cooling fluid. Some of the further complications for the bipolar plate are considered in the next section.

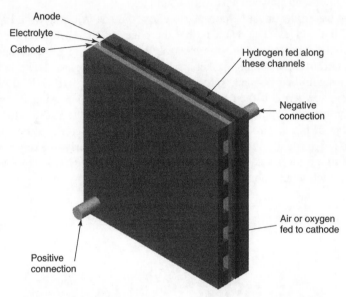

Figure 1.8   Single cell, with end plates for taking current from all over the face of the electrodes, and also supplying gas to the whole electrode.

Figure 1.9   Two bipolar plates of very simple design. There are horizontal grooves on one side and vertical grooves on the other.

1.4 Gas Supply and Cooling

The arrangement shown in Figure 1.10 has been simplified to show the basic principle of the bipolar plate. However, the problem of gas supply and of preventing leaks means that in reality the design is somewhat more complex.

Because the electrodes must be porous (to allow the gas in), they would allow the gas to leak out of their edges. The result is that the edges of the electrodes must be sealed. Sometimes this is done by making the electrolyte somewhat larger than one or both of the electrodes and fitting a sealing gasket around each electrode, as shown in Figure 1.11. Such assemblies can then be made into a stack, as in Figures 1.10 and 1.12.

The fuel and oxygen can then be supplied to the electrodes using the manifolds as shown disassembled in Figure 1.12 and assembled in Figure 1.13. Because of the seals around the edge of the electrodes, the hydrogen should only come into contact with the anodes as it is fed vertically through the fuel cell stack. Similarly, the oxygen (or air) fed horizontally through the stack should only contact the cathodes, and not even the edges of the anodes. Such would not be the case in Figure 1.10.

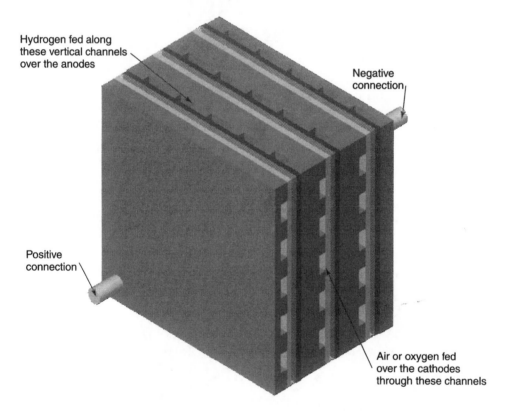

Figure 1.10   A three-cell stack showing how bipolar plates connect the anode of one cell to the cathode of its neighbour.

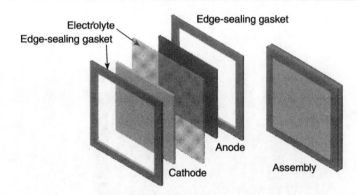

Figure 1.11 The construction of anode/electrolyte/cathode assemblies with edge seals. These prevent the gases leaking in or out through the edges of the porous electrodes.

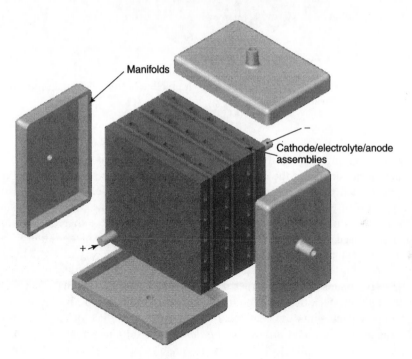

Figure 1.12 Three-cell stack, with external manifolds. Unlike Figure 1.10, the electrodes now have edge seals.

The arrangement of Figures 1.12 and 1.13 is used in some systems. It is called *external manifolding*. It has the advantage of simplicity. However, it has two major disadvantages. The first is that it is difficult to cool the system. Fuel cells are far from 100% efficient, and considerable quantities of heat energy as well as electrical power are generated. (Chapter 3

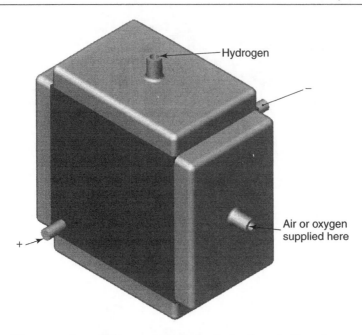

Hydrogen

Air or oxygen
supplied here

+

Figure 1.13   The external manifolds are fitted to the fuel cell stack. Note that no provision has
been made for cooling – see Section 1.4.

gives the reasons.) It is clear from Figures 1.12 and 1.13 that it would be hard to supply
a cooling fluid running through the cells. In practice, this type of cell has to be cooled by
the reactant air passing over the cathodes. This means air has to be supplied at a higher
rate than demanded by the cell chemistry; sometimes this is sufficient to cool the cell,
but it is a waste of energy. The second disadvantage is that the gasket round the edge
of the electrodes is not evenly pressed down – at the point where there is a channel, the
gasket is not pressed firmly onto the electrode. This results in an increased probability of
leakage of the reactant gases.

A more common arrangement requires a more complex bipolar plate and is shown in
Figure 1.14. The plates are made larger relative to the electrodes and have extra channels
running through the stack that feed the fuel and oxygen to the electrodes. Carefully placed
holes feed the reactants into the channels that run over the surface of the electrodes. This
type of arrangement is called *internal manifolding*. It results in a fuel cell stack that has
the appearance of the solid block with the reactant gases fed in at the ends where the
positive and negative connections are also made.

Such a fuel cell stack is shown under test in Figure 1.15. The end plate is quite
complex, with several connections. The stack is a solid block. Electrical connections
have been made to each of the approximately 60 cells in the stack for testing purposes.
The typical form of a fuel cell as a solid block with connections at each end is also
illustrated in Figure 4.1.

The bipolar plate with internal manifolding can be cooled in various ways. The simplest way is to make narrow channels up through the plates and to drive cooling air or

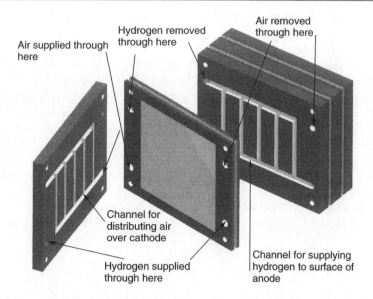

Air supplied through here

Hydrogen removed through here

Air removed through here

Channel for distributing air over cathode

Hydrogen supplied through here

Channel for supplying hydrogen to surface of anode

Figure 1.14 Internal manifolding. A more complex bipolar plate allows reactant gases to be fed to electrodes through internal tubes. (Picture by kind permission of Ballard Power Systems.)

Figure 1.15 Photograph of a fuel cell stack under test. The voltage of each of the approximately 60 cells in the stack is being measured. Note the carefully engineered end plates. (Photograph reproduced by kind permission of Ballard Power Systems.)

water through them. Such an approach is used in several systems shown in Chapter 4. Alternatively, channels can be provided along the length of the cell, and there is provision for this in the system shown in Figure 1.14. The preferred cooling method varies greatly with the different fuel cell types, and is addressed in Chapters 4 to 7.

It should now be clear that the bipolar plate is usually quite a complex item in a fuel cell stack. In addition to being a fairly complex item to make, the question of its material is often difficult. Graphite, for example, is often used, but this is difficult to work with and is brittle. Stainless steel can also be used, but this will corrode in some types of fuel cells. Ceramic materials have been used in the very high temperature fuel cells. The bipolar plate nearly always makes a major contribution to the cost of a fuel cell.

Anyone who has made fuel cells knows that **leaks** are a major problem. If the path of hydrogen through a stack using internal manifolding (as in Figure 1.14) is imagined, the possibilities for the gas to escape are many. The gas must reach the edge of every porous electrode – so the entire edge of every electrode is a possible escape route, both under and over the edge gasket. Other likely trouble spots are the joints between each and every bipolar plate. In addition, if there is the smallest hole in any of the electrolytes, a serious leak is certain. In Chapters 4 to 7, when the different fuel cell types are described in more detail, attention is given to the different ways in which this problem is solved.

1.5 Fuel Cell Types

Leaving aside practical issues such as manufacturing and materials costs, the two fundamental technical problems with fuel cells are

* the slow reaction rate, leading to low currents and power, discussed briefly in Section 1.2, and
* that hydrogen is not a readily available fuel.

To solve these problems, many different fuel cell types have been tried. The different fuel cell types are usually distinguished by the electrolyte that is used, though there are always other important differences as well. The situation now is that six classes of fuel cell have emerged as viable systems for the present and near future. Basic information about these systems is given in Table 1.1.

In addition to facing different problems, the various fuel types also try to play to the strengths of fuel cells in different ways. The **proton exchange membrane (PEM) fuel cell** capitalises on the essential simplicity of the fuel cell. The electrolyte is a solid polymer in which protons are mobile. The chemistry is the same as the acid electrolyte fuel cell of Figure 1.3. With a solid and immobile electrolyte, this type of cell is inherently very simple.

These cells run at quite low temperatures, so the problem of slow reaction rates is addressed by using sophisticated catalysts and electrodes. Platinum is the catalyst, but developments in recent years mean that only minute amounts are used, and the cost of the platinum is a small part of the total price of a PEM fuel cell. The problem of hydrogen supply is not really addressed – quite pure hydrogen **must** be used, though various ways of supplying this are possible, as is discussed in Chapter 8.

One theoretically very attractive solution to the hydrogen supply problem is to use methanol[1] as a fuel instead. This can be done in the PEM fuel cell, and such cells are

[1] A fairly readily available liquid fuel, formula CH_3OH.

Table 1.1 Data for different types of fuel cell

Fuel cell type	Mobile ion	Operating temperature	Applications and notes
Alkaline (AFC)	OH^-	50–200°C	Used in space vehicles, e.g. Apollo, Shuttle.
Proton exchange membrane (PEMFC)	H^+	30–100°C	Vehicles and mobile applications, and for lower power CHP systems
Direct methanol (DMFC)	H^+	20–90°C	Suitable for portable electronic systems of low power, running for long times
Phosphoric acid (PAFC)	H^+	~220°C	Large numbers of 200-kW CHP systems in use.
Molten carbonate (MCFC)	CO_3^{2-}	~650°C	Suitable for medium- to large-scale CHP systems, up to MW capacity
Solid oxide (SOFC)	O^{2-}	500–1000°C	Suitable for all sizes of CHP systems, 2 kW to multi-MW.

called *direct methanol fuel cells*. 'Direct' because they use the methanol as the fuel, as it is in liquid form, as opposed to extracting the hydrogen from the methanol using one of the methods described in Chapter 8. Unfortunately, these cells have very low powers, but nevertheless, even at low power, there are many potential applications in the rapidly growing area of portable electronics equipment. Such cells, in the foreseeable future at least, are going to be of very low power, and used in applications requiring slow and steady consumption of electricity over long periods.

Although PEM fuel cells were used on the first manned spacecraft, the **alkaline fuel cell** was used on the Apollo and Shuttle Orbiter craft. The problem of slow reaction rate is overcome by using highly porous electrodes with a platinum catalyst, and sometimes by operating at quite high pressures. Although some historically important alkaline fuel cells have operated at about 200°C, they more usually operate below 100°C. Alkaline fuel cells are discussed in Chapter 6, where their main problem is described – that the air and fuel supplies must be free from CO_2, or else pure oxygen and hydrogen must be used.

The **phosphoric acid fuel cell (PAFC)** was the first to be produced in commercial quantities and enjoys widespread terrestrial use. Many 200-kW systems, manufactured by the International Fuel Cells Corporation (now trading as UTC Fuel Cells Inc.), are installed in the USA and Europe, as well as systems produced by Japanese companies. Porous electrodes, platinum catalysts, and a fairly high temperature (~220°C) are used to boost the reaction rate to a reasonable level. The problem of fuelling with hydrogen is solved by 'reforming' natural gas (predominantly methane) to hydrogen and carbon dioxide, but the equipment needed to do this adds considerably to the cost, complexity, and size of the fuel cell system. Nevertheless, PAFC systems use the inherent simplicity of a fuel cell to provide an extraordinarily reliable and maintenance-free power system. Several PAFC systems have run continuously for periods of one year or more with little maintenance requiring shutdown or human intervention. (See Figure 1.16.)

Figure 1.16   Phosphoric acid fuel cell. In addition to providing 200 kW of electricity, it also provides about 200 kW of heat energy in the form of steam. Such units are called *combined heat and power* or *CHP systems*. (Reproduced by kind permission of ONSI Corporation.)

As is the way of things, each fuel cell type solves some problems, but brings new difficulties of its own. The **solid oxide fuel cell** (**SOFC**) operates in the region of 600 to 1000°C. This means that high reaction rates can be achieved without expensive catalysts, and that gases such as natural gas can be used directly, or 'internally reformed' within the fuel cell, without the need for a separate unit. This fuel cell type thus addresses all the problems and takes full advantage of the inherent simplicity of the fuel cell concept. Nevertheless, the ceramic materials that these cells are made from are difficult to handle, so they are expensive to manufacture, and there is still quite a large amount of extra equipment needed to make a full fuel cell system. This extra plant includes air and fuel pre-heaters; also, the cooling system is more complex, and they are not easy to start up.

Despite operating at temperatures of up to 1000°C, the SOFC always stays in the solid state. This is not true for the **molten carbonate fuel cell** (**MCFC**), which has the interesting feature that it needs the carbon dioxide in the air to work. The high temperature means that a good reaction rate is achieved by using a comparatively inexpensive catalyst – nickel. The nickel also forms the electrical basis of the electrode. Like the SOFC it can use gases such as methane and coal gas (H_2 and CO) directly, without an external reformer. However, this simplicity is somewhat offset by the nature of the electrolyte, a hot and corrosive mixture of lithium, potassium, and sodium carbonates.

1.6 Other Cells – Some Fuel Cells, Some Not

In addition to the major fuel cell types described above, there are other fuel cells that are mentioned in scientific journals from time to time, and also cells that are described as 'fuel cells', but are not really so.

A fuel cell is usually defined as an electrochemical device that converts a supplied fuel to electrical energy (and heat) continuously, so long as reactants are supplied to its electrodes. The implication is that neither the electrodes nor the electrolyte are consumed by the operation of the cell. Of course, in all fuel cells the electrodes and electrolytes are degraded and subject to 'wear and tear' in use, but they are not entirely consumed in the way that happens with two of the three types of cells briefly described below, both of which are sometimes described as 'fuel cells'.

1.6.1 Biological fuel cells

One type of genuine fuel cell that does hold promise in the very long term is the biological fuel cell. These would normally use an organic fuel, such as methanol or ethanol. However, the distinctive 'biological' aspect is that enzymes, rather than conventional 'chemical' catalysts such as platinum, promote the electrode reactions. Such cells replicate nature in the way that energy is derived from organic fuels. However, this type of cell is not yet anywhere near commercial application, and is not yet suitable for detailed consideration in an application-oriented book such as this.

The biological fuel cell should be distinguished from biological methods for generating hydrogen, which is then used in an ordinary fuel cell. This is discussed in Chapter 8.

1.6.2 Metal/air cells

The most common type of cell in this category is the zinc air battery, though aluminium/air and magnesium/air cells have been commercially produced. In all cases the basis of operation is the same. Such cells are sometimes called *zinc fuel cells*.

At the negative electrode, the metal reacts with an alkaline electrolyte to form the metal oxide or hydroxide. For example, in the case of zinc the reaction is

$$Zn + 2OH^- \rightarrow ZnO + H_2O + 2e^-$$

The electrons thus released pass around the external electric circuit to the air cathode where they are available for the reaction between water and oxygen to form more hydroxyl ions. *The cathode reaction is exactly the same as for the alkaline fuel cell shown in Figure* 1.4. The metal oxide or hydroxide should remain dissolved in the electrolyte. Cells that use a salt-water electrolyte work reasonably well when they use aluminium or magnesium as the 'fuel'.

Such cells have a very good energy density. Zinc/air batteries are very widely used in applications that require long running times at low currents, such as hearing aids. Several companies are also developing higher power units for applications such as electric vehicles. This is because they can also be 'refuelled' by adding more metal to the anode – which is why they are sometimes called *fuel cells*. The fact that the cathode reaction is exactly the same as for a fuel cell, and that the same electrodes can be used, is another reason. However, the electrolyte also has to be renewed to remove the metal oxide. Thus, they consume both the anode and the electrolyte, and cannot really be described as fuel cells. They are mechanically rechargeable primary batteries.

1.6.3 Redox flow cells or regenerative fuel cells

Another type of cell sometimes called a *fuel cell* is the redox flow cell (Vincent and Scrosati, 1997). In this type of cell, the reactants are removed from the electrodes during charging and are stored in tanks. The capacity of such cells can thus be very large. They are discharged by resupplying the reactants to the electrodes. Because the operation of the cell involves supply of chemicals to the electrodes, these devices are sometimes called *fuel cells*. However, this is a misnomer, as will become clear.

This type of cell is used to make very large capacity rechargeable batteries and may be used by electricity utilities to balance peaks in supply and demand. There are a number of different chemistries that can be used. Cells based on vanadium have been made (Shibata and Sato, 1999), as have zinc/bromine systems (Lex and Jonshagen, 1999). This type of system is perhaps best exemplified in the so-called Regenesys™ fuel cell (Zito, 1997 or Price et al., 1999).

The operating principles of the Regenesys™ system is shown in Figure 1.17. Two fluids ('fuels') are involved. When fully charged, a solution of sodium sulphide (Na_2S_2) in water is fed to the negative electrode, and a sodium tribromide ($NaBr_3$) solution is fed to the positive electrode. The reaction at the negative electrode is

$$2Na_2S_2 \rightarrow Na_2S_4 + 2Na^+ + 2e^-$$

The electrons flow around the external circuit, and the sodium ions pass through the membrane to the positive electrolyte. Here the reaction is

$$NaBr_3 + 2Na^+ + 2e^- \rightarrow 3NaBr$$

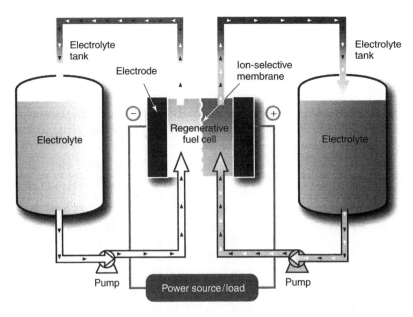

Figure 1.17 Diagram showing the principle of operation of a redox flow cell. See text for explanation. (Reproduced by kind permission of Regenesys Technology Ltd.)

Figure 1.18   Picture of the 100-MWh electrical energy storage facility being installed in Cambridgeshire, England. The storage tanks for the electrolyte solutions ('fuel') can be clearly seen. (Reproduced by kind permission of Regenesys Technology Ltd.)

So, as the system discharges, the sodium sulphide solution gradually changes to sodium polysulphide, and the sodium tribromide solution changes to sodium bromide. Figure 1.18 shows such a system under construction in Cambridgeshire, England. The two tanks to hold the solutions can be seen, together with a building that will hold the cells. This particular system has a storage capacity of 100 MWh, which is equivalent to the energy held in about 240,000 typical lead acid car batteries, and is believed to be the largest electrochemical electrical energy storage system in the world.

The rationale for calling this system a *fuel cell* is presumably that the electrodes are simply a surface where reactions take place and are not consumed. Furthermore, the electrodes are fed an energy-containing liquid. However, the electrolyte most certainly changes during operation, and the system cannot work indefinitely. Also, the electrolyte solutions are not fuels in any conventional sense. Indeed, this is a rather unusual and very high capacity rechargeable battery. Exactly the same arguments apply to the other cells of this type.

1.7 Other Parts of a Fuel Cell System

The core of a fuel cell power system is the electrodes, the electrolyte, and the bipolar plate that we have already considered. However, other parts frequently make up a large proportion of the engineering of the fuel cell system. These 'extras' are sometimes called the *balance of plant* (BOP). In the higher-temperature fuel cells used in CHP systems, the fuel cell stack often appears to be quite a small and insignificant part of the whole system, as is shown in Figure 1.19. The extra components required depend greatly on the

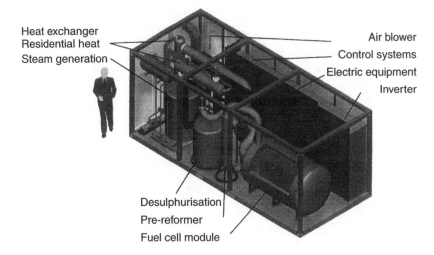

Heat exchanger
Residential heat
Steam generation

Air blower
Control systems
Electric equipment
Inverter

Desulphurisation
Pre-reformer
Fuel cell module

Figure 1.19 Design of a 100-kW fuel cell–based combined heat and power system. (Reproduced by kind permission of Siemens Power Generation.)

type of fuel cell, and crucially on the fuel used. These vitally important subsystem issues are described in much more detail in Chapters 8 to 10, but a summary is given here.

On all but the smallest fuel cells the air and fuel will need to be circulated through the stack using *pumps* or *blowers*. Often *compressors* will be used, which will sometimes be accompanied by the use of *intercoolers*, as in internal combustion engines.[2]

The direct current (DC) output of a fuel cell stack will rarely be suitable for direct connection to an electrical load, and so some kind of *power conditioning* is nearly always needed. This may be as simple as a voltage regulator, or a *DC/DC converter*. In CHP systems, a DC to AC *inverter* is needed, which is a significant part of the cost of the whole system.[3] *Electric motors*, which drive the pumps, blowers, and compressors mentioned above, will nearly always be a vital part of a fuel cell system. Frequently also, the electrical power generated will be destined for an electric motor – for example, in motor vehicles.

The supply and storage of hydrogen is a very critical problem for fuel cells. *Fuel storage* will clearly be a part of many systems. If the fuel cell does not use hydrogen, then some form of *fuel processing system* will be needed. These are often very large and complex, for example, when obtaining hydrogen from petrol in a car. In many cases *desulphurisation* of the fuel will be necessary. These vitally important subsystems will be discussed in detail in Chapter 8.

Various *control valves*, as well as *pressure regulators*, will usually be needed. In most cases a *controller* will be needed to coordinate the parts of the system. A special problem the controller has to deal with is the start-up and shutdown of the fuel cell system, as this can be a complex process, especially for high-temperature cells.

---

[2] These components are discussed in some detail in Chapter 9.

[3] Electrical subsystems are covered in Chapter 10.

Figure 1.20 The 75-kW (approx.) fuel cell system for a prototype Mercedes Benz A-class car. (Photograph reproduced by kind permission of Ballard Power Systems.)

For all but the smallest fuel cells a *cooling system* will be needed. In the case of CHP systems, this will usually be called a *heat exchanger*, as the idea is not to lose the heat but to use it somewhere else. Sometimes, in the case of the higher-temperature cells, some of the heat generated in the fuel cell will be used in fuel and/or air *pre-heaters*. In the case of the PEM fuel cell, to be described in detail in Chapter 4, there is often the need to *humidify* one or both of the reactant gases.

This very important idea of the balance of plant is illustrated in Figures 1.19 and 1.20. In Figure 1.19, we see that the fuel cell module is, in terms of size, a small part of the overall system, which is dominated by the fuel and heat processing systems. This will nearly always be the case for combined heat and power systems running on ordinary fuels such as natural gas. Figure 1.20 is the fuel cell engine from a car. It uses hydrogen fuel, and the waste heat is only used to warm the car interior. The fuel cell stack is the rectangular block to the left of the picture. The rest of the unit (pumps, humidifier, power electronics, compressor) is much less bulky than that of Figure 1.19, but still takes up over half the volume of the whole system.

1.8 Figures Used to Compare Systems

When comparing fuel cells with each other, and with other electric power generators, certain standard key figures are used. For comparing fuel cell electrodes and electrolytes, the key figure is the current per unit area, always known as the current density. This is usually given in $mA\,cm^{-2}$ though some Americans use $A\,ft^{-2}$. (Both figures are in fact quite similar: $1.0\,mA\,cm^{-2} = 0.8\,A\,ft^{-2}$.)

This figure should be given at a specific operating voltage, typically about 0.6 or 0.7 V. These two numbers can then be multiplied to give the power per unit area, typically given in $mW\,cm^{-2}$.

A note of warning should be given here. Electrodes frequently do not 'scale up' properly. That is, if the area is doubled, the current will often *not* double. The reasons for this are varied and often not well understood, but relate to issues such as the even delivery of reactants and removal of products from all over the face of the electrode.

Bipolar plates will be used to connect many cells in series. To the fuel cell stack will be added the 'balance of plant' components mentioned in Section 1.7. This will give a system of a certain power, mass, and volume. These figures give the key figures of merit for comparing electrical generators – specific power and power density.

$$\text{Power Density} = \frac{\text{Power}}{\text{Volume}}$$

The most common unit is kW m^{-3}, though kW L^{-1} is also used.

The measure of power per unit mass is called the *specific power*

$$\text{Specific Power} = \frac{\text{Power}}{\text{Mass}}$$

The straightforward SI unit of $W\ kg^{-1}$ is used for specific power.

The cost of a fuel cell system is obviously vital, and this is usually quoted in *US dollars per kilowatt*, for ease of comparison.

The lifetime of a fuel cell is rather difficult to specify. Standard engineering measures such as MTBF (mean time between failures) do not really apply well, as a fuel cell's performance always gradually deteriorates, and their power drops fairly steadily with time as the electrodes and electrolyte age. This is sometimes given as the *'percentage deterioration per hour'*. The gradual decline in voltage is also sometimes given in units of millivolts per 1000 hours. Formally, the life of a fuel cell is over when it can no longer deliver the rated power, that is, when a '10-kW fuel cell' can no longer deliver 10 kW. When new, the system may have been capable of, say, 25% more than the rated power −12.5 kW in this case.

The final figure of key importance is the efficiency, though as is explained in the next chapter, this is not at all a straightforward figure to give, and any information needs to be treated with caution.

In the automotive industry, the two key figures are the cost per kilowatt and the power density. In round figures, current internal combustion engine technology is about 1 kW L^{-1} and $10 per kW. Such a system should last about 4000 h (i.e. about 1 h use each day for over 10 years). For combined heat and power systems the cost is still important, but a much higher figure of $1000 per kW is the target. The cost is raised by the extra heat exchanger and mains grid connection systems, which are also needed by rival technologies, and because the system must withstand much more constant usage −40,000 h use would be a minimum.

1.9 Advantages and Applications

The most important disadvantage of fuel cells at the present time is the same for all types – the cost. However, there are varied advantages, which feature more or less strongly for different types and lead to different applications (Figure 1.21). These include the following:

- *Efficiency*. As is explained in the following chapter, fuel cells are generally more efficient than combustion engines whether piston or turbine based. A further feature of this is that small systems can be just as efficient as large ones. This is very important in the case of the small local power generating systems needed for combined heat and power systems.

- *Simplicity*. The essentials of a fuel cell are very simple, with few if any moving parts. This can lead to highly reliable and long-lasting systems.

- *Low emissions*. The by-product of the main fuel cell reaction, when hydrogen is the fuel, is pure water, which means a fuel cell can be essentially 'zero emission'. This is their main advantage when used in vehicles, as there is a requirement to reduce vehicle emissions, and even eliminate them within cities. However, it should be noted that, at present, emissions of CO_2 are nearly always involved in the production of hydrogen that is needed as the fuel.

- *Silence*. Fuel cells are very quiet, even those with extensive extra fuel processing equipment. This is very important in both portable power applications and for local power generation in combined heat and power schemes.

The fact that hydrogen is the preferred fuel in fuel cells is, in the main, one of their principal disadvantages. However, there are those who hold that this is a major advantage. It is envisaged that as fossil fuels run out, hydrogen will become the major world fuel and energy vector. It would be generated, for example, by massive arrays of solar cells electrolysing water. This may be true, but is unlikely to come to pass within the lifetime of this book.

The advantages of fuel cells impact particularly strongly on **combined heat and power** systems (for both large- and small-scale applications), and on **mobile power systems**,

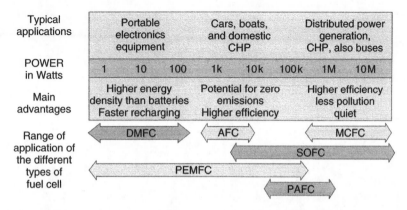

Figure 1.21 Chart to summarize the applications and main advantages of fuel cells of different types, and in different applications.

especially for **vehicles** and electronic equipment such as **portable computers, mobile telephones, and military communications equipment**. These areas are the major fields in which fuel cells are being used. Several example applications are given in the chapters in which the specific fuel cell types are described – especially Chapters 4 to 7. A key point is the wide range of applications of fuel cell power, from systems of a few watts up to megawatts. In this respect, fuel cells are quite unique as energy converters – their range of application far exceeds all other types.

References

Lex P. and Jonshagen B. (1999) "The zinc/bromine battery system for utility and remote area applications", *Power Engineering Journal*, **13**(3), 142–148.
Price A., Bartley S., Male S., and Cooley G. (1999) "A novel approach to utility scale energy storage", *Power Engineering Journal*, **13**(3), 122–129.
Shibata A. and Sato K. (1999) "Development of vanadium redox flow battery for electricity storage", *Power Engineering Journal*, **13**(3), 130–135.
Vincent C.A. and Scrosati B. (1997) *Modern Batteries*, 2nd ed., Arnold, London.
Zito R. (1997) "Process for energy storage and/or power delivery with means for restoring electrolyte balance", US patent 5612148.

2

Efficiency and Open Circuit Voltage

In this chapter we consider the **efficiency of fuel cells** – how it is defined and calculated, and what the limits are. The energy considerations give us information about the open circuit voltage (OCV) of a fuel cell, and the formulas produced also give important information about how factors such as pressure, gas concentration, and temperature affect the voltage.

2.1 Energy and the EMF of the Hydrogen Fuel Cell

In some electrical power-generating devices, it is very clear what form of energy is being converted into electricity. A good example is a wind-driven generator, as in Figure 2.1. The energy source is clearly the kinetic energy of the air moving over the blades.

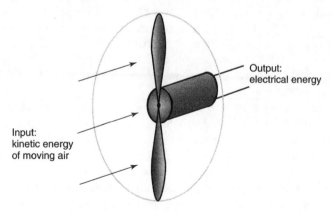

Figure 2.1 A wind-driven turbine. The input and output powers are simple to understand and calculate.

Fuel Cell Systems Explained, Second Edition James Larminie and Andrew Dicks
© 2003 John Wiley & Sons, Ltd   ISBN: 0-470-84857-X

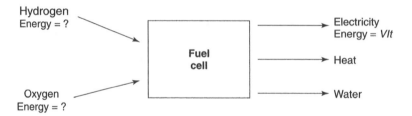

Figure 2.2   Fuel cell inputs and outputs.

With a fuel cell, such energy considerations are much more difficult to visualise. The basic operation has already been explained, and the input and outputs are shown in Figure 2.2.

The electrical power and energy output are easily calculated from the well-known formulas

$$\text{Power} = VI \qquad \text{and} \qquad \text{Energy} = VIt$$

However, the energy of the chemical input and output is not so easily defined. At a simple level we could say that it is the 'chemical energy' of the H_2, O_2, and H_2O that is in question. The problem is that 'chemical energy' is not simply defined – and terms such as enthalpy, Helmholtz function, and Gibbs free energy are used. In recent years the useful term 'exergy' has become quite widely used, and the concept is particularly useful in high-temperature fuel cells. There are also older (but still useful) terms such as calorific value.

In the case of fuel cells, it is the 'Gibbs free energy' that is important. This can be defined as the 'energy available to do external work, neglecting any work done by changes in pressure and/or volume' In a fuel cell, the 'external work' involves moving electrons round an external circuit – any work done by a change in volume between the input and output is not harnessed by the fuel cell.[1] Exergy is *all* the external work that can be extracted, including that due to volume and pressure changes. Enthalpy, simply put, is the Gibbs free energy plus the energy connected with the entropy (see Appendix 1).

All these forms of 'chemical energy' are rather like ordinary mechanical 'potential energy' in two important ways.

The **first** is that the *point of zero energy can be defined as almost anywhere.* When working with chemical reactions, the zero energy point is normally defined as pure elements, in the normal state, at standard temperature and pressure (25°C, 0.1 MPa). The term 'Gibbs free energy of formation', G_f, rather than the 'Gibbs free energy' is used when adopting this convention. Similarly, we also use 'enthalpy of formation' rather than just 'enthalpy'. For an ordinary hydrogen fuel cell operating at standard temperature and pressure (STP),[2] this means that the 'Gibbs free energy of formation' of the input is zero – a useful simplification.

---

[1] Though it may be harnessed by some kind of turbine in a combined cycle system, as discussed in Chapter 6.

[2] Standard temperature and pressure, or standard reference state, that is, 100 kPa and 25°C or 298.15 K.

The **second** parallel with mechanical potential energy is that it is the *change* in energy that is important. In a fuel cell, it is the change in this Gibbs free energy of formation, ΔG_f, that gives us the energy released. This change is the difference between the Gibbs free energy of the products and the Gibbs free energy of the inputs or reactants.

$$\Delta G_f = G_f \text{ of products} - G_f \text{ of reactants}$$

To make comparisons easier, it is nearly always most convenient to consider these quantities in their 'per mole' form. These are indicated by $^-$ over the lower case letter, for example, $(\bar{g}_f)_{H_2O}$ is the molar specific Gibbs free energy of formation for water.

Moles – g mole and kg mole

The *mole* is a measure of the 'amount' of a substance that takes into account its molar mass. The molar mass of H_2 is 2.0 amu, so one *gmole* is 2.0 g and one *kgmole* is 2.0 kg. Similarly, the molecular weight of H_2O is 18 amu, so 18 g is one *gmole*, or 18 kg is one *kgmole*. The *gmole*, despite the SI preference for kg, is still the most commonly used, and the 'unprefixed' *mole* means *gmole*.

A *mole* of any substance always has the same number of entities (e.g. molecules) – 6.022×10^{23} – called *Avogadro's number*. This is represented by the letter N or N_a.

A 'mole of electrons' is 6.022×10^{23} electrons. The charge is $N \cdot e$, where e is 1.602×10^{-19} C – the charge on one electron. This quantity is called the *Faraday constant*, and is designated by the letter F.

$$F = N \cdot e = 96485\,C$$

Consider the basic reaction for the hydrogen/oxygen fuel cell:

$$2H_2 + O_2 \rightarrow 2H_2O$$

which is equivalent to

$$H_2 + \tfrac{1}{2}O_2 \rightarrow H_2O$$

The 'product' is one mole of H_2O and the 'reactants' are one mole of H_2 and half a mole of O_2. Thus,

$$\Delta\bar{g}_f = \bar{g}_f \text{ of products} - \bar{g}_f \text{ of reactants}$$

So we have

$$\Delta\bar{g}_f = (\bar{g}_f)_{H_2O} - (\bar{g}_f)_{H_2} - \tfrac{1}{2}(\bar{g}_f)_{O_2}$$

This equation seems straightforward and simple enough. However, the Gibbs free energy of formation is *not constant*; it changes with temperature and state (liquid or gas). Table 2.1 below shows $\Delta\bar{g}_f$ for the basic hydrogen fuel cell reaction

$$H_2 + \tfrac{1}{2}O_2 \rightarrow H_2O$$

Table 2.1 $\Delta\bar{g}_f$ for the reaction $H_2 + \frac{1}{2}O_2 \rightarrow H_2O$ at various temperatures

Form of water product	Temperature (°C)	$\Delta\bar{g}_f$ (kJ mol^{-1})
Liquid	25	−237.2
Liquid	80	−228.2
Gas	80	−226.1
Gas	100	−225.2
Gas	200	−220.4
Gas	400	−210.3
Gas	600	−199.6
Gas	800	−188.6
Gas	1000	−177.4

for a number of different conditions. The method for calculating these values is given in Appendix 1. Note that the values are negative, which means that energy is released.

If there are no losses in the fuel cell, or as we should more properly say, if the process is 'reversible', then **all** this Gibbs free energy is converted into electrical energy. (In practice, some is also released as heat.) We will use this to find the reversible OCV of a fuel cell.

The basic operation of a fuel cell was explained in Chapter 1. A review of this chapter will remind you that, for the hydrogen fuel cell, **two** electrons pass round the external circuit for each water molecule produced and each molecule of hydrogen used. So, for one *mole* of hydrogen used, $2N$ electrons pass round the external circuit – where N is Avogadro's number. If $-e$ is the charge on one electron, then the charge that flows is

$$-2Ne = -2F \quad \text{coulombs}$$

F being the Faraday constant, or the charge on one mole of electrons.

If E is the voltage of the fuel cell, then the electrical work done moving this charge round the circuit is

$$\text{Electrical work done} = \text{charge} \times \text{voltage} = -2FE \quad \text{joules}$$

If the system is reversible (or has no losses), then this electrical work done will be equal to the Gibbs free energy released $\Delta\bar{g}_f$. So

$$\Delta\bar{g}_f = -2F \cdot E$$

Thus

$$E = \frac{-\Delta\bar{g}_f}{2F} \qquad [2.1]$$

This fundamental equation gives the electromotive force (EMF) or reversible open circuit voltage of the hydrogen fuel cell.

For example, a hydrogen fuel cell operating at 200°C has $\Delta \overline{g}_f = -220\,kJ$, so

$$E = \frac{220,000}{2 \times 96,485} = 1.14\,V$$

Note that this figure assumes no 'irreversibilities', and as we shall see later, assumes pure hydrogen and oxygen at standard pressure (0.1 MPa). In practice the voltage would be lower than this because of the voltage drops discussed in Chapter 3. Some of these irreversibilities apply a little even when no current is drawn, so even the OCV of a fuel cell will usually be lower than the figure given by equation 2.1.

Reversible processes, irreversibilities, and losses – an explanation of terms

An example of a simple reversible process is that shown in Figure 2.3. In position **A**, the ball has no kinetic energy, but potential energy *mgh*. In position **B**, this PE has been converted into kinetic energy. If there is no rolling resistance or wind resistance the process is *reversible*, in that the ball can roll up the other side and recover its potential energy.

In practice, some of the potential energy will be converted into heat because of friction and wind resistance. This is an *irreversible* process, as the heat cannot be converted back into kinetic or potential energy. We might be tempted to describe this as a 'loss' of energy, but that would not be very precise. In a sense, the potential energy is no more 'lost' to heat than it is 'lost' to kinetic energy. The difference is that in one you can get it back – it's reversible, and in the other you cannot – it's irreversible. So, the term 'irreversible energy loss' or 'irreversibility' is a rather more precise description of situations that many would describe as simply 'a loss'.

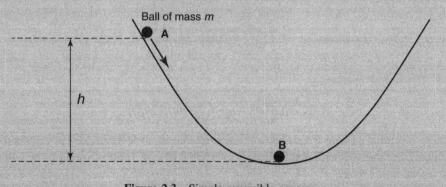

Figure 2.3 Simple reversible process.

2.2 The Open Circuit Voltage of Other Fuel Cells and Batteries

The equation that we have derived for the OCV of the hydrogen fuel cell

$$E = \frac{-\Delta \bar{g}_f}{2F}$$

can be applied to other reactions too. The only step in the derivation that was specific to the hydrogen fuel cell was the '2' electrons for each molecule of fuel, which led to the 2 in the equation. If we generalise it to any number of electrons per molecule, we have the formula

$$E = \frac{-\Delta \bar{g}_f}{zF} \qquad [2.2]$$

where z is the number of electrons transferred for each molecule of fuel.

The derivation was also not specific to fuel cells and applies just as well to other electrochemical power sources, particularly primary and secondary batteries. For example, the reaction in the familiar alkali battery used in radios and other portable appliances can be expressed (Bockris, 1981) by the equation

$$2MnO_2 + Zn \rightarrow ZnO + Mn_2O_3$$

for which $\Delta \bar{g}_f$ is $-277\,kJ\,mol^{-1}$. At the anode the reaction is

$$Zn + 2OH^- \rightarrow ZnO + H_2O + 2e^-$$

And at the cathode we have | Electrons flow from anode to cathode

$$2MnO_2 + H_2O + 2e^- \rightarrow Mn_2O_3 + 2OH^-$$

Thus, two electrons are passed round the circuit, and so the equation for the OCV is exactly the same as equation 2.1. This gives

$$E = \frac{2.77 \times 10^5}{2 \times 96485} = 1.44\,V$$

Another useful example is the methanol fuel cell, which we will look at later in Chapter 5. The overall reaction is

$$2CH_3OH + 3O_2 \rightarrow 4H_2O + 2CO_2$$

with 12 electrons passing from anode to cathode, that is, 6 electrons for each molecule of methanol. For the methanol reaction, $\Delta \bar{g}_f$ is $-698.2\,kJ\,mol^{-1}$. Substituting these numbers into equation 2.2 gives

$$E = \frac{6.98 \times 10^5}{6 \times 96485} = 1.21\,V$$

We note that this is similar to the open circuit reversible voltage for the hydrogen fuel cell.

2.3 Efficiency and Efficiency Limits

It is not straightforward to define the efficiency of a fuel cell, and efficiency claims cannot usually be taken at face value. In addition, the question of the maximum possible efficiency of a fuel cell is not without its complications.

The wind-driven generator of Figure 2.1 is an example of a system where the efficiency is fairly simple to define, and it is also clear that there must be some limit to the efficiency. Clearly, the air passing through the circle defined by the turbine blades cannot lose all its kinetic energy. If it did, it would stop dead and there would be an accumulation of air behind the turbine blades. As is explained in books on wind power, it can be shown, using fluid flow theory, that

Maximum energy from generator $= 0.58 \times$ kinetic energy of the wind

The figure of 0.58 is known as the 'Betz Coefficient'.

A more well-known example of an efficiency limit is that for heat engines – such as steam and gas turbines. If the maximum temperature of the heat engine is T_1, and the heated fluid is released at temperature T_2, then Carnot showed that the maximum efficiency possible is

$$\text{Carnot limit} = \frac{T_1 - T_2}{T_1}$$

The temperatures are in Kelvin, where 'room temperature' is about 290 K, and so T_2 is never likely to be small. As an example, for a steam turbine operating at 400°C (675 K), with the water exhausted through a condenser at 50°C (325 K), the Carnot efficiency limit is

$$\frac{675 - 325}{675} = 0.52 = 52\%$$

The reason for this efficiency limit for heat engines is not particularly mysterious. Essentially there must be some heat energy, proportional to the lower temperature T_2, which is always 'thrown away' or wasted. This is similar to the kinetic energy that must be retained by the air that passes the blades of the wind-driven generator.

With the fuel cell the situation is not so clear. It is quite well known that fuel cells are not subject to the Carnot efficiency limit. Indeed, it is commonly supposed that if there were no 'irreversibilities' then the efficiency could be 100%, and if we define efficiency in a particular (not very helpful) way, then this is true.

In Section 2.1 we saw that it was the 'Gibbs free energy' that is converted into electrical energy. If it were not for the irreversibilities to be discussed in Chapter 3, all this energy would be converted into electrical energy, and the efficiency could be said to be 100%. However, this is the result of choosing one among several types of 'chemical energy'. We have also noted, in Table 2.1, that the Gibbs free energy changes with temperature, and we will see in the section following that it also changes with pressure and other factors. All in all, to define efficiency as

$$\frac{\text{electrical energy produced}}{\text{Gibbs free energy change}}$$

is not very useful, and is rarely done, as whatever conditions are used the efficiency limit is always 100%.

Since a fuel cell uses materials that are usually burnt to release their energy, it would make sense to compare the electrical energy produced with the heat that would be produced by burning the fuel. This is sometimes called the *calorific value*, though a more precise description is the change in ' enthalpy of formation'. Its symbol is $\Delta \overline{h}_f$. As with the Gibbs free energy, the convention is that $\Delta \overline{h}_f$ is negative when energy is released. So to get a good comparison with other fuel-using technologies, the efficiency of the fuel cell is usually defined as

$$\frac{\text{electrical energy produced per mole of fuel}}{-\Delta \overline{h}_f} \qquad [2.3]$$

However, even this is not without its ambiguities, as there are two different values that we can use for $\Delta \overline{h}_f$. For the 'burning' of hydrogen

$$H_2 + \tfrac{1}{2}O_2 \rightarrow H_2O \text{ (steam)}$$

$$\Delta \overline{h}_f = -241.83 \, \text{kJ mol}^{-1}$$

whereas if the product water is condensed back to liquid, the reaction is

$$H_2 + \tfrac{1}{2}O_2 \rightarrow H_2O \text{ (liquid)}$$

$$\Delta \overline{h}_f = -285.84 \, \text{kJ mol}^{-1}$$

The difference between these two values for $\Delta \overline{h}_f$ (44.01 kJ mol^{-1}) is the molar enthalpy of vaporisation[3] of water. The higher figure is called the *higher heating value* (HHV), and the lower, quite logically, the 'lower heating value' (LHV). Any statement of efficiency should say whether it relates to the higher or lower heating value. If this information is not given, the LHV has probably been used, since this will give a higher efficiency figure.

We can now see that there is a limit to the efficiency, if we define it as in equation 2.3. The maximum electrical energy available is equal to the change in Gibbs free energy, so

$$\text{Maximum efficiency possible} = \frac{\Delta \overline{g}_f}{\Delta \overline{h}_f} \times 100\% \qquad [2.4]$$

This maximum efficiency limit is sometimes known as the 'thermodynamic efficiency'. Table 2.2 gives the values of the efficiency limit, relative to the HHV, for a hydrogen fuel cell. The maximum voltage, from equation 2.1, is also given.

---

[3] This used to be known as the molar 'latent heat'.

Table 2.2 $\Delta \bar{g}_f$, maximum EMF (or reversible open circuit voltage), and efficiency limit (HHV basis) for hydrogen fuel cells

Form of water product	Temp °C	$\Delta \bar{g}_f$, kJ mol^{-1}	Max EMF V	Efficiency limit %
Liquid	25	−237.2	1.23	83
Liquid	80	−228.2	1.18	80
Gas	100	−225.2	1.17	79
Gas	200	−220.4	1.14	77
Gas	400	−210.3	1.09	74
Gas	600	−199.6	1.04	70
Gas	800	−188.6	0.98	66
Gas	1000	−177.4	0.92	62

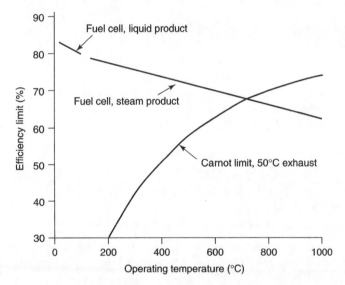

Figure 2.4 Maximum H_2 fuel cell efficiency at standard pressure, with reference to higher heating value. The Carnot limit is shown for comparison, with a 50°C exhaust temperature.

The graphs in Figure 2.4 show how these values vary with temperature, and how they compare with the 'Carnot limit'. Three important points should be noted:

- Although the graph and table would suggest that lower temperatures are better, the voltage losses discussed in Chapter 3 are nearly always less at higher temperatures. So *in practice* fuel cell voltages are usually *higher* at higher temperatures.
- The waste heat from the higher-temperature cells is more useful than that from lower-temperature cells.

- Contrary to statements often made by their supporters, fuel cells do NOT always have a higher efficiency limit than heat engines.[4]

This decline in maximum possible efficiency with temperature associated with the hydrogen fuel cell does not occur in exactly the same way with other types of fuel cells. For example, when using carbon monoxide we get

$$CO + \tfrac{1}{2}O_2 \rightarrow CO_2$$

$\Delta \bar{g}_f$ changes even more quickly with temperature, and the maximum possible efficiency falls from about 82% at 100°C to 52% at 1000°C. On the other hand, for the reaction

$$CH_4 + 2O_2 \rightarrow CO_2 + 2H_2O$$

$\Delta \bar{g}_f$ is fairly constant with temperature, and the maximum possible efficiency hardly changes.

2.4 Efficiency and the Fuel Cell Voltage

It is clear from Table 2.2 that there is a connection between the maximum EMF of a cell and its maximum efficiency. The *operating* voltage of a fuel cell can also be very easily related to its efficiency. This can be shown by adapting equation 2.1. If all the energy from the hydrogen fuel, its 'calorific value', heating value, or enthalpy of formation, were transformed into electrical energy, then the EMF would be given by

$$E = \frac{-\Delta \bar{h}_f}{2F}$$
$$= 1.48 \, \text{V} \ \text{if using the HHV}$$
$$\text{or} = 1.25 \, \text{V} \ \text{if using the LHV}$$

These are the voltages that would be obtained from a 100% efficient system, with reference to the HHV or LHV. The actual efficiency of the cell is then the actual voltage divided by these values, or

$$\text{Cell efficiency} = \frac{V_c}{1.48} 100\% \ (\text{with reference to HHV})$$

However, in practice it is found that not all the fuel that is fed to a fuel cell can be used, for reasons discussed later. Some fuel usually has to pass through unreacted. A *fuel utilisation coefficient* can be defined as

$$\mu_f = \frac{\text{mass of fuel reacted in cell}}{\text{mass of fuel input to cell}}$$

---

[4] In Chapter 7 we see how a heat engine and a high-temperature fuel cell can be combined into a particularly efficient system.

This is equivalent to the ratio of fuel cell current and the current that would be obtained if all the fuel were reacted. The fuel cell efficiency is therefore given by

$$\text{Efficiency, } \eta = \mu_f \frac{V_c}{1.48} 100\% \qquad [2.5]$$

If a figure relative to the LHV is required, use 1.25 instead of 1.48. A good estimate for μ_f is 0.95, which allows the efficiency of a fuel cell to be accurately estimated from the very simple measurement of its voltage. However, it can be a great deal less in some circumstances, as is discussed in Section 2.5.3, and in Chapter 6.

2.5 The Effect of Pressure and Gas Concentration

2.5.1 The Nernst equation

In Section 2.1 we noted that the Gibbs free energy changes in a chemical reaction vary with temperature. Equally important, though more complex, are the changes in Gibbs free energy with reactant pressure and concentration.

Consider a general reaction such as

$$j\text{J} + k\text{K} \rightarrow m\text{M} \qquad [2.6]$$

where j moles of J react with k moles of K to produce m moles of M. Each of the reactants, and the products, has an associated 'activity'. This 'activity' is designated by a, a_J, and a_K being the activity of the reactants, and a_M the activity of the product. It is beyond the scope of this book to give a thorough description of 'activity'. However, in the case of gases behaving as 'ideal gases', it can be shown that

$$\text{activity } a = \frac{P}{P^0}$$

where P is the pressure or partial pressure of the gas and P^0 is standard pressure, 0.1 MPa. Since fuel cells are generally gas reactors, this simple equation is very useful. We can say that activity is proportional to partial pressure. In the case of dissolved chemicals, the activity is linked to the molarity (strength) of the solution. The case of the water produced in fuel cells is somewhat difficult, since this can be either as steam or as liquid. For steam, we can say that

$$a_{H_2O} = \frac{P_{H_2O}}{P^0_{H_2O}}$$

where $P^0_{H_2O}$ is the vapour pressure of the steam at the temperature concerned. This has to be found from steam tables. In the case of liquid water product, it is a reasonable approximation to assume that $a_{H_2O} = 1$.

The activities of the reactants and products modify the Gibbs free energy change of a reaction. Using thermodynamic arguments (Balmer, 1990), it can be shown that in a chemical reaction such as that given in equation 2.6

$$\Delta \bar{g}_f = \Delta \bar{g}_f{}^0 - RT \ln \left(\frac{a_J^j \cdot a_K^k}{a_M^m} \right)$$

where $\Delta \bar{g}_f{}^0$ is the change in molar Gibbs free energy of formation at standard pressure.

Despite not looking very 'friendly', this equation is useful, and not very difficult. In the case of the hydrogen fuel cell reaction

$$H_2 + \tfrac{1}{2}O_2 \rightarrow H_2O$$

the equation becomes

$$\Delta \bar{g}_f = \Delta \bar{g}_f{}^0 - RT \ln \left(\frac{a_{H_2} \cdot a_{O_2}^{\frac{1}{2}}}{a_{H_2O}} \right)$$

$\Delta \bar{g}_f{}^0$ is the quantity given in Tables 2.1 and 2.2. We can see that if the activity of the reactants increases, $\Delta \bar{g}_f$ becomes more negative, that is, more energy is released. On the other hand, if the activity of the product increases, $\Delta \bar{g}_f$ increases, so becomes less negative, and less energy is released. To see how this equation affects voltage, we can substitute it into equation 2.1 and obtain

$$E = \frac{-\Delta \bar{g}_f{}^0}{2F} + \frac{RT}{2F} \ln \left(\frac{a_{H_2} \cdot a_{O_2}^{\frac{1}{2}}}{a_{H_2O}} \right)$$

$$= E^0 + \frac{RT}{2F} \ln \left(\frac{a_{H_2} \cdot a_{O_2}^{\frac{1}{2}}}{a_{H_2O}} \right) \qquad [2.7]$$

where E^0 is the EMF at standard pressure and is the number given in column 4 of Table 2.2. The equation shows precisely how raising the activity of the reactants increases the voltage.

Equation 2.7, and its variants below, which give an EMF in terms of product and/or reactant activity, are called *Nernst equations*. The EMF calculated from such equations is known as the 'Nernst voltage' and is the reversible cell voltage that would exist at a given temperature and pressure. The logarithmic function involving the reactants allows us to use the regular rules of the logarithmic functions such as

$$\ln \left(\frac{a}{b} \right) = \ln(a) - \ln(b) \quad \text{and} \quad \ln \left(\frac{c^2}{d^{\frac{1}{2}}} \right) = 2 \ln(c) - \frac{1}{2} \ln(d)$$

This makes it straightforward to manipulate equation 2.7 to get at the effect of different parameters. For example, in the reaction

$$H_2 + \tfrac{1}{2}O_2 \rightarrow H_2O \quad \text{(steam)}$$

at high temperature (e.g. in a solid oxide fuel cell (SOFC) at $1000°C$) we can assume that the steam behaves as an ideal gas, and so

$$a_{H_2} = \frac{P_{H_2}}{P^0}, \qquad a_{O_2} = \frac{P_{O_2}}{P^0}, \qquad a_{H_2O} = \frac{P_{H_2O}}{P^0}$$

Then equation 2.7 will become

$$E = E^0 + \frac{RT}{2F} \ln \left(\frac{\dfrac{P_{H_2}}{P^o} \cdot \left(\dfrac{P_{O_2}}{P^o} \right)^{\frac{1}{2}}}{\dfrac{P_{H_2O}}{P^o}} \right)$$

If all the pressures are given in bar, then $P^0 = 1$ and the equation simplifies to

$$E = E^0 + \frac{RT}{2F} \ln \left(\frac{P_{H_2} \cdot P_{O_2}^{\frac{1}{2}}}{P_{H_2O}} \right) \qquad [2.8]$$

In nearly all cases the pressures in equation 2.8 will be partial pressures, that is, the gases will be part of a mixture. For example, the hydrogen gas might be part of a mixture of H_2 and CO_2 from a fuel reformer, together with product steam. The oxygen will nearly always be part of air. It is also often the case that the pressure on both the cathode and the anode is approximately the same – this simplifies the design. If this system pressure is P, then we can say that

$$P_{H_2} = \alpha P$$

$$P_{O_2} = \beta P$$

$$\text{and} \quad P_{H_2O} = \delta P$$

where α, β, and δ are constants depending on the molar masses and concentrations of H_2, O_2, and H_2O. Equation 2.8 then becomes

$$E = E^0 + \frac{RT}{2F} \ln \left(\frac{\alpha \cdot \beta^{\frac{1}{2}}}{\delta} \cdot P^{\frac{1}{2}} \right)$$

$$= E^0 + \frac{RT}{2F} \ln \left(\frac{\alpha \cdot \beta^{\frac{1}{2}}}{\delta} \right) + \frac{RT}{4F} \ln(P) \qquad [2.9]$$

The two equations 2.8 and 2.9 are forms of the Nernst equation. They provide a theoretical basis and a quantitative indication for a large number of variables in fuel cell design and operation. Some of these are discussed in later chapters, but some points are considered briefly here.

Partial Pressures

In a mixture of gases, the total pressure is the sum of all the 'partial pressures' of the components of the mixture. For example, in air at 0.1 MPa, the partial pressures are as shown in Table 2.3

Table 2.3 Partial pressures of atmospheric gases

Gas	Partial pressure MPa
Nitrogen	0.07809
Oxygen	0.02095
Argon	0.00093
Others (including CO_2)	0.00003
Total	0.10000

In fuel cells, the partial pressure of oxygen is important. On the fuel side the partial pressure of hydrogen and/or carbon dioxide is important if a hydrocarbon (e.g. CH_4) is used as the fuel source. In such cases the partial pressure will vary according to the reformation method used – see Chapter 7.

However, if the mixture of the gases is known, the partial pressures can be easily found. It can be readily shown, using the gas law equation $PV = NRT$, that the volume fraction, molar fraction, and pressure fraction of a gas mixture are all equal. For example, consider the reaction

$$CH_4 + H_2O \rightarrow 3H_2 + CO_2$$

The product gas stream contains three parts H_2 and one part CO_2 by moles and volume. So, if the reaction takes place at 0.10 MPa

$$P_{H_2} = \tfrac{3}{4} \times 0.1 = 0.075\,\text{MPa} \quad \text{and} \quad P_{CO_2} = \tfrac{1}{4} \times 0.1 = 0.025\,\text{MPa}$$

2.5.2 Hydrogen partial pressure

Hydrogen can either be supplied pure or as part of a mixture. If we isolate the pressure of hydrogen term in equation 2.8, we have

$$E = E^0 + \frac{RT}{2F} \ln \left(\frac{P_{O_2}^{\frac{1}{2}}}{P_{H_2O}} \cdot \right) + \frac{RT}{2F} \ln(P_{H_2})$$

So, if the hydrogen partial pressure changes, say, from P_1 to P_2 bar, with P_{O_2} and P_{H_2O} unchanged, then the voltage will change by

$$\Delta V = \frac{RT}{2F} \ln(P_2) - \frac{RT}{2F} \ln(P_1)$$

$$= \frac{RT}{2F} \ln \left(\frac{P_2}{P_1} \right) \qquad\qquad [2.10]$$

The use of H_2 mixed with CO_2 occurs particularly in phosphoric acid fuel cells, operating at about 200°C. Substituting the values for R, T, and F gives

$$\Delta V = 0.02 \ln \left(\frac{P_2}{P_1} \right) \text{V}$$

This agrees well with experimental results, which correlate best with a factor of 0.024 instead of 0.020 (Parsons Inc., 2000).[5] As an example, changing from pure hydrogen to 50% hydrogen/carbon dioxide mixture will reduce the voltage by 0.015 V per cell.

2.5.3 Fuel and oxidant utilisation

As air passes through a fuel cell, the oxygen is used, and so the partial pressure will reduce. Similarly, the fuel partial pressure will often decline, as the proportion of fuel reduces and reaction products increase. Referring to equation 2.9 we can see that α and β decrease, whereas δ increases. All these changes make the term

$$\frac{RT}{2F} \ln \left(\frac{\alpha \cdot \beta^{\frac{1}{2}}}{\delta} \right)$$

from equation 2.9 smaller, and so the EMF will fall. This will vary within the cell – it will be worst near the fuel outlet as the fuel is used. Because of the low-resistance bipolar plates on the electrode, it is not actually possible for different parts of one cell to have different voltages, so the current varies. The current density will be lower nearer the exit where the fuel concentration is lower. The RT term in the equation also shows us that this drop in Nernst voltage due to fuel utilisation will be greater in high-temperature fuel cells.

We have seen in Section 2.4 above that for a high system efficiency the fuel utilisation should be as high as possible. However, this equation shows us that cell voltage, and hence the cell efficiency, will *fall* with higher utilisation. So we see fuel and oxygen utilisation need careful optimising, especially in higher-temperature cells. The selection of utilisation is an important aspect of system design and is especially important when reformed fuels are used. It is given further consideration in Chapter 7.

---

[5] Parsons Inc (2000) uses base 10 logarithms, and so the coefficient given there is 0.055.

2.5.4 System pressure

The Nernst equation in the form of equation 2.9 shows us that the EMF of a fuel cell is increased by the system pressure according to the term

$$\frac{RT}{4F}\ln(P)$$

So, if the pressure changes from P_1 to P_2 there will be a change of voltage

$$\Delta V = \frac{RT}{4F}\ln\left(\frac{P_2}{P_1}\right)$$

For an SOFC operating at 1000°C, this would give

$$\Delta V = 0.027\ln\left(\frac{P_2}{P_1}\right)$$

This agrees well with reported results (Bevc, 1997 and Parsons Inc., 2000) for high-temperature cells. However, for other fuel cells, working at lower temperatures, the agreement is not as good. For example, a phosphoric acid fuel cell working at 200°C should be affected by system pressure by the equation

$$\Delta V = \frac{RT}{4F}\ln\left(\frac{P_2}{P_1}\right) = 0.010\ln\left(\frac{P_2}{P_1}\right)$$

whereas reported results (Parsons, 2000) give a correlation to the equation

$$\Delta V = 0.063\ln\left(\frac{P_2}{P_1}\right)$$

In other words, at lower temperatures, the benefits of raising system pressure are much greater than the Nernst equation predicts. This is because, except for very high temperature cells, increasing the pressure also reduces the losses at the electrodes, especially the cathode. (This is considered further in Chapter 3.)

A similar effect occurs when studying the change from air to oxygen. This effectively changes β in equation 2.9 from 0.21 to 1.0. Isolating β in equation 2.9 gives

$$E = E^0 + \frac{RT}{4F}\ln(\beta) + \frac{RT}{2F}\ln\left(\frac{\alpha}{\delta}\right) + \frac{RT}{4F}\ln(P)$$

For the change in β from 0.21 to 1.0, with all other factors remaining constant, we have

$$\Delta V = \frac{RT}{4F}\ln\left(\frac{1.0}{0.21}\right)$$

For a proton exchange membrane (PEM) fuel cell at 80°C this would give

$$\Delta V = 0.012 \, \text{V}$$

In fact, reported results (Prater, 1990) give a much larger change, 0.05 V being a typical result. This is also due to the improved performance of the cathode when using oxygen, reducing the voltage losses there.

2.5.5 An application – blood alcohol measurement

In addition to generating electrical power, fuel cells are also the basis of some types of sensors. One of the most successful is the fuel cell–based alcohol sensor – the 'breathalyser' (Figure 2.5). This measures the concentration of alcohol in the air that someone breathes out of his or her lungs. It has been shown that this is directly proportional to the concentration of alcohol in the blood. The basic chemistry is that the alcohol (ethanol) reacts in a simple fuel cell to give a (very small) voltage. In theory the ethanol could be fully oxidised to CO_2 and water. However, the ethanol is probably not fully reacted and is only partially oxidised to ethanal.

The anode and cathode reactions are probably

$$C_2H_5OH \text{ (ethanol)} \rightarrow CH_3CHO \text{ (ethanal)} + 2H^+ + 2e^-$$

and

Electrons flow round external circuit

$$\tfrac{1}{2}O_2 + 2H^+ + 2e^- \rightarrow H_2O$$

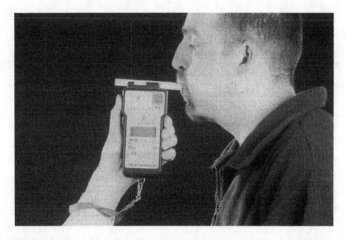

Figure 2.5 Fuel cell–based breathalyser as used by police forces in the USA. (Reproduced by kind permission of Lion Laboratories Ltd.)

We have seen that the voltage is affected by the concentration of the reactants, and so the voltage of the cell will be affected by the concentration of the alcohol in the gas blown into the cell. Thus, we can get a measure of the alcohol concentration in the blood. This type of fuel cell is the basis of the majority of roadside blood alcohol measurement instruments used by police forces throughout the world.

2.6 Summary

The reversible OCV for a hydrogen fuel cell is given by the equation

$$E = \frac{-\Delta \bar{g}_f}{2F} \qquad [2.1]$$

In general, for a reaction where z electrons are transferred for each molecule of fuel the reversible OCV is

$$E = \frac{-\Delta \bar{g}_f}{zF} \qquad [2.2]$$

However, $\Delta \bar{g}_f$ changes with temperature and other factors. The maximum efficiency is given by the expression

$$\eta_{max} = \frac{\Delta \bar{g}_f}{\Delta \bar{h}_f} \times 100\% \qquad [2.4]$$

The efficiency of a working hydrogen fuel cell can be found from the simple formula

$$\text{Efficiency, } \eta = \mu_f \frac{V}{1.48} 100\% \qquad [2.5]$$

where μ_f is the fuel utilisation (typically about 0.95) and V is the voltage of a single cell within the fuel cell stack. This gives the efficiency relative to the HHV of hydrogen.

The pressure and concentration of the reactants affects the Gibbs free energy, and thus the voltage. This is expressed in the Nernst equation, which can be given in many forms. For example, if the pressures of the reactants and products are given in bar and the water product is in the form of steam, then

$$E = E^0 + \frac{RT}{2F} \ln \left(\frac{P_{H_2} \cdot P_{O_2}^{\frac{1}{2}}}{P_{H_2O}} \right) \qquad [2.8]$$

where E^0 is the cell EMF at standard pressure.

Now, in most parts of this chapter we have referred to or given equations for the EMF of a cell or its reversible open circuit voltage. In practice the operating voltage is less than these equations give, and in some cases much less. This is the result of losses or irreversibilities, to which we give careful consideration in the following chapter.

References

Balmer R. (1990) *Thermodynamics*, West, St Paul, Minnesota.

Bevc F. (1997) "Advances in solid oxide fuel cells and integrated power plants", *Proceedings of the Institution of Mechanical Engineers*, **211**(Part A), 359–366.

Bockris J.O'M., Conway B.E., Yeager E. and White R.E. (eds) (1981) *A Comprehensive Treatment of Electrochemistry*, Vol. 3, Plenum Press, New York, p. 220.

Parsons Inc., EG&G Services (2000) *Fuel Cells: A Handbook*, 5th ed., US Department of Energy, p. 5-19 and p. 8-22.

Prater K. (1990) "The renaissance of the solid polymer fuel cell", *Journal of Power Sources*, **29**, 239–250.

3

Operational Fuel Cell Voltages

3.1 Introduction

We have seen in Chapter 2 that the theoretical value of the open circuit voltage (OCV) of a hydrogen fuel cell is given by the formula

$$E = \frac{-\Delta \overline{g}_f}{2F} \qquad\qquad [3.1]$$

This gives a value of about 1.2 V for a cell operating below 100°C. However, when a fuel cell is made and put to use, it is found that the voltage is less than this, often considerably less. Figure 3.1 shows the performance of a typical single cell operating at about 70°C, at normal air pressure. The key points to notice about this graph of the cell voltage against current density[1] are as follows:

- Even the open circuit voltage is less than the theoretical value.
- There is a rapid initial fall in voltage.
- The voltage then falls less rapidly, and more linearly.
- There is sometimes a higher current density at which the voltage falls rapidly.

If a fuel cell is operated at higher temperatures, the shape of the voltage/current density graph changes. As we have seen in Chapter 2, the reversible 'no loss' voltage falls. However, the difference between the actual operating voltage and the 'no loss' value usually becomes less. In particular, the initial fall in voltage as current is drawn from the cell is markedly less. Figure 3.2 shows the situation for a typical solid oxide fuel cell (SOFC) operating at about 800°C. The key points here are as follows:

- The open circuit voltage is equal to or only a little less than the theoretical value.
- The initial fall in voltage is very small, and the graph is more linear.

---

[1] Note that we usually refer to current **density**, or current per unit area, rather than to just current. This is to make the comparison between cells of different sizes easier.

Fuel Cell Systems Explained, Second Edition James Larminie and Andrew Dicks
© 2003 John Wiley & Sons, Ltd   ISBN: 0-470-84857-X

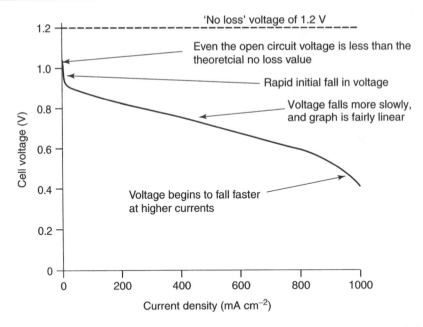

Figure 3.1 Graph showing the voltage for a typical low temperature, air pressure, fuel cell.

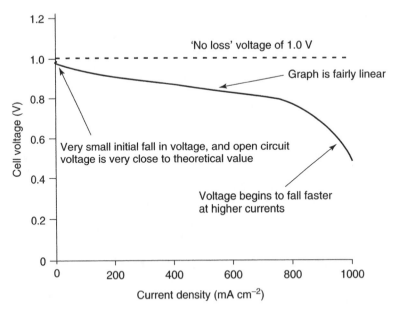

Figure 3.2 Graph showing the voltage of a typical air pressure fuel cell operating at about 800°C.

- There may be a higher current density at which the voltage falls rapidly, as with lower-temperature cells.

Comparing Figures 3.1 and 3.2, we see that although the reversible or 'no loss' voltage is lower for the higher temperature, the operating voltage is generally *higher*, because the voltage drop or irreversibilities are smaller.

In this chapter, we consider what causes the voltage to fall below the reversible value, and what can be done to improve the situation.

3.2 Terminology

Fuel cell power systems are highly interdisciplinary – they require the skills of chemists, electrochemists, materials scientists, thermodynamicists, electrical and chemical engineers, control and instrumentation engineers, and others, to make them work well. One problem with this is that there are occasions when these various disciplines have their own names for what is often essentially the same thing. The main topic of this chapter is a case in point.

The graphs of Figures 3.1 and 3.2 show the difference between the voltage we would expect from a fuel cell operating ideally (reversibly) and the actual voltage. This difference is the main focus of this chapter. However, a problem is that there are at least *five* commonly used names for this voltage difference!

- *Overvoltage* or *overpotential* is the term often used by electrochemists. This is because it is a voltage superimposed over the reversible or ideal voltage. Strictly, this term only applies to differences generated at an electrode interface. A disadvantage of it is that it tends to imply making the voltage larger – whereas in fuel cells the overvoltage opposes and reduces the reversible ideal voltage.
- *Polarisation* is another term much used by some electrochemists and others. For many engineers and scientists this has connotations of static electricity, which is not helpful.
- *Irreversibility* is the best term from a thermodynamics point of view. However, it is perhaps too general and does not connect well with its main effect; in this case – reducing voltage.
- *Losses* is a term that can be used, but is rather too vague. See also the section on 'reversibility, irreversibility, and losses' in Chapter 2.
- *Voltage drop* is not a very scientifically precise term, but it conveys well the effect observed, and is a term readily understood by electrical engineers.

The English language is rich, frequently having many words for similar ideas. This is an example of how this richness can develop, and most of these terms will be used in this book.

3.3 Fuel Cell Irreversibilities – Causes of Voltage Drop

The characteristic shape of the voltage/current density graphs of Figures 3.1 and 3.2 results from four major irreversibilities. These will be outlined very briefly here and then considered in more detail in the sections that follow.

1. *Activation losses*. These are caused by the slowness of the reactions taking place on the surface of the electrodes. A proportion of the voltage generated is lost in driving the chemical reaction that transfers the electrons to or from the electrode. As we shall see in Section 3.4 below, this voltage drop is highly non-linear.
2. *Fuel crossover and internal currents*. This energy loss results from the waste of fuel passing through the electrolyte, and, to a lesser extent, from electron conduction through the electrolyte. The electrolyte should only transport ions through the cell, as in Figures 1.3 and 1.4. However, a certain amount of fuel diffusion and electron flow will always be possible. Except in the case of direct methanol cells the fuel loss and current is small, and its effect is usually not very important. However, it does have a marked effect on the OCV of low-temperature cells, as we shall see in Section 3.5.
3. *Ohmic losses*. This voltage drop is the straightforward resistance to the flow of electrons through the material of the electrodes and the various interconnections, as well as the resistance to the flow of ions through the electrolyte. This voltage drop is essentially proportional to current density, linear, and so is called *ohmic losses*, or sometimes as *resistive losses*.
4. *Mass transport or concentration losses*. These result from the change in concentration of the reactants at the surface of the electrodes as the fuel is used. We have seen in Chapter 2 that concentration affects voltage, and so this type of irreversibility is sometimes called *concentration loss*. Because the reduction in concentration is the result of a failure to transport sufficient reactant to the electrode surface, this type of loss is also often called *mass transport loss*. This type of loss has a third name – 'Nernstian'. This is because of its connections with concentration, and the effects of concentration are modelled by the Nernst equation.

These four categories of irreversibility are considered one by one in the sections that follow.

3.4 Activation Losses

3.4.1 The Tafel equation

As a result of experiments, rather than theoretical considerations, Tafel observed and reported in 1905 that the overvoltage at the surface of an electrode followed a similar pattern in a great variety of electrochemical reactions. This general pattern is shown in Figure 3.3. It shows that if a graph of overvoltage against *log* of current density is plotted, then, for most values of overvoltage, the graph approximates to a straight line. Such plots of overvoltage against *log* of current density are known as 'Tafel Plots'. The diagram shows two typical plots.

For most values of overvoltage its value is given by the equation

$$\Delta V_{act} = a \log \left(\frac{i}{i_0} \right)$$

This equation is known as the Tafel equation. It can be expressed in many forms. One simple variation is to use natural logarithms instead of base 10, which is preferred since

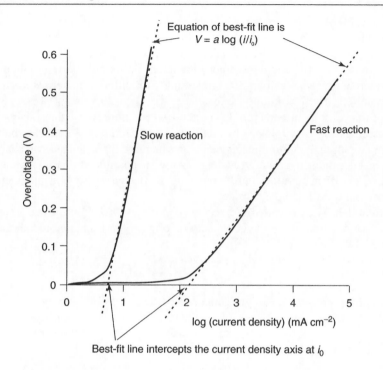

Figure 3.3 Tafel plots for slow and fast electrochemical reactions.

they feature in the equations used in Chapter 2. This gives

$$\Delta V_{act} = A \ln \left(\frac{i}{i_0} \right) \qquad [3.2]$$

The constant A is higher for an electrochemical reaction that is slow. The constant i_0 is higher if the reaction is faster. The current density i_0 can be considered as the current density at which the overvoltage begins to move from zero. It is important to remember that the Tafel equation only holds true when $i > i_0$. This current density i_0 is usually called the *exchange current density*, as we shall see in Section 3.4.2.

3.4.2 The constants in the Tafel equation

Although the Tafel equation was originally deduced from experimental results, it also has a theoretical basis. It can be shown (McDougall, 1976) that for a hydrogen fuel cell with two electrons transferred per mole, the constant A in equation 3.2 above is given by

$$A = \frac{RT}{2\alpha F} \qquad [3.3]$$

The constant α is called the *charge transfer coefficient* and is the proportion of the electrical energy applied that is harnessed in changing the rate of an electrochemical reaction. Its value depends on the reaction involved and the material the electrode is made from, but it must be in the range 0 to 1.0. For the hydrogen electrode, its value is about 0.5 for a great variety of electrode materials (Davies, 1967). At the oxygen electrode the charge transfer coefficient shows more variation, but is still between about 0.1 and 0.5 in most circumstances. In short, experimenting with different materials to get the best possible value for A will make little impact.

The appearance of T in equation 3.3 might give the impression that raising the temperature increases the overvoltage. In fact this is very rarely the case, as the effect of increases in i_0 with temperature far outweigh any increase in A. Indeed, we shall see that the key to making the activation overvoltage as low as possible is this i_0, which can vary by several orders of magnitude. Furthermore, it is affected by several parameters other than the material used for the electrode.

The current density i_0 is called the *exchange current density*, and it can be visualised as follows. The reaction at the oxygen electrode of a proton exchange membrane (PEM) or acid electrolyte fuel cell is

$$O_2 + 4e^- + 4H^+ \longrightarrow 2H_2O$$

At zero current density, we might suppose that there was no activity at the electrode and that this reaction does not take place. In fact this is not so; the reaction is taking place all the time, but the reverse reaction is also taking place at the same rate. There is an equilibrium expressed as

$$O_2 + 4e^- + 4H^+ \leftrightarrow 2H_2O$$

Thus, there is a continual backwards and forwards flow of electrons from and to the electrolyte. This current density is i_0, the 'exchange' current density. It is self-evident that if this current density is high, then the surface of the electrode is more 'active' and a current in one particular direction is more likely to flow. We are simply shifting in one particular direction something already going on, rather than starting something new.

This exchange current density i_0 is crucial in controlling the performance of a fuel cell electrode. It is vital to make its value as high as possible.

We should note that it is possible to change around the Tafel equation 3.2 so that it gives the current, rather than the voltage. This is done by rearranging, and converting from the logarithmic to the exponential form. It is thus possible to show that equation 3.2, with 3.3, can be rearranged to give

$$i = i_0 \exp\left(\frac{2\alpha F \Delta V_{act}}{RT}\right)$$

The equation is called the *Butler–Vollmer equation* and is quite often used as an equivalent alternative to the Tafel equation.

Imagine a fuel cell that has no losses at all except for this activation overvoltage on *one* electrode. Its voltage would then be given by the equation

$$V = E - A \ln\left(\frac{i}{i_0}\right) \tag{3.4}$$

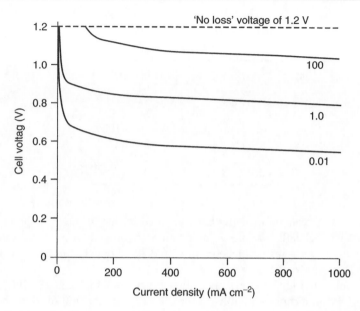

Figure 3.4 Graphs of cell voltage against current density, assuming losses are due only to the activation overvoltage at one electrode, for exchange current density i_0 values of 0.01, 1.0, and 100 mA cm^{-2}.

where E is the reversible OCV given by equation 3.1. If we plot graphs of this equation using values of i_0 of 0.01, 1.0, and 100 mA cm^{-2}, using a typical value for A of 0.06 V, we get the curves shown in Figure 3.4.

The importance of i_0 can be clearly seen. The effect, for most values of current density, is to reduce the cell voltage by a fairly fixed amount, as we could predict from the Tafel equation. The smaller the i_0, the greater is this voltage drop. Note that when i_0 is 100 mA cm^{-2}, there is no voltage drop until the current density i is greater than 100 mA cm^{-2}.

It is possible to measure this overvoltage at each electrode, either using reference electrodes within a working fuel cell or using half-cells. Table 3.1 below gives the values of i_0 for the hydrogen electrode at 25°C, for various metals. The measurements are for flat smooth electrodes.

The most striking thing about these figures is their great variation, indicating a strong catalytic effect. The figures for the oxygen electrode also vary greatly and are generally lower by a factor of about 10^5, that is, they are much smaller (Appleby and Foulkes, 1993). This would give a figure that is about 10^{-8} A cm^{-2}, even using Pt catalyst, far worse than even the lowest curve on Figure 3.4. However, the value of i_0 for a real fuel cell electrode is much higher than the figures in Table 3.1, because of the roughness of the electrode. This makes the real surface area many times bigger, typically at least 10^3 times larger than the nominal length × width.

We have noted that i_0 at the oxygen electrode (the cathode) is much smaller than that at the hydrogen anode, sometimes 10^5 times smaller. Indeed, it is generally reckoned that

Table 3.1 i_0 for the hydrogen electrode for various metals for an acid electrolyte. (Bloom, 1981)

Metal	i_0 $(A\,cm^{-2})$
Pb	2.5×10^{-13}
Zn	3×10^{-11}
Ag	4×10^{-7}
Ni	6×10^{-6}
Pt	5×10^{-4}
Pd	4×10^{-3}

the overvoltage at the anode is negligible compared to that of the cathode, at least in the case of hydrogen fuel cells. For a low temperature, hydrogen-fed fuel cell running on air at ambient pressure, a typical value for i_0 would be about $0.1\,mA\,cm^{-2}$ at the cathode and about $200\,mA\,cm^{-2}$ at the anode.

In other fuel cells, for example, the direct methanol fuel cell (DMFC), the anode overvoltage is by no means negligible. In these cases, the equation for the total activation overvoltage would combine the overvoltages at both anode and cathode, giving

$$\text{Activation Voltage drop} = A_a \ln\left(\frac{i}{i_{0a}}\right) + A_c \ln\left(\frac{i}{i_{0c}}\right)$$

However, it is readily proved that this equation can be expressed as

$$\Delta V_{act} = A \ln\left(\frac{i}{b}\right)$$

where

$$A = A_a + A_c \quad \text{and} \quad b = i_{0a}^{\frac{A_a}{A}} + i_{0c}^{\frac{A_c}{A}} \qquad [3.5]$$

This is exactly the same form as equation 3.2, the overvoltage for one electrode. So, whether the activation overvoltage arises mainly at one electrode only or at both electrodes, the equation that models the voltage is of the same form. Furthermore, in all cases, the item in the equation that shows the most variation is the exchange current density i_0 rather than A.

3.4.3 Reducing the activation overvoltage

We have seen that the exchange current density i_0 is the crucial factor in reducing the activation overvoltage. A crucial factor in improving fuel cell performance is, therefore, to increase the value of i_0, especially at the cathode. This can be done in the following ways:

- *Raising the cell temperature.* This fully explains the different shape of the voltage/current density graphs of low- and high-temperature fuel cells illustrated in

Figures 3.1 and 3.2. For a low-temperature cell, i_0 at the cathode will be about $0.1\,\text{mA}\,\text{cm}^{-2}$, whereas for a typical $800°C$ cell, it will be about $10\,\text{mA}\,\text{cm}^{-2}$, a 100-fold improvement.

- *Using more effective catalysts.* The effect of different metals in the electrode is shown clearly by the figures in Table 3.1.
- *Increasing the roughness of the electrodes.* This increases the real surface area of each nominal $1\,\text{cm}^2$, and this increases i_0.
- *Increasing reactant concentration, for example, using pure O_2 instead of air.* This works because the catalyst sites are more effectively occupied by reactants. (As we have seen in Chapter 2, this also increases the reversible open circuit voltage.)
- *Increasing the pressure.* This is also presumed to work by increasing catalyst site occupancy. (This also increases the reversible open circuit voltage, and so brings a 'double benefit'.)

Increasing the value of i_0 has the effect of raising the cell voltage by a constant amount at most currents, and so mimics raising the open circuit voltage (OCV). (See Figure 3.4 above.) The last two points in the above list explain the discrepancy between theoretical OCV and actual OCV noted in Section 2.5.4 in the previous chapter.

3.4.4 Summary of activation overvoltage

In low- and medium-temperature fuel cells, activation overvoltage is the most important irreversibility and cause of voltage drop, and occurs mainly at the cathode. Activation overvoltage at *both* electrodes is important in cells using fuels other than hydrogen, such as methanol. At higher temperatures and pressures the activation overvoltage becomes less important.

Whether the voltage drop is significant at both electrodes or just the cathode, the size of the voltage drop is related to the current density i by the equation

$$\Delta V_{\text{act}} = A \ln\left(\frac{i}{b}\right)$$

where A and b are constants depending on the electrode and cell conditions. This equation is only valid for $i > b$.

3.5 Fuel Crossover and Internal Currents

Although the electrolyte of a fuel cell would have been chosen for its ion conducting properties, it will always be able to support very small amounts of electron conduction. The situation is akin to minority carrier conduction in semiconductors. Probably more important in a practical fuel cell is that some fuel will diffuse from the anode through the electrolyte to the cathode. Here, because of the catalyst, it will react directly with the oxygen, producing no current from the cell. This small amount of wasted fuel that migrates through the electrolyte is known as *fuel crossover*.

These effects – fuel crossover and internal currents – are essentially equivalent. The crossing over of one hydrogen molecule from anode to cathode where it reacts, wasting two electrons, amounts to exactly the same as two electrons crossing from anode to cathode internally, rather than as an external current. Furthermore, if the major loss in the cell is the transfer of electrons at the cathode interface, which is the case in hydrogen fuel cells, then the effect of both these phenomena on the cell voltage is also the same.

Although internal currents and fuel crossover are essentially equivalent, and the fuel crossover is probably more important, the effect of these two phenomena on the cell voltage is easier to understand if we just consider the internal current. We, as it were, assign the fuel crossover as 'equivalent to' an internal current. This is done in the explanation that follows.

The flow of fuel and electrons will be small, typically the equivalent of only a few mA cm^{-2}. In terms of energy loss this irreversibility is not very important. However, in low-temperature cells it does cause a very noticeable voltage drop at open circuit. Users of fuel cells can readily accept that the working voltage of a cell will be less than the theoretical 'no loss' reversible voltage. However, at open circuit, when no work is being done, surely it should be the same! With low-temperature cells, such as PEM cells, if operating on air at ambient pressure, the voltage will usually be at least 0.2 V less than the $\sim$1.2 V reversible voltage that might be expected.

If, as in the last section, we suppose that we have a fuel cell that only has losses caused by the 'activation overvoltage' on the cathode, then the voltage will be as in equation 3.4

$$V = E - A \ln\left(\frac{i}{i_0}\right)$$

For the case in point, a PEM fuel cell using air, at normal pressure, at about 30°C, reasonable values for the constants in this equation are

$$E = 1.2\,\text{V} \qquad A = 0.06\,\text{V} \quad \text{and} \quad i_0 = 0.04\,\text{mA cm}^{-2}$$

If we draw up a table of the values of V at low values of current density, we get the following values given in Table 3.2.

Now, because of the internal current density, the cell current density is *not zero*, even if the cell is open circuit. So, for example, if the internal current density is $1\,\text{mA cm}^{-2}$ then the open circuit would be 0.97 V, over 0.2 V (or 20%) less than the theoretical OCV. This large deviation from the reversible voltage is caused by the very steep initial fall in voltage that we can see in the curves of Figure 3.4. The steepness of the curve also explains another observation about low-temperature fuel cells, which is that the OCV is highly variable. The graphs and Table 3.2 tell us that a small change in fuel crossover and/or internal current, caused, for example, by a change in humidity of the electrolyte, can cause a large change in OCV.

The equivalence of the fuel crossover and the internal currents on the open circuits is an approximation, but is quite a fair one in the case of hydrogen fuel cells where the cathode activation overvoltage dominates. However, the term 'mixed potential' is often used to describe the situation that arises with fuel crossover.

The fuel crossover and internal current are obviously not easy to measure – an ammeter cannot be inserted in the circuit! One way of measuring it is to measure the consumption

Table 3.2 Cell voltages at low current densities

Current density (mA cm^{-2})	Voltage (V)
0	1.2
0.25	1.05
0.5	1.01
1.0	0.97
2.0	0.92
3.0	0.90
4.0	0.88
5.0	0.87
6.0	0.86
7.0	0.85
8.0	0.84
9.0	0.83

If the internal current density is 1.0 mA cm^{-2}, then the open circuit voltage will drop to 0.97 V

of reactant gases at open circuit. For single cells and small stacks, the very low gas usage rates cannot be measured using normal gas flow meters, and it will normally have to be done using bubble counting, gas syringes, or similar techniques. For example, a small PEM cell of area 10 cm^2 might have an open circuit hydrogen consumption of 0.0034 cm^3 sec^{-1}, at normal temperature and pressure (NTP) (author's measurement of a commercial cell). We know, from Avogadro's law that at standard temperature and pressure (STP) the volume of one mole of any gas is 2.43×10^4 cm^3. So the gas usage is 1.40×10^{-7} mol s^{-1}.

In Appendix 2 it is shown in equation A2.6 that the rate of hydrogen fuel usage in a single cell ($n = 1$) is related to current by the formula

$$\text{Gas usage} = \frac{I}{2F} \text{ moles s}^{-1}$$

so

$$I = \text{Gas usage} \times 2F$$

So, in this case the losses correspond to a current I of $1.40 \times 10^{-7} \times 2 \times 9.65 \times 10^4 = 27$ mA. The cell area is 10 cm^2, so this corresponds to a current density of 2.7 mA cm^{-2}. This current density gives the total of the current density equivalent of fuel lost because of fuel crossover and the actual internal current density.

If i_n is the value of this internal current density, then the equation for cell voltage that we have been using, equation 3.4, can be refined to

$$V = E - A \ln\left(\frac{i + i_n}{i_0}\right) \qquad [3.6]$$

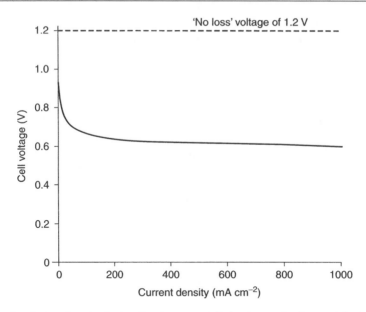

Figure 3.5 Graph showing the fuel cell voltage modelled using activation and fuel crossover/internal current losses only.

Using typical values for a low-temperature cell, that is, $E = 1.2\,\text{V}$, $A = 0.06\,\text{V}$, $i_0 = 0.04\,\text{mA cm}^{-2}$, and $i_n = 3\,\text{mA cm}^{-2}$, we get a graph of voltage against current density as shown in Figure 3.5.

The reader will observe that we are already getting a curve that is quite similar to Figure 3.1. The importance of this internal current is much less in the case of higher-temperature cells, because the exchange current density i_0 is so much higher, and so the initial fall in voltage is not nearly so marked.

So, to sum up, the internal current and/or diffusion of hydrogen through the electrolyte of a fuel cell is not usually of great importance in terms of operating efficiency. However, in the case of low-temperature cells, it has a very marked effect on the open circuit voltage.

3.6 Ohmic Losses

The losses due to the electrical resistance of the electrodes, and the resistance to the flow of ions in the electrolyte, are the simplest to understand and to model. The size of the voltage drop is simply proportional to the current, that is,

$$V = IR$$

In most fuel cells the resistance is mainly caused by the electrolyte, though the cell interconnects or bipolar plates (see Section 1.3) can also be important.

To be consistent with the other equations for voltage loss, the equation should be expressed in terms of current density. To do this we need to bring in the idea of the resistance corresponding to $1\,cm^2$ of the cell, for which we use the symbol r. (This quantity is called the *area-specific resistance* or *ASR*.) The equation for the voltage drop then becomes

$$\Delta V_{ohm} = ir \qquad\qquad [3.7]$$

where i is, as usual, the current density. If i is given in $mA\,cm^{-2}$, then the area-specific resistance, r, should be given in $k\Omega\,cm^2$.

Using the methods described below in Section 3.10, it is possible to distinguish this particular irreversibility from the others. Using such techniques it is possible to show that this 'ohmic' voltage loss is important in all types of cell, and especially important in the case of the solid oxide fuel cell (SOFC). Three ways of reducing the internal resistance of the cell are as follows:

- The use of electrodes with the highest possible conductivity.
- Good design and use of appropriate materials for the bipolar plates or cell interconnects. This issue has already been addressed in Section 1.3.
- Making the electrolyte as thin as possible. However, this is often difficult, as the electrolyte sometimes needs to be fairly thick as it is the support onto which the electrodes are built, or it needs to be wide enough to allow a circulating flow of electrolyte. In any case, it must certainly be thick enough to prevent any shorting of one electrode to another through the electrolyte, which implies a certain level of physical robustness.

3.7 Mass Transport or Concentration Losses

If the oxygen at the cathode of a fuel cell is supplied in the form of air, then it is self-evident that during fuel cell operation there will be a slight reduction in the concentration of the oxygen in the region of the electrode, as the oxygen is extracted. The extent of this change in concentration will depend on the current being taken from the fuel cell, and on physical factors relating to how well the air around the cathode can circulate, and how quickly the oxygen can be replenished. This change in concentration will cause a reduction in the partial pressure of the oxygen.

Similarly, if the anode of a fuel cell is supplied with hydrogen, then there will be a slight drop in pressure if the hydrogen is consumed as a result of a current being drawn from the cell. This reduction in pressure results from the fact that there will be a flow of hydrogen down the supply ducts and tubes, and this flow will result in a pressure drop due to their fluid resistance. This reduction in pressure will depend on the electric current from the cell (and hence H_2 consumption) and the physical characteristics of the hydrogen supply system.

In both cases, the reduction in gas pressure will result in a reduction in voltage. However, it is generally agreed among fuel cell researchers that there is no analytical solution to the problem of modelling the changes in voltage that works satisfactorily in all cases (Kim et al., 1995). One approach that does yield an equation that has some value and use is to see the effect of this reduction in pressure (or partial pressure) by revisiting equations 2.8 and 2.10. These give the change in OCV caused by a change in pressure of the reactants. In equation 2.10 we saw that the change in voltage caused by a change in hydrogen pressure only is

$$\Delta V = \frac{RT}{2F} \ln \left(\frac{P_2}{P_1} \right)$$

Now, the change in pressure caused by the use of the fuel gas can be estimated as follows. We postulate a limiting current density i_1 at which the fuel is used up at a rate equal to its maximum supply speed. The current density cannot rise above this value, because the fuel gas cannot be supplied at a greater rate. At this current density the pressure would have just reached zero. If P_1 is the pressure when the current density is zero, and we assume that the pressure falls linearly down to zero at the current density i_1, then the pressure P_2 at any current density i is given by the formula

$$P_2 = P_1 \left(1 - \frac{i}{i_1} \right)$$

If we substitute this into equation 2.10 (given above), we obtain

$$\Delta V = \frac{RT}{2F} \ln \left(1 - \frac{i}{i_1} \right) \qquad [3.8]$$

This gives us the voltage change due to the mass transport losses. We have to be careful with signs here; equation 2.10 and 3.8 are written in terms of a voltage *gain*, and the term inside the brackets is always less than 1. So if we want an equation for voltage *drop*, we should write it as

$$\Delta V_{\text{trans}} = -\frac{RT}{2F} \ln \left(1 - \frac{i}{i_1} \right)$$

Now the term that in this case is $RT/2F$ will be different for different reactants, as should be evident from equation 2.8. For example, for oxygen it will be $RT/4F$. In general, we may say that the concentration or mass transport losses are given by the equation

$$\Delta V_{\text{trans}} = -B \ln \left(1 - \frac{i}{i_1} \right) \qquad [3.9]$$

where B is a constant that depends on the fuel cell and its operating state. For example, if B is set to 0.05 V and i_1 to 1000 mA cm^{-2}, then quite a good fit is made to curves such as those of Figures 3.1 and 3.2. However, this theoretical approach has many weaknesses, especially in the case of fuel cells supplied with air rather than pure oxygen – which is the vast majority. There are also problems with lower-temperature cells, and those supplied with hydrogen mixed with other gases such as carbon dioxide for the fuel. No account is taken for the production and removal of reaction products, such as water, and neither is any account taken of the build-up of nitrogen in air systems.

Another approach that has no claim for a theoretical basis, but is entirely empirical, has become more favoured lately, and yields an equation that fits the results very well

(Kim et al., 1995 and Laurencelle et al., 2001). This approach uses equation 3.10 below because it gives a very good fit to the results, provided the constants m and n are chosen properly.

$$\Delta V_{trans} = m \exp(ni) \qquad [3.10]$$

The value of m will typically be about 3×10^{-5} V, and n about 8×10^{-3} cm^2 mA^{-1}. Although the equations 3.9 and 3.10 look very different, if the constants are chosen carefully the results can be quite similar. However, equation 3.10 can be used to give a better fit to measured results, and so this will be used in the rest of this chapter, and appears to be quite widely used in the fuel cell community.

The mass transport or concentration overvoltage is particularly important in cases where the hydrogen is supplied from some kind of reformer, as there might be a difficulty in increasing the rate of supply of hydrogen quickly to respond to demand. Another important case is at the air cathode, if the air supply is not well circulated. A particular problem is that the nitrogen that is left behind after the oxygen is consumed can cause a mass transport problem at high currents – it effectively blocks the oxygen supply. In proton exchange membrane fuel cells (PEMFCs), the removal of water can also be a cause of mass transport or concentration overvoltage.

3.8 Combining the Irreversibilities

It is useful to construct an equation that brings together all these irreversibilities. We can do so and arrive at the following equation for the operating voltage of a fuel cell at a current density i.

$$V = E - \Delta V_{ohm} - \Delta V_{act} - \Delta V_{trans}$$

$$V = E - ir - A \ln\left(\frac{i + i_n}{i_0}\right) + m \exp(ni) \qquad [3.11]$$

In this equation,

   E is the reversible OCV given by equations 2.1 and 3.1

   i_n is the internal and fuel crossover equivalent current density described in Section 3.5

   A is the slope of the Tafel line as described in Section 3.4.2

   i_0 is either the exchange current density at the cathode if the cathodic overvoltage is much greater than the anodic or it is a function of both exchange current densities as given in equation 3.5

m and n are the constants in the mass-transfer overvoltage equation 3.10 as discussed in Section 3.7

   r is the area-specific resistance, as described in Section 3.6.

Although correct, this equation is often simplified in a useful and practical way. The crossover current i_n is usually very small, and although useful for explaining the initial fall in voltage, it has little impact on operating losses of fuel cells at working currents. It is also very difficult to measure. We can also largely account for the term resulting from i_0 if we assume that the current is always greater than this exchange current. Because of the crossover current, this is nearly always the case. The equation for the activation overvoltage is rearranged to

$$\Delta V_{\text{act}} = A \ln \left(\frac{i}{i_0} \right) = A \ln(i) - A \ln(i_0) \qquad [3.12]$$

Because the second half of this equation is a constant, we can deal with this by postulating a real, practical, open circuit voltage E_{oc} that is given by the equation

$$E_{\text{oc}} = E + A \ln(i_0) \qquad [3.13]$$

E is the theoretical, reversible, open circuit voltage given by equations 2.1 and 3.1. Note that E_{oc} will always be *less* than E because i_0, being small, will generate negative logarithms. If we substitute equations 3.12 and 3.13 into 3.11 and remove i_n, we obtain

$$V = E_{\text{oc}} - ir - A \ln(i) + m \exp(ni) \qquad [3.14]$$

This equation is simple, yet has been found to give an excellent fit with the results of real fuel cells. Example values of the constants are given in Table 3.3 for two different types of fuel cell. Those for one cell from a Ballard Mark V fuel cell stack are taken from Laurencelle et al. (2001).

It is simple to model this equation using a spreadsheet (such as EXCEL), or programs such as MATLAB, or graphics calculators. However, it must be borne in mind that the logarithmic model does not work at very low currents, especially at zero. It is best to start the plots with a current of $1.0\,\text{mA}\,\text{cm}^{-2}$. As an example, we have given below the MATLAB script file that was used to produce the graph in Figure 3.1.

```
Eoc=1.031, A=0.03, r=0.000245, m=2.11E-5, n=0.008
i=linspace(1,1000,200)
v= Eoc - r*i - A*log(i) - m*exp(n*i)

plot (i,v)
```

Table 3.3   Example constants for equation 3.14

Constant	Ballard Mark V PEMFC at 70°C	High temperature, e.g. SOFC
E_{oc} (V)	1.031	1.01
r (kΩ cm^2)	2.45×10^{-4}	2.0×10^{-3}
A (V)	0.03	0.002
m (V)	2.11×10^{-5}	1.0×10^{-4}
n (cm^2 mA^{-1})	8×10^{-3}	8×10^{-3}

The approach we have taken in considering these different losses has been moderately rigorous and mathematical, suitable for an initial understanding of the issues. It is of course possible to be much more rigorous. For such a rigorous theoretical approach, yet soundly based upon the performance of a real fuel cell stack, the reader is referred to Amphlett et al. (1995).

3.9 The Charge Double Layer

The 'charge double layer' is a complex and interesting electrode phenomenon, and whole books have been written on the topic (Bokins et al., 1975). However, a much briefer account will suffice in this context. The charge double layer is important in understanding the dynamic electrical behaviour of fuel cells.

Whenever two different materials are in contact, there is a build-up of charge on the surfaces or a charge transfer from one to the other. For example, in semiconductor materials, there is a diffusion of 'holes' and electrons across junctions between n-type and p-type materials. This forms a 'charge double layer' at the junction, of electrons in the p-type region and 'holes' in the n-type, which has a strong impact on the behaviour of semiconductor devices. In electrochemical systems, the charge double layer forms in part due to diffusion effects, as in semiconductors, and also because of the reactions between the electrons in the electrodes and the ions in the electrolyte, and also as a result of applied voltages. For example, the situation in Figure 3.6 might arise at the cathode of an acid electrolyte fuel cell. Electrons will collect at the surface of the electrode and H^+ ions will be attracted to the surface of the electrolyte. These electrons and ions, together with the O_2 supplied to the cathode will take part in the cathode reaction

$$O_2 + 4e^- + 4H^+ \longrightarrow 2H_2O$$

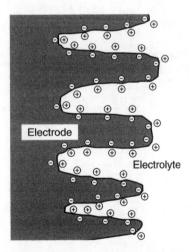

Figure 3.6   The charge double layer at the surface of a fuel cell cathode.

The probability of the reaction taking place obviously depends on the density of the charges, electrons, and H^+ ions on the electrode and electrolyte surfaces. The more the charge, the greater is the current. However, any collection of charge, such as of these electrons and H^+ ions at the electrode/electrolyte interface will generate an electrical voltage. The voltage in this case is the 'activation overvoltage' we have been considering in Section 3.4. So the charge double layer gives an explanation of why the activation overvoltage occurs. It shows that a charge double layer needs to be present for a reaction to occur, that more charge is needed if the current is higher, and so the overvoltage is higher if the current is greater. We can also see that the catalytic effect of the electrode is important, as an effective catalyst will also increase the probability of a reaction – so that a higher current can flow without such a build-up of charge.

The layer of charge on or near the electrode–electrolyte interface is a store of electrical charge and energy, and as such behaves much like an electrical capacitor. If the current changes, it will take some time for this charge (and its associated voltage) to dissipate (if the current reduces) or to build up (if there is a current increase). So, the activation overvoltage does not immediately follow the current in the way that the ohmic voltage drop does. The result is that if the current suddenly changes, the operating voltage shows an immediate change due to the internal resistance, but moves fairly slowly to its final equilibrium value. One way of modelling this is by using an equivalent circuit, with the charge double layer represented by an electrical capacitor. The capacitance of a capacitor is given by the formula

$$C = \varepsilon \frac{A}{d}$$

where ε is the electrical permittivity, A is the surface area, and d is the separation of the plates. In this case, A is the real surface area of the electrode, which is several thousand times greater than its length × width. Also d, the separation, is very small, typically only a few nanometres. The result is that, in some fuel cells, the capacitance will be of the order of a few Farads, which is high in terms of capacitance values. (In electrical circuits, a 1 μF capacitor is on the large size of average.) The connection between this capacitance, the charge stored in it, and the resulting activation overvoltage, leads to an equivalent circuit as shown in Figure 3.7.

The resistor Rr models the ohmic losses. A change in current gives an immediate change in the voltage drop across this resistor. The resistor Ra models the activation

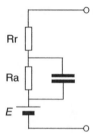

Figure 3.7   Simple equivalent circuit model of a fuel cell.

overvoltage and the capacitor 'smoothes' any voltage drop across this resistor. If we were to include the concentration overvoltage this would be incorporated into this resistor too.

Generally speaking, the effect of this capacitance resulting from the charge double layer gives the fuel cell a 'good' dynamic performance in that the voltage moves gently and smoothly to a new value in response to a change in current demand. It also permits a simple and effective way to distinguish between the main types of voltage drop, and hence to analyse the performance of a fuel cell, which is described in the next section.

3.10 Distinguishing the Different Irreversibilities

At various points in this chapter it has been asserted that 'such and such an overvoltage is important in so and so conditions'. For example, it has been said that in SOFC the ohmic voltage drop is more important than activation losses. What is the evidence for these claims?

Some of the evidence is derived from experiments using specialised electrochemical test equipment such as half-cells, which is beyond the scope of this book to describe. (For such information, see, for example, Greef et al. (1985).) A method that is fairly straightforward to understand is that of *electrical impedance spectroscopy*. A variable frequency alternating current is driven through the cell, the voltage is measured, and the impedance calculated. At higher frequencies, the capacitors in the circuits will have less impedance. By plotting graphs of impedance against frequency, it is possible to find the values of the equivalent circuit of Figure 3.7. It is sometimes even possible to distinguish between the losses at the cathode and the anode, and certainly between mass transport and activation-type losses. Wagner (1998) gives a particularly good example of this type of experiment as applied to fuel cells. However, because the capacitances are large and the impedances are small, special signal generators and measurement systems are needed. Frequencies as low as 10 mHz may be used, so the experiments are often rather slow.

The *current interrupt technique* is an alternative that can be used to give accurate quantitative results (Lee, 1998), but is also used to give quick qualitative indications. It can be performed using standard low-cost electronic equipment. Suppose a cell is providing a current at which the concentration (or mass transport) overvoltage is negligible. The ohmic losses and the activation overvoltage will in this case cause the voltage drop. Suppose now that the current is suddenly cut off. The charge double layer will take some time to disperse, and so will the associated overvoltage. However, the ohmic losses will *immediately* reduce to zero. We would therefore expect the voltage to change as in Figure 3.8 if the load was suddenly disconnected from the cell.

The simple circuit needed to perform this current interrupt test is shown in Figure 3.9. The switch is closed and the load resistor adjusted until the desired test current is flowing. The storage oscilloscope is set to a suitable timebase, and the load current is then switched off. The oscilloscope triggering will need to be set so that the oscilloscope moves into 'hold' mode – though with some cells the system is so slow that this can be done by hand. The two voltages Vr and Va are then read off the screen. Although the method is simple, when obtaining quantitative results, care must be taken, as it is possible to overestimate Vr by missing the point where the vertical transition ends. The oscilloscope timebase setting

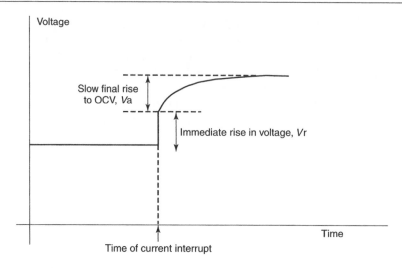

Figure 3.8   Sketch graph of voltage against time for a fuel cell after a current interrupt.

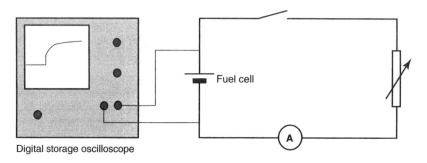

Figure 3.9   Simple circuit for performing a current interrupt test.

needed will vary for different fuel cell types, depending on the capacitance, as is done in the three example interrupt tests overleaf. These issues are addressed, for example, by Büchi et al. (1995).[2]

The current interrupt test is particularly easy to perform with single cells and small fuel cell stacks. With larger cells the switching of the higher currents can be problematic. Current interrupts and electrical impedance spectroscopy give us two powerful methods of finding the causes of fuel cell irreversibilities, and both methods are widely used.

Typical results from three current interrupt tests are shown in Figures 3.10, 3.11, and 3.12. These three examples are shown because of the clear *qualitative* indication they give of the importance of the different types of voltage drop we have been describing. Because oscilloscopes do not show vertical lines, the appearance is slightly different

---

[2] This paper also outlines an interesting variation on the current interrupt test, in which a pulse of current is applied to the cell.

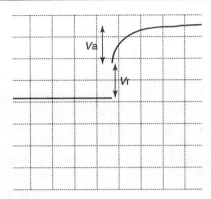

Figure 3.10 Current interrupt test for a low-temperature, ambient pressure, hydrogen fuel cell. The ohmic and activation voltage drops are similar. (Time scale $0.2\,\text{s/div}^{-1}$, $i = 100\,\text{mA cm}^{-2}$.)

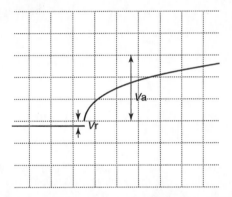

Figure 3.11 Current interrupt test for a direct methanol fuel cell. There is a large activation overvoltage at **both** electrodes. As a result, the activation overvoltage is much greater than the ohmic, which is barely discernible. (Time scale $2\,\text{s/div}$. $i = 10\,\text{mA cm}^{-2}$.)

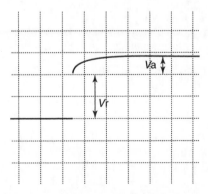

Figure 3.12 Current interrupt test for a small solid oxide fuel cell working at about $700°C$. The large immediate rise in voltage shows that most of the voltage drop is caused by ohmic losses. (Time scale $0.02\,\text{s/div}$. $I = 100\,\text{mA cm}^{-2}$.)

from Figure 3.8, as there is no vertical line corresponding to Vr. The tests were done on three different types of fuel cell, a PEM hydrogen fuel cell, a direct methanol fuel cell, and a solid oxide fuel cell. In each case the *total* voltage drop was about the same, though the current density certainly was not.

These three examples   give a good summary of the causes of voltage losses in fuel cells. Concentration or mass transport losses are important only at higher currents, and in a well-designed system, with good fuel and oxygen supply, they should be very small at rated currents. In low-temperature hydrogen fuel cells, the activation overvoltage (at the cathode) is important, especially at low currents, but the ohmic losses play an important part too, and the activation and ohmic loses are similar (Figure 3.10). In fuel cells using fuels such as methanol, there is a considerable activation overvoltage at *both* the anode and cathode, and so the activation overvoltage dominates at all times (Figure 3.11). On the other hand, in higher-temperature cells the activation overvoltage becomes much less important and ohmic losses is the main problem (Figure 3.12).

We now have a sufficient understanding of the principles of fuel cell operation, and in the following chapters we look much more closely at the practical details of different types of fuel cell systems.

References

Amphlett J.C., Baument R.M., Mann R.E., Peppley B.A., Roberge P.R., and Harris T.J. (1995) "Performance modelling of the Ballard Mark V solid polymer electrolyte fuel cell", *Journal of the Electrochemical Society*, **142**(1), 1–5.

Appleby A.J and Foulkes F.R. (1993) *A Fuel Cell Handbook*, 2nd ed., Kreiger Publishing Co., p. 22.

Bloom H. and Cutman F. (eds) (1981) *Electrochemistry*, Plenum Press, New York, p. 121.

Bokins J.O'M., Conway B.E., and Yeager E.(eds) (1975) *Comprehensive Treatment of Electrochemistry*, Vol. 1, Plenum Press, New York.

Büchi F.N., Marek A., and Schere G.G. (1995) "In-situ membrane resistance measurements in polymer electrolyte fuel cells by fast auxiliary current pulses", *Journal of the Electrochemical Society*, **142**(6), 1895–1901.

Davies C.W. (1967) *Electrochemistry*, Newnes, London, p. 188.

Greef R., Peat R., Peter L.M., Pletcher D., and Robinson J. (1985) *Instrumental Methods in Electrochemistry*, Ellis Horwood/John Wiley & Sons.

Kim J., Lee S-M., Srinivasan S., and Chamberlin C.E. (1995) "Modeling of proton exchange membrane fuel cell performance with an empirical equation", *Journal of the Electrochemical Society*, **142**(8), 2670–2674.

Laurencelle F., Chahine R., Hamelin J., Fournier M., Bose T.K., and Laperriere A. (2001) "Characterization of a Ballard MK5-E proton exchange membrane stack", *Fuel Cells*, **1**(1), 66–71.

Lee C.G., Nakano H., Nishina T., Uchida I., and Kuroe S. (1998) "Characterisation of a $100 \, cm^2$ class molten carbonate fuel cell with current interruption", *Journal of the Electrochemical Society*, **145**(8), 2747–2751.

McDougall A. (1976) *Fuel Cells*, Macmillan, London, pp. 37–41.

Wagner N., Schnarnburger N., Mueller B., and Lang M. (1998) "Electrochemical impedance spectra of solid-oxide fuel cells and polymer membrane fuel cells", *Electrochimica Acta*, **43**(24), 3785–3790.

4

Proton Exchange Membrane Fuel Cells

4.1 Overview

The proton exchange membrane fuel cell (PEMFC), also called the *solid polymer fuel cell* (SPFC), was first developed by General Electric in the United States in the 1960s for use by NASA on their first manned space vehicles.

The electrolyte is an ion conduction polymer, described in more detail in Section 4.2. Onto each side is bonded a catalysed porous electrode. The anode–electrolyte–cathode assembly is thus one item, and is very thin, as shown in Figure 4.9. These 'membrane electrode assemblies' (or MEAs) are connected in series, usually using bipolar plates, as in Figure 1.8.

The mobile ion in the polymers used is an H^+ ion or proton, so the basic operation of the cell is essentially the same as for the acid electrolyte fuel cell, as shown in Figure 1.3.

The polymer electrolytes work at low temperatures, which has the advantage that a PEMFC can start quickly. The thinness of the MEAs means that compact fuel cells can be made. Further advantages are that there are no corrosive fluid hazards and that the cell can work in any orientation. This means that the PEMFC is particularly suitable for use in vehicles and in portable applications.

The early versions of the PEMFC, as used in the NASA Gemini spacecraft, had a lifetime of only about 500 h, but that was sufficient for those limited early missions. The development program continued with the incorporation of a new polymer membrane in 1967 called *Nafion*, a registered trademark of Dupont. This type of membrane, outlined in Section 4.2, became standard for the PEMFC, as it still is today.

However, the problem of water management in the electrolyte, which we consider in some detail in Section 4.4 below, was judged too difficult to manage reliably, and for the Apollo vehicles, NASA selected the 'rival' alkaline fuel cell (Warshay, 1990). General Electric also chose not to pursue commercial development of the PEMFC, probably because the costs were seen as higher than other fuel cells, such as the phosphoric acid fuel cell then being developed. At that time catalyst technology was such that 28 mg of

Fuel Cell Systems Explained, Second Edition James Larminie and Andrew Dicks
© 2003 John Wiley & Sons, Ltd   ISBN: 0-470-84857-X

Figure 4.1 Four PEM fuel cell stacks illustrating developments through the 1990s. The 1989 model on the left has a power density of $100\,W\,L^{-1}$. The 1996 model on the right has a power density of $1.1\,kW\,L^{-1}$. (Reproduced by kind permission of Ballard Power Systems.)

platinum was needed for each square centimeter of electrode, compared to $0.2\,mg\,cm^{-2}$ or less now.

The development of proton exchange membrane (PEM) cells went more or less into abeyance in the 1970s and early 1980s. However, in the latter half of the 1980s and the early 1990s, there was a renaissance of interest in this type of cell (Prater, 1990). A good deal of the credit for this must go to Ballard Power Systems of Vancouver, Canada and to the Los Alamos National Laboratory in the United States.[1]

The developments over recent years have brought the current densities up to around $1\,A\,cm^{-2}$ or more, while at the same time reducing the use of platinum by a factor of over 100. These improvements have led to huge reduction in cost per kilowatt of power, and much improved power density, as can be seen from the fuel cell stacks in Figure 4.1.

PEMFCs are being actively developed for use in cars and buses, as well as for a very wide range of portable applications, and also for combined heat and power (CHP) systems. A sign of the dominance of this type of cell is that they are again the preferred option for NASA, and the new Space Shuttle Orbiter will use PEM cells (Warshay et al.,

---

[1] The story of the Ballard fuel cell company is told in Koppel (1999).

1997). It could be argued that PEMFCs exceed all other electrical energy generating technologies with respect to the scope of their possible applications. They are a possible power source at a few watts for powering mobile phones and other electronic equipment such as computers, right through to a few kilowatt for boats and domestic systems, to tens of kilowatt for cars, to hundreds of kilowatt for buses and industrial CHP systems.

Within this huge range of applications, two aspects of PEM fuel cells are more or less similar. These are

- the electrolyte used, which is described in Section 4.2,
- the electrode structure and catalyst, which we cover in Section 4.3.

However, other important aspects of fuel cell design vary greatly depending on the application and the outlook of the designer. The most important of these are as follows:

- Water management – a vital topic for PEMFCs dealt with in Section 4.4.
- The method of cooling the fuel cell, which we discuss in Section 4.5.
- The method of connecting cells in series. The bipolar plate designs vary greatly, and some fuel cells use altogether different methods. We discuss these in Section 4.6.
- The question of at what pressure to operate the fuel cell, which we consider in Section 4.7.
- The reactants used is also an important issue – pure hydrogen is not the only possible fuel, and oxygen can be used instead of air. This is briefly discussed in Section 4.8.

Finally, in Section 4.9 we look at some sample PEM fuel cell systems. There are, of course, other important questions, such as 'where does the hydrogen come from?' but these questions are large and apply to all fuel cell types, and so are dealt with in separate chapters.

4.2 How the Polymer Electrolyte Works

The different companies producing polymer electrolyte membranes have their own special tricks, mostly proprietary. However, a common theme is the use of sulphonated fluoro-polymers, usually fluoroethylene. The most well known and well established of these is Nafion ($^{®}$Dupont), which has been developed through several variants since the 1960s. This material is still the electrolyte against which others are judged, and is in a sense an 'industry standard'. Other polymer electrolytes function in a similar way.[2]

The construction of the electrolyte material is as follows. The starting point is the basic, simplest to understand, man-made polymer – polyethylene. Its molecular structure based on ethylene is shown in Figure 4.2.

This basic polymer is modified by substituting fluorine for the hydrogen. This process is applied to many other compounds and is called *perfluorination*. The 'mer' is called *tetrafluoroethylene*.[3] The modified polymer, shown in Figure 4.3, is polytetrafluoroethylene, or PTFE. It is also sold as Teflon, the registered trademark of ICI. This remarkable

---

[2] For a review of work with other types of proton exchange membrane, see Rozière and Jones, 2001.

[3] 'Tetra' indicates that all four hydrogens in each ethylene group have been replaced by fluorine.

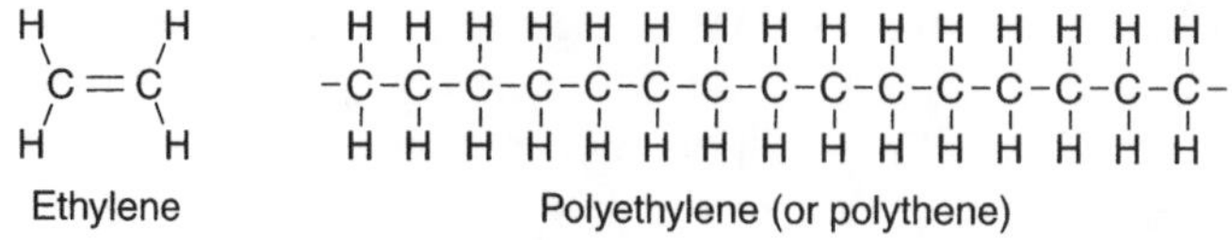

Figure 4.2 Structure of polyethylene.

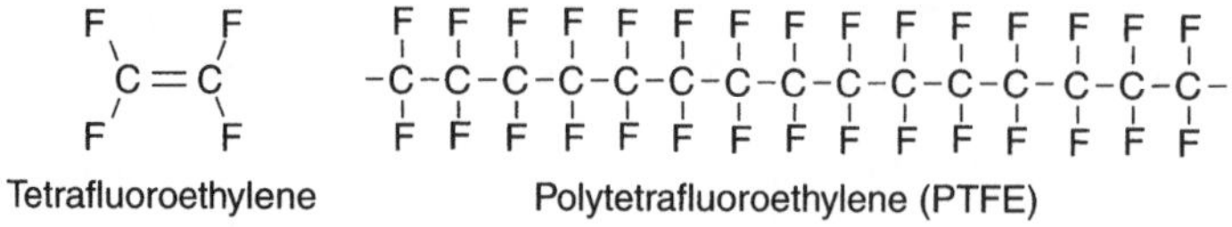

Figure 4.3 Structure of PTFE.

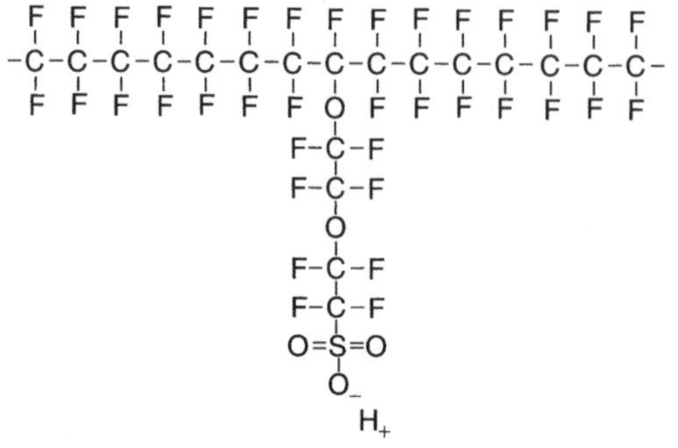

Figure 4.4 Example structure of a sulphonated fluoroethylene (also called *perfluorosulphonic acid PTFE copolymer*).

material has been very important in the development of fuel cells. The strong bonds between the fluorine and the carbon make it durable and resistant to chemical attack. Another important property is that it is strongly hydrophobic, and so it is used in fuel cell electrodes to drive the product water out of the electrode, and thus it prevents flooding. It is used in this way in phosphoric acid and alkali fuel cells, as well as in PEMFCs. (The same property gives it a host of uses in outdoor clothing and footwear.)

However, to make an electrolyte, a further stage is needed. The basic PTFE polymer is 'sulphonated' – a side chain is added, ending with sulphonic acid HSO_3. Sulphonation of complex molecules is a widely used technique in chemical processing. It is used, for example, in the manufacture of detergent. One possible side chain structure is shown in Figure 4.4 – the details vary for different types of Nafion, and with different manufacturers of these membranes. The methods of creating and adding the side chains are proprietary, though one modern method is discussed in Kiefer et al. (1999).

The HSO_3 group added is ionically bonded, and so the end of the side chain is actually an SO_3^- ion. For this reason, the resulting structure is called an *ionomer*. The result of the presence of these SO_3^- and H^+ ions is that there is a strong mutual attraction between the + and − ions from each molecule. The result is that the side chain molecules tend to 'cluster' within the overall structure of the material. Now, a key property of sulphonic acid is that it is highly hydrophyllic – it attracts water. (This is why it is used in detergents; it makes one end of the molecule mix readily with water, while the other end attaches to the dirt.) In Nafion, this means we are creating hydrophyllic regions within a generally hydrophobic substance, which is bound to create interesting results.

The hydrophyllic regions around the clusters of sulphonated side chains can lead to the absorption of large quantities of water, increasing the dry weight of the material by up to 50%. Within these hydrated regions, the H^+ ions are relatively weakly attracted to the SO_3^- group and are able to move. This creates what is essentially a dilute acid. The resulting material has different phases – dilute acid regions within a tough and strong hydrophobic structure. This 'micro-phase separated morphology' is illustrated in Figure 4.5. Although the hydrated regions are somewhat separate, it is still possible for the H^+ ions to move through the supporting long molecule structure. However, it is easy to see that for this to happen the hydrated regions must be as large as possible. In a well-hydrated electrolyte, there will be about 20 water molecules for each SO_3^- side chain. This will typically give a conductivity of about $0.1\,S\,cm^{-1}$. As the water content falls, the conductivity falls in a more or less linear fashion.

From the point of view of fuel cell use, the main features of Nafion and other fluorosulphonate ionomers are that

- they are chemically highly resistant,
- they are strong (mechanically), and so can be made into very thin films, down to $50\,\mu m$,
- they are acidic,

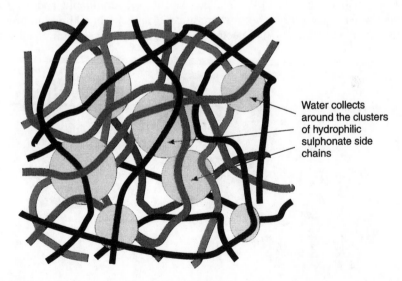

Water collects around the clusters of hydrophilic sulphonate side chains

Figure 4.5 The structure of Nafion-type membrane materials. Long chain molecules containing hydrated regions around the sulphonated side chains.

- they can absorb large quantities of water,
- if they are well hydrated, the H^+ ions can move quite freely within the material – they are good proton conductors.

4.3 Electrodes and Electrode Structure

The best catalyst for both the anode and the cathode is platinum. In the early days of PEMFC development, this catalyst was used at the rate of $28\,mg\,cm^{-2}$ of platinum. This high rate of usage led to the myth, still widely held, that platinum is a major factor in the cost of a PEMFC. In recent years the usage has been reduced to around $0.2\,mg\,cm^{-2}$, yet with power increasing. At such 'loadings' the basic raw material cost of the platinum metal in a 1-kW PEMFC would be about \$10 – a small proportion of the total cost.

The basic structure of the electrode in different designs of PEMFC is similar, though of course details vary. The anodes and the cathodes are essentially the same too – indeed, in many PEMFCs they are identical.

The platinum catalyst is formed into very small particles on the surface of somewhat larger particles of finely divided carbon powders. A carbon-based powder, XC72 (®Cabot), is widely used. The result, in a somewhat idealised form, is shown in Figure 4.6. A real picture of this type of supported catalyst is shown in Figure 1.6. The platinum is highly divided and spread out, so that a very high proportion of the surface area will be in contact with the reactants.

For the next stage, one of two alternative routes is used, though the end result is essentially the same in both cases.

In the *separate electrode method*, the carbon-supported catalyst is fixed, using proprietary techniques, to a porous and conductive material such as carbon cloth or carbon paper. PTFE will often be added also, because it is hydrophobic and so will expel the product water to the surface from where it can evaporate. In addition to providing the basic mechanical structure for the electrode, the carbon paper or cloth also diffuses the

Figure 4.6   The structure (idealised) of carbon-supported catalyst.

gas onto the catalyst and so is often called the *gas diffusion layer*. An electrode is then fixed to each side of a piece of polymer electrolyte membrane. A fairly standard procedure for doing this is described in several papers (Lee et al., 1998a). First, the electrolyte membrane is cleaned by immersing it in boiling 3% hydrogen peroxide in water for 1 h, and then in boiling sulphuric acid for the same time, to ensure as full protonation of the sulphonate group as possible. The membrane is then rinsed in boiling deionised water for 1 h to remove any remaining acid. The electrodes are then put onto the electrolyte membrane and the assembly is hot pressed at 140°C at high pressure for 3 min. The result is a complete membrane electrode assembly, or MEA.

The alternative method involves *building the electrode directly onto the electrolyte*. The platinum on carbon catalyst is fixed directly to the electrolyte, thus manufacturing the electrode directly onto the membrane, rather than separately. The catalyst, which will often (but not always) be mixed with hydrophobic PTFE, is applied to the electrolyte membrane using rolling methods (Bever et al., 1998), or spraying (Giorgi et al., 1998) or an adapted printing process (Ralph et al., 1997). With the exception of the first paper describing this idea (Wilson and Gottesfeld, 1992), the literature generally gives very little detail of the method used, usually referring to 'proprietary techniques'. Once the catalyst is fixed to the membrane, a gas diffusion layer must be applied. This will be carbon cloth or paper, and about 0.2 to 0.5-mm thick, as is used in the separate electrodes. 'Gas diffusion layer' is a slightly misleading name for this part of the electrode, as it does much more than diffuse the gas. It also forms an electrical connection between the carbon-supported catalyst and the bipolar plate, or other current collector. In addition, it carries the product water away from the electrolyte surface and also forms a protective layer over the very thin (typically ~30 m) layer of catalyst. This gas diffusion layer may or may not be an integral part of the membrane electrode assembly.

Whichever of these two methods is chosen, the result is a structure as shown, in idealised form, in Figure 4.7. The carbon-supported catalyst particles are joined to the electrolyte on one side, and the gas diffusion (+ current collecting, water removing, physical support) layer on the other. The hydrophobic PTFE that is needed to remove water from the catalyst is not shown explicitly, but will almost always be present.

Two further points need to be discussed. The **first** relates to the impregnation of the electrode with electrolyte material. In Figure 4.8, a portion of the catalyst/electrode region is shown enlarged. It can be seen that the electrolyte material spreads out over the catalyst. It does not cover the catalyst, but makes a direct connection between catalyst and electrolyte. This increases the performance of the MEA markedly (Lee, 1998b), by promoting the important 'three-phase contact' between reactant gas, electrolyte, and electrode catalyst. This light covering of the catalyst with the electrolyte is achieved by brushing the electrode with a solubilised form of the electrolyte. In the case of the 'separate electrode' method, this is done before the electrode is hot pressed onto the membrane. In the case of the integral membrane/electrode method, it is done before the gas diffusion layer is added.

The **second** relates to the selection of the gas diffusion layer. We have seen that this is generally either a carbon paper or carbon cloth material. Carbon paper (e.g. ELAT™, TORAY™, and CARBEL™ are widely used brands) is chosen when it is required to make the cell as thin as possible in compact designs. In an interesting paper, Lee et al. (1999)

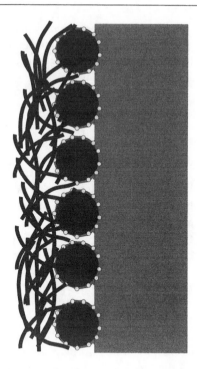

Figure 4.7   Simplified and idealised structure of a PEM fuel cell electrode.

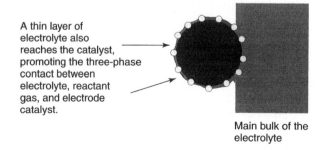

A thin layer of
electrolyte also
reaches the catalyst,
promoting the three-phase
contact between
electrolyte, reactant
gas, and electrode
catalyst.

Main bulk of the
electrolyte

Figure 4.8   Enlargement of part of Figure 4.7, showing that the electrolyte reaches out to the
catalyst particles.

compare the performance of cells using these three materials under different levels of
compression. Carbon cloths are thicker, and so will absorb a little more water, and also
simplify mechanical assembly, since they will fill small gaps and irregularities in bipolar
plate manufacture and assembly. On the other hand, they will slightly expand out into
the gas diffusion channels on the bipolar plates, which may be significant if these have
been made very shallow, as will be the case in a very thin design. In the case of the very
low power small PEMFCs to be discussed in Sections 4.4 and 4.5, even thicker materials

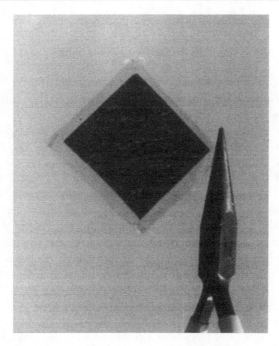

Figure 4.9 An example of a membrane electrode assembly (MEA). The membrane is a little larger than the electrodes that are attached. These electrodes have the gas diffusion layer attached, which gives it a 'grainy' texture. The membrane is typically 0.05 to 0.1 mm thick, the electrodes are about 0.03 mm thick, and the gas diffusion layer is between 0.2 and 0.5-mm thick.

such as carbon felt may be used, as a comparatively wide cell separation is needed for air circulation.

So, we have the heart of our proton exchange membrane fuel cell, which is the **membrane electrode assembly** shown in Figure 4.9. No matter how this is made, or which company made it, the MEA will look similar, work in essentially the same way, and require similar care in use. However, the way in which working fuel cells stacks are made around these MEAs varies enormously. In the following four sections, we look at some of these different approaches, taking as our themes the ways in which the major engineering problems of the PEMFC are solved.

4.4 Water Management in the PEMFC

4.4.1 Overview of the problem

It will be clear from the description of a proton exchange membrane given in Section 4.2 that there must be sufficient water content in the polymer electrolyte. The proton conductivity is directly proportional to the water content. However, there must not be so much water that the electrodes that are bonded to the electrolyte flood, blocking the pores in the electrodes or the gas diffusion layer. A balance is therefore needed, which takes care to achieve.

In the PEMFC, water forms at the cathode – revisit Figure 1.3 if you are not sure why. In an ideal world, this water would keep the electrolyte at the correct level of hydration. Air would be blown over the cathode, and apart from supplying the necessary oxygen it would dry out any excess water. Because the membrane electrolyte is so thin, water would diffuse from the cathode side to the anode, and throughout the whole electrolyte a suitable state of hydration would be achieved without any special difficulty. This happy situation can sometimes be achieved, but needs a good engineering design to bring to pass.

There are several complications. One is that during the operation of the cell the H^+ ions moving from the anode to the cathode (see Figure 1.3) pull water molecules with them. This process is sometimes called *electro-osmotic drag*. Typically, between one and five water molecules are 'dragged' for each proton (Zawodzinski et al., 1993 and Ren and Gottesfeld, 2001). This means that, especially at high current densities, the anode side of the electrolyte can become dried out – even if the cathode is well hydrated. Another major problem is the drying effect of air at high temperatures. We will show this quantitatively in Section 4.4.2, but it suffices to say at this stage that at temperatures of over approximately 60°C, the air will *always* dry out the electrodes faster than the water is produced by the H_2/O_2 reaction. One common way to solve these problems is to humidify the air, the hydrogen, or both, before they enter the fuel cell. This may seem bizarre, as it effectively adds a by-product to the inputs in the process, and there cannot be many other processes where this is done. However, we will see that this is sometimes needed and greatly improves the fuel cell performance.

Yet another complication is that the water balance in the electrolyte must be correct throughout the cell. In practice, some parts may be just right, others too dry, and others flooded. An obvious example of this can be seen if we think about the air as it passes through the fuel cell. It may enter the cell quite dry, but by the time it has passed over some of the electrodes it may be about right. However, by the time it has reached the exit it may be so saturated that it cannot dry off any more excess water. Obviously, this is more of a problem when designing larger cells and stacks.

The different water movements are shown in Figure 4.10. Fortunately, all these water movements are predictable and controllable. Starting from the top of Figure 4.10, the water production and the water drag are both directly proportional to the current. The water evaporation can be predicted with care, using the theory outlined below in Section 4.4.2. The back diffusion of water from cathode to anode depends on the thickness of the electrolyte membrane and the relative humidity of each side. Finally, if external humidification of the reactant gases is used prior to entry into the fuel cell, this is a process that can be controlled.

4.4.2 Airflow and water evaporation

Except in the special case of PEM fuel cells supplied with pure oxygen, it is universally the practice to remove the product water using the air that flows through the cell. The air will also always be fed through the cell at a rate faster than that needed just to supply the necessary oxygen. If it were fed at exactly the 'stoichiometric' rate, then there would

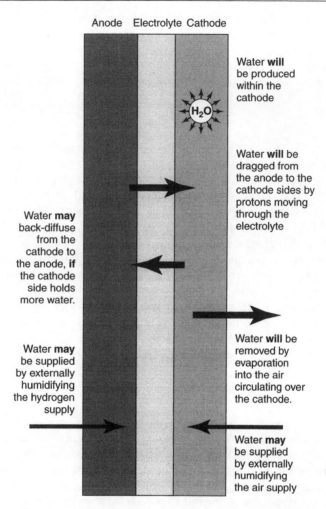

Anode   Electrolyte Cathode

Water **will** be produced within the cathode

Water **will** be dragged from the anode to the cathode sides by protons moving through the electrolyte

Water **may** back-diffuse from the cathode to the anode, **if** the cathode side holds more water.

Water **may** be supplied by externally humidifying the hydrogen supply

Water **will** be removed by evaporation into the air circulating over the cathode.

Water **may** be supplied by externally humidifying the air supply

Figure 4.10   The different water movements to, within, and from the electrolyte of a PEM fuel cell.

be very great 'concentration losses', as described in the previous chapter, Section 3.7. This is because the exit air would be completely depleted of oxygen. In practice, the stoichiometry (λ) will be at least 2. In Appendix 2, Section A2.2, a very useful equation is derived connecting the air flowrate, the power of a fuel cell and the stoichiometry (equation A2.4). Problems arise because the drying effect of air is so non-linear in its relationship to temperature. To understand this, we have to consider the precise meaning and quantitative effects of terms such as *relative humidity, water content,* and *saturated vapour pressure.*

When considering the effect of oxygen concentration in Section 2.5, the partial pressures of the various gases that make up air were given. At that point we ignored the fact that air also contains water vapour. We did this because the amount of water vapour in air varies greatly, depending on the temperature, location, weather conditions, and other factors. A straightforward way of measuring and describing the amount of water vapour in air is to give the ratio of water to other gases – the other gases being nitrogen, oxygen, argon, carbon dioxide, and others that make up 'dry air'. This quantity is variously known as the *humidity ratio*, *absolute humidity*, or *specific humidity* and is defined as

$$\text{humidity ratio}, \omega = \frac{m_w}{m_a} \qquad [4.1]$$

where m_w is the mass of water present in the sample of the mixture and m_a is the mass of dry air. The total mass of the air is $m_w + m_a$.

However, this does not give a very good idea of the drying effect or the 'feel' of the air. Warm air, with quite a high water content, can feel very dry, and indeed have a very strong drying effect. On the other hand, cold air, with a low water content, can feel very damp. This is due to the change in the *saturated vapour pressure* of the water vapour. The saturated vapour pressure is the *partial pressure* of the water when a mixture of air and liquid water is at equilibrium – the rate of evaporation is equal to the rate of condensation. When the air cannot hold any more water vapour, it is 'saturated'. This is illustrated in Figure 4.11.

Air that has no 'drying effect', that will not be able to hold any more water, could reasonably be said to be 'fully humidified'. This state is achieved when $P_w = P_{sat}$, where P_w is the partial pressure of the water and P_{sat} is the saturated vapour pressure of the water. We define the *relative humidity* as the ratio of these two pressures:

$$\text{relative humidity}, \phi = \frac{P_w}{P_{sat}} \qquad [4.2]$$

Typical relative humidities vary from about 0.3 (or 30%) in the ultra-dry conditions of the Sahara desert to about 0.7 (or 70%) in New York on an 'average day'. Very important for us is the fact that the drying effect of air, or the rate of evaporation of water, is directly proportional to the difference between the water partial pressure P_w and the saturated vapour pressure P_{sat}.

The cause of the complication for PEM fuel cells is that the saturated vapour pressure varies with temperature in a highly non-linear way – P_{sat} increases more and more rapidly at higher temperatures. The saturated vapour pressure for a range of temperatures is given in Table 4.1.

The result of the rapid rise in P_{sat} is that air that might be only moderately drying, say 70% relative humidity, at ambient temperature, can be fiercely drying when heated to about 60°C. For example, for air at 20°C, relative humidity 70%, the pressure of the water vapour in the mixture is

$$P_w = 0.70 \times P_{sat} = 0.70 \times 2.338 = 1.64 \, \text{kPa}$$

If this air is then heated to 60°C, at constant pressure, without adding water, then P_W will not change, and so the new relative humidity will be

$$\text{relative humidity}, \phi = \frac{P_w}{P_{sat}} = \frac{1.64}{19.94} = 0.08 = 8\%$$

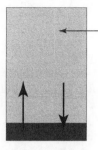 Water/air mixture. The partial pressure of the water vapour is **equal** to the *saturated vapour pressure*. There is thus an equilibrium. The water content of the air does not change. It is 'saturated' and can take no more.

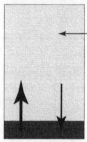 Water/air mixture. The partial pressure of the water vapour is **less than** the *saturated vapour pressure*. The water will evaporate faster than any condensation. The rate of evaporation is proportional to the difference between the saturated vapour pressure P_{sat} and the partial pressure of the water P_w.

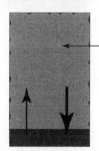 Water/air mixture. The partial pressure of the water vapour is **greater than** the *saturated vapour pressure*. In this case the rate of condensation exceeds that rate of evaporation. The condensation will normally be observed on the wall of the vessel. The rate of condensation is proportional to the difference between the partial pressure of the water vapour P_w and the saturated vapour pressure P_{sat}.

Figure 4.11   Diagram to explain saturated vapour pressure.

Table 4.1   The saturated vapour pressure of water at selected temperatures

$T°C$	Saturated vapour pressure (kPa)
15	1.705
20	2.338
30	4.246
40	7.383
50	12.35
60	19.94
70	31.19
80	47.39
90	70.13

This is very dry, far more drying than the Sahara desert for example, for which ϕ is typically about 30% (data for the Dakhla Oasis in Daly, 1979). This would have a catastrophic effect upon polymer electrolyte membranes, which not only rely totally on a high water content but are also very thin, and so prone to rapid drying out.

Another way of describing the water content that should be mentioned is the *dew point*. This is the temperature to which the air should be cooled to reach saturation. For example, if the partial pressure of the water in a sample of air is 12.35 kPa, then, referring to Table 4.1, the *dew point* would be 50°C.

We shall see in the sections that follow that it is sometimes necessary to humidify the gases going into a fuel cell. To do this in a controlled way it might be important to calculate the mass of water to be added to the air to achieve a set humidity at any pressure and temperature. To derive such a formula, we need to note that the mass of any species in a mixture is proportional to the product of the molecular mass and the partial pressure. The molecular mass of water is 18. The molecular mass of air is usually taken to be 28.97. So, using equation 4.1 we can say that

$$\omega = \frac{m_w}{m_a} = \frac{18 \times P_w}{28.97 \times P_a} = 0.622 \frac{P_w}{P_a}$$

The partial pressure of the dry air P_a is not usually known, only the total pressure P. And

$$P = P_a + P_w \quad \therefore \ P_a = P - P_w$$

$$\text{and so} \quad m_w = 0.622 \frac{P_w}{P - P_w} m_a \qquad\qquad [4.3]$$

The water vapour pressure P_W will be found using the desired humidity, the temperature, and figures from Table 4.1. This equation will normally be used in the 'mass per second' form. The mass of dry air per second needed by a fuel cell can be found in Appendix 2, equation A2.4. Note the very important point that the mass of water needed is inversely proportional to the total air pressure P. Higher pressure systems require less added water to achieve the same humidity.

A worked example using this formula is given in Section 4.4.5.

4.4.3 Humidity of PEMFC air

The humidity of the air in a PEMFC must be carefully controlled. The air must be dry enough to evaporate the product water, but not so dry that it dries too much – it is essential that the electrolyte membrane retains a high water content. The humidity should be above about 80% to prevent excess drying, but must be below 100%, or liquid water would collect in the electrodes.

What conditions are required to bring this about? Higher airflow would obviously reduce humidity, as would raising the temperature. But is there some theory that can help us put some detail into the picture? Fortunately, it is not too difficult to derive a simple formula for the humidity of the exit air of a fuel cell.

The partial pressure of a gas is proportional to the number of molecules of that gas in the mixture – the 'molar fraction'. If we consider the exit gas of a fuel cell, then we can say that

$$\frac{P_W}{P_{exit}} = \frac{\text{number of water molecules}}{\text{total number of molecules}}$$

$$\frac{P_W}{P_{exit}} = \frac{\dot{n}_W}{\dot{n}_W + \dot{n}_{O_2} + \dot{n}_{rest}} \qquad [4.4]$$

Where $\dot{n}_W$ is the number of moles of water leaving the cell per second,
   $\dot{n}_{O_2}$ is the number of moles of oxygen leaving the cell per second,
   $\dot{n}_{rest}$ is the number of moles of the 'non-oxygen' component of air per second,
   P_W is the vapour pressure of the water, and
   P_{exit} is the total air pressure at the fuel cell exit.

We can assume that all the product water is removed by the cathode air supply – there is usually no other way of removing it. Thus we can use equation A2.9, in Appendix 2, and write

$$\dot{n}_W = \frac{P_e}{2V_c F}$$

And from equation A2.2, which gives the rate of use of oxygen, we can say that

$$\dot{n}_{O_2} = \text{rate of supply of } O_2 - \text{rate of use of } O_2$$

$$\text{So} \quad \dot{n}_{O_2} = (\lambda - 1)\frac{P_e}{4V_c F}$$

Here λ is the air stoichiometry.[4] The exit flow rate of the 'non-oxygen' component of air is the same as the entry flow rate – these gases just pass through the cell. These non-oxygen components amount to 79% of the air. This will be greater than the oxygen molar flow rate by the ratio $0.79/0.21 = 3.76$. So

$$\dot{n}_{rest} = 3.76 \times \lambda \frac{P_e}{4V_c F}$$

Substituting these three equations into equation 4.4 gives

$$\frac{P_W}{P_{exit}} = \frac{\dfrac{P_e}{2V_c F}}{\dfrac{P_e}{2V_c F} + (\lambda - 1)\dfrac{P_e}{4V_c F} + 3.76\lambda\dfrac{P_e}{4V_c F}}$$

$$= \frac{2}{2 + (\lambda - 1) + 3.76\lambda} = \frac{2}{1 + 4.76\lambda}$$

---

[4] If you are not familiar with this term, please read the beginning of Appendix 2 NOW.

And so we have

$$P_W = \frac{0.420 P_{exit}}{\lambda + 0.210} \qquad [4.5]$$

This impressively simple formula shows that the water vapour pressure at the outlet depends, quite simply, on the air stoichiometry and the air pressure at the exit P_{exit}. In this derivation, we ignored any water vapour in the inlet air – thus this formula gives us the 'worst case' situation, with dry inlet air. We will consider the effect of inlet water vapour later.

Let us take an example. Suppose we have a fuel cell operating with an exit air pressure of 1.1 bar (= 110 kPa), slightly above air pressure. The cell temperature is 70°C, and the air stoichiometry is 2. The inlet air is of low humidity, and cold, so any inlet water can be ignored. Equation 4.5 tells us that

$$P_W = \frac{0.420 \times 110}{2 + 0.210} = 20.91 \text{ kPa}$$

Referring to Table 4.1, and using equation 4.2, we can see that the relative humidity of the exit air is

$$\phi = \frac{20.91}{31.19} = 0.67 = 67\%$$

This would be judged too dry and would require attention. The cell could be made more humid by

- lowering the temperature, which would increase losses,
- lowering the air flowrate and hence λ, which could be done a little, but would reduce cathode performance
- increasing the pressure, which would take more energy to run the compressor

None of these options is particularly attractive. Another alternative is to condense or otherwise acquire the water from the exit gas and use it to humidify the inlet air. This has an obvious penalty of extra equipment, weight, size, and cost – but is often justified by the increase in performance that is possible. If the water content of the inlet air is not negligible, it can be shown that the outlet water vapour pressure is given by the less simple formula

$$P_W = \frac{(0.420 + \Psi\lambda) P_{exit}}{(1 + \Psi)\lambda + 0.210} \qquad [4.6]$$

Where Ψ is a coefficient whose value is given by the simple equation

$$\Psi = \frac{P_{Win}}{P_{in} - P_{Win}}$$

Here P_{in} is the total inlet air pressure, which will usually be a little larger than P_{exit}, and P_{Win} is the water vapour pressure at the inlet.

With equations 4.5 and 4.6 we now have the means to properly consider the different ways of running a PEM fuel cell so that the air is not too wet or too dry.

4.4.4 Running PEM fuel cells without extra humidification

In the example given in the previous section, the exit air from the fuel cell was too dry. However, by choosing suitable operating temperatures and air flowrates, it is possible to run a PEM fuel cell that does not get too dry without using any extra humidification. We can find these conditions using equation 4.5 (or 4.6), together with the saturated vapour pressure taken from Table 4.1. This gives us the exit air humidity at different temperatures and air flowrates. Graphs of the humidity at air stoichiometries of 2 and 4 are shown in Figure 4.12 for a cell operating at 100 kPa (1 bar). Some selected figures are also given in Table 4.2. It can be readily seen that for most operating conditions the fuel cell will either be too wet or too dry. However, correct conditions are not impossible to find.

As one would expect, the humidities are lower at greater airflow. Also, at higher temperatures the relative humidity falls sharply. If the relative humidity of the exit air is much less than 100%, then the effect will be for the cell to dry out, and the PEM to cease working. This may not be obvious, but remember that these figures are calculated assuming that *all* the product water is evaporated. In these circumstances, if the exit air is still less than 100% relative humidity then the fuel cell has given all it has got to give, and yet the air still wants more! It must also be remembered that the conditions at the entry will be *even more drying*.

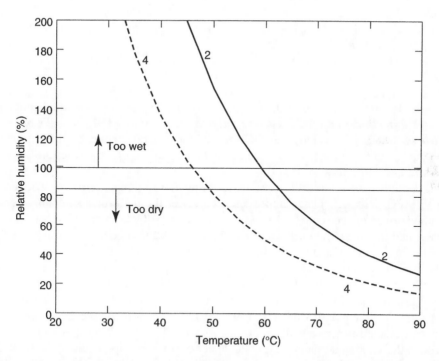

Figure 4.12 A graph of relative humidity versus temperature for the exit air of a PEM fuel cell with air stoichiometry of 2 and 4. The entry air is assumed to be dry, and the total pressure is 1 bar.

Table 4.2 Exit air relative humidities at selected temperatures and stoichiometries. In this case the inlet air is assumed to be at 20°C and 70% relative humidity. The gaps are where the relative humidity is absurdly high or low. (Figures are for exit air pressure of 1.0 bar)

Temp (°C)	$\lambda = 1.5$	$\lambda = 2$	$\lambda = 3$	$\lambda = 6$	$\lambda = 12$	$\lambda = 24$
20					213	142
30				194	117	78
40		273	195	112	68	45
50	208	164	118	67	40	26
60	129	101	72	41		
70	82	65	46			
80	54	43	30			
90	37	28				

On the other hand, relative humidity greater than 100% is basically impossible, and the airstream will contain condensed water droplets. This can be dealt with in moderation, with the water blown out by the air. However, if the theoretical humidity is greater than 100% the electrodes will become flooded. The result of this two-way constraint is a fairly narrow band of operating conditions. Nevertheless, if we stay below about 60°C, then there will be an air flowrate that will give a suitable humidity, that is, around 100%. Some of these conditions are shown in Table 4.2.

A very important point from Figure 4.12 and Table 4.2 is that *at temperatures above about 60°C, the relative humidity of the exit air is below or well below 100% at all reasonable values of stoichiometry*. (If λ is less than 2, then the oxygen concentration will be too low for the cells near the exit airflow.) This leads to the important conclusion that *extra humidification of the reactant gases is essential in PEM fuel cells operating at above about 60.°C*. This has been confirmed by the general experience of PEM fuel cell users (Büchi and Srinivasan, 1997). The methods for humidifying the gases are discussed in the sections that follow.

This feature makes for difficulties in choosing the optimum operating temperature for a PEMFC. The higher the temperature, the better the performance, mainly because the cathode overvoltage reduces. However, once over 60°C the humidification problems increase, and the extra weight and cost of the humidification equipment can exceed the savings coming from a smaller and lighter fuel cell.

The key to running a fuel cell without external humidification is to set the air stoichiometry so that the relative humidity of the exit air is about 100% *and* to ensure that the cell design is such that the water is balanced within the cell. One way of doing this is described by Büchi and Srinivasan (1997) and is shown in Figure 4.13. The air and hydrogen flows are in opposite directions across the MEA. The water flow from anode to cathode is the same in all parts, as it is the 'electro-osmotic drag', and is directly proportional to the current. The back diffusion from cathode to anode varies, but is compensated for by the gas circulation. Other aids to an even spread of humidity are narrow electrodes and thicker gas diffusion layers, which hold more water.

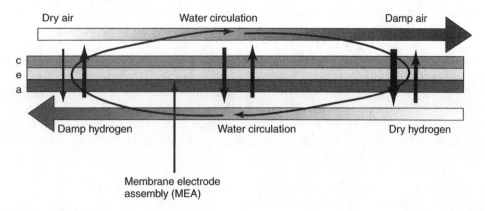

Dry air   Water circulation   Damp air

c
e
a

Damp hydrogen   Water circulation   Dry hydrogen

Membrane electrode
assembly (MEA)

Figure 4.13   Contra flow of reactant gases to spread humidification (Büchi and Srinivasan, 1997).

A key to correct PEM fuel cell water balance is clearly the air flowrate, and hence stoichiometry. At lower temperatures, say 30°C, this will be about $\lambda = 24$. This may seem a wildly extravagant airflow, but in practice it is not. Let us take the example of the small type of fuel cell to which this sort of technique applies. The smallest fan from the catalogue of a major manufacturer has a flow rate of 4.7 cfm ($=$ ft^3 min^{-1}, the manufacturer is in the United States). If we assume that only 2.0 cfm of this air actually flows over the cathodes, and that the stack operates with an average cell voltage of 0.6 V, then, using equation A2.4 (adapted for cfm), we find that this airflow is suitable for a fuel cell of power

$$P_e = \frac{0.6 \times 2.0}{3.57 \times 10^{-7} \times 24 \times 1795} = 78 \text{ watts}$$

The electrical power of the fan in question is just 0.7 W, so the very high stoichiometry for air circulation is supplied using less than 1% of the fuel cell power. This is a very modest parasitic power loss.

One of the best sources of data and further discussion of the issues of running a PEM fuel cell without humidification of the gases is Büchi and Srinivasan (1997). Here it is shown that even if the air flowrate and temperature are correctly set, then the unhumidified cell will still perform about 40% worse than an identical cell with humidified reactants. This is because, although the overall water balance may be correct, there will be areas within the cell, especially near the gas entry points, where the humidity will be too low. Thus, although not having any humidification greatly reduces system complexity, size, and cost – it is quite rare for PEM fuel cells not to use one of the techniques in the section that follows.

4.4.5 External humidification – principles

Although we have seen that small fuel cells can be operated without additional or external humidification, in larger cells this is rarely done. Operating temperatures of over 60°C are desirable to reduce losses, especially the cathode activation voltage drops described in

Section 3.4. Also, it makes economic sense to operate the fuel cell at maximum possible power density, even if the extra weight, volume, cost, and complexity of the humidification system are taken into account. With larger cells, all these are proportionally less important.

The need and the value of external humidification is shown by revisiting equation 4.6, and taking an example

Let us take a fuel cell operating at 90°C, moderately pressurised, with an inlet pressure of 220 kPa and an exit pressure of 200 kPa. We will assume that the air stoichiometry is 2, a very typical value. If we also assume that the inlet is 'normal air' at 20°C and 70% relative humidity, then putting these numbers into equation 4.6, with values from Table 4.1, we find the following:

- The inlet water vapour pressure $P_{win} = 1.64$ kPa (Table 4.1 & equation 4.2).
- Thus the Ψ term in equation 4.6 is 0.00751.
- So, from equation 4.6, the vapour pressure of the water at the exit is 39.1 kPa.
- And so, using Table 4.1 and equation 4.2, the exit air humidity is 56%.

This is far too low a figure, and these conditions would dry out the PEM very quickly. However, if the inlet air is warm and damp, say 80°C and 90% relative humidity, then the figures become as follows:

- The inlet water vapour pressure $P_{win} = 42.65$ kPa (Table 4.1 & equation 4.2).
- Thus the Ψ term in equation 4.6 is 0.2405.
- So, from equation 4.6, the vapour pressure of the water at the exit is 66.96 kPa.
- And so, using Table 4.1 and equation 4.2, the exit air humidity is 95%.

This is a very reasonable figure, and shows that it is possible to run a PEM fuel cell at a higher temperature (90°C in this case) in a way that does not cause the PEM to dry out

Water is added to achieve the necessary humidification of the reactant air. We can use equation 4.3 to calculate how much water is needed. For example, if the fuel cell of the previous paragraph was a 10-kW cell, then using equation A2.4 from Appendix 2 we see that the mass flow rate of dry air into the cell is

$$\dot{m}_a = 3.57 \times 10^{-7} \times 2 \times \frac{10 \times 10^3}{0.65} = 0.0110 \,\text{kg s}^{-1}$$

The desired water vapour pressure is 42.7 kPa (equation 4.2 and Table 4.1). Equation 4.3 then becomes

$$\dot{m}_w = 0.622 \times \frac{42.7}{220 - 42.7} \times 0.0110 = 0.0016 \,\text{kg s}^{-1} = 1.6 \,\text{g s}^{-1}$$

This water needs to be added to the air. The rate is approximately equivalent to 100 ml min^{-1}, or 6 L h^{-1}, and is definitely a considerable throughput of water to deal with. The question that needs to be asked is – where is this water to come from?

Having the water as an extra input to the fuel cell system is obviously not desirable – they are complicated enough without having to supply water in the manner of steam engines. The water must be extracted from the exit gas of the fuel cell. All this water added to the reactant air stays in the air and comes out at the other side. In addition, the exit air carries the product water from the reaction. Equation A2.10 in Appendix 2 tells us that for this 10-kW cell the rate of water production is

$$9.34 \times 10^{-8} \times \frac{10 \times 10^3}{0.65} = 0.0014 \, \text{kg s}^{-1}$$

This water, plus the water added to the reactant air, will leave the cell, making a total water output flow rate of $0.0016 + 0.0014 = 0.003 \, \text{kg s}^{-1}$. This will all be leaving as water vapour, and it means that the condensation or water separation system in the exit path must extract rather more that half the water present in the exit gas, so that it can be recycled for use in a humidifier. This method of water supply can also ensure the purity of the water, provided the system is kept clean. The water used will essentially be distilled water.

Before we look at some of the practicalities of HOW the reactants are humidified, three simple points need to be made.

- Firstly, it is often not the case that only the air is humidified. To spread the humidity more evenly, sometimes the hydrogen fuel is humidified as well.
- Secondly, the humidification process involves evaporating water in the incoming gas. This will cool the gas, as the energy to make the water evaporate will come from the air. In pressurised systems this is positively helpful. It will be seen in Chapter 9, when compressors are considered, that a result of compressing the air is considerable heating, and the humidification process can be an ideal way of bringing the air temperature to a suitable value.
- Thirdly, we should note that the quantities of water to be added to the air, and the benefits in terms of humidity increase, are all much improved by operating at higher pressure. The reader could recalculate all the values in the example used in this section, with pressures of 140 kPa (inlet) and 120 kPa (outlet), and it will be found that the mass of water to be added to the inlet airstream rises to $0.003 \, \text{kg s}^{-1}$, and yet the exit humidity is a hopelessly inadequate 34%. This provides a link to the important question of operating pressure, which we consider later.

4.4.6 External humidification – methods

The method used to humidify the reactant gases of a fuel cell is one of the features of the PEM fuel cell where no standard has yet emerged, and different designers are experimenting with different methods. Fuel cells are not the only systems where gas streams have to be humidified, and so the technology used elsewhere can often be adapted. One useful area is air conditioning technology. Another analogous technology is the 'humidification' of the air entering a four-stroke engine with gasoline. The methods used in these areas, together with others created especially for fuel cells, are used. Eight different methods will be outlined below. These illustrate the breadth of the possibilities and the ingenuity of PEM fuel cell designers.

In laboratory test systems the reactant gases of fuel cells are humidified by bubbling them through water, whose temperature is controlled. This process can be called *sparging*.

It is generally assumed that the dew point of the air is the same as the temperature of the water it has bubbled through, which makes control straightforward. This is good for experimental work, but this will rarely be practical in the field.

One of the easiest to control is the direct injection of water as a spray. This has the further advantage that it will cool the gas, which will be necessary if it has been compressed or if the fuel gas has been formed by reforming some other fuel and is still hot, as in Chapter 8. The method involves the use of pumps to pressurise the water, and also a solenoid valve to open and close the injector. It is therefore fairly expensive in terms of equipment and parasitic energy use. Nevertheless, it is a mature technology and is widely used, especially on larger fuel cell systems.

Another method recently described (Floyd D.E., 2001) uses metal foam to make a kind of fine water spray to humidify the water. This system has the advantage that only a pump is needed to move the water – the water is finally introduced to the air in a passive way.

A method that has been used in the past, but not so much lately, has been to humidify the PEM directly, rather than humidifying the reactant gases. Wicks are constructed into the PEM, usually as part of the gas diffusion layer. These wicks dip into the water and draw it directly to the PEM. The system is somewhat self-regulating, as no water will be drawn up if the wicks are saturated. The difficulty is that they add very greatly to the problem of sealing the cell – the wick will also be an easy exit route for reactant gases. The possibility of cooling the incoming air is also lost with this method.

Another approach is to directly inject liquid water into the fuel cell. Normally this would lead to the electrode flooding and the cell ceasing to work. However, the technique is combined with a bipolar plate and 'flow field' design that forces the reactant gases to blow the water through the cell and over the entire electrode. This is well described by Wood et al. (1998), who have coined the term 'interdigitated flow field' for the method. The principle is shown in Figure 4.14. The 'flow field' is the name of the channel cut in the bipolar plate that is the route for the reactant gas. The flow field shown in Figure 4.14 is like a maze with no exit. The gas is forced under the bipolar plate and into the electrode, driving the water with it. If the flow field is well designed, this will happen all over the electrode. Good results are reported for this method, though the reactant gases must be driven at pressure through the cell and the energy used to do this is not clear, neither is the wear and tear on the electrodes and the effect on long-term performance. The possibility of cooling the incoming air is also lost with this method.

All the systems described above require liquid water to operate. This means that the exit air must be treated so that a good deal of the water content is condensed out as liquid water, stored, and then pumped to where it is needed for the humidification process. This clearly adds to the system size, cost, and complexity. There are some systems that use the water in the exit gas to humidify the inlet air, but without actually condensing it out as liquid water. One way to do this is to use a rotating piece of water-absorbing material. It is put into the path of the exit air, and it absorbs water. It is then rotated so that it is in the path of the dry entry air, which will at least partly dry it out. If the piece is made circular this can be done on a continuous basis – constantly transferring water from exit to entry gases. This method has the disadvantage that it will still be fairly bulky and will require power and control to operate.

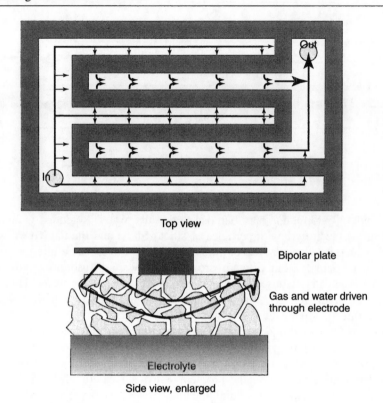

Top view

Bipolar plate

Gas and water driven through electrode

Electrolyte

Side view, enlarged

Figure 4.14 Diagrams to show the principle of humidification using interdigitated flow fields, after Wood et al. (1998).

Engineers at the Paul Scherrer Institute in Switzerland have demonstrated a more ingenious system that also uses the exit water to humidify without a separate condenser, pump, and tank. The principle is shown in Figure 4.15. The warm, damp air leaving the cell passes over one side of a membrane, where it is cooled. Some of the water condenses on the membrane. The liquid water passes through the membrane and is evaporated by the drier gas going into the cell on the other side. Such a humidifier unit can be seen on the top of the fuel cell system shown in Figure 4.29. The membrane used can be exactly the same as a PEM electrolyte. This has led to a modification of this idea, where a smaller version is included in each and every cell. The bipolar plate is so designed that extra space is provided for a miniature version of Figure 4.15 by each and every cell. This has the great advantage that the humidification of the air is evenly spread through the cell. Whether only at the fuel cell entry, or by each cell, this system involves no moving parts. However, everything has its downside, and this simplicity means there is no way of controlling it.

All seven humidification methods described so far could use the water leaving the cell to do the humidification. One method that does not do this should be mentioned. Watanabe (1996) has described a system called *self-humidification*, where the electrolyte

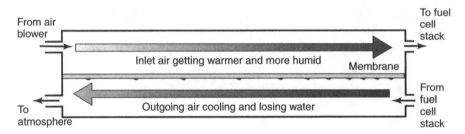

Figure 4.15 Humidification of reactant air using exit air, as demonstrated by the Paul Scherrer Institute (1999). See also Figure 4.29.

is modified, not only to retain water but also to produce water. Retention is increased by impregnating the electrolyte with particles of silica (SiO_2) and titania (TiO_2), which are hygroscopic materials. Nanocrystals of platinum are also impregnated into the electrolyte, which is made particularly thin. Some hydrogen and oxygen diffuse through the electrode and, because of the catalytic effect of the platinum, react, producing water. This of course uses up valuable hydrogen gas, but it is claimed that the improved performance of the electrolyte justifies this parasitic fuel loss.

The actual method used and the extent of humidification will vary depending on the size of the cell, the operating pressure (to be considered below in Section 4.7), the balance sought between optimum performance and simplicity, and the fuel source – among other considerations. In any case the problem will need careful thought at the design stage. In larger systems, the extra humidification of the reactant gases will need to be actively controlled.

4.5 PEM Fuel Cell Cooling and Air Supply

4.5.1 Cooling using the cathode air supply

PEM fuel cells are, of course, not 100% efficient. In converting the hydrogen energy into electricity, efficiencies are normally about 50%. This means that a fuel cell of power X watts will also have to dispose of about X watts of heat. More precisely, it is shown in Appendix 2, Section A2.6, that the heat produced by a fuel cell, if the product water is evaporated within the cell, is

$$\text{Heating rate} = P_e \left(\frac{1.25}{V_c} - 1 \right) \text{ watts}$$

The way this heat is removed depends greatly on the size of the fuel cell. With fuel cells below 100 W it is possible to use purely convected air to cool the cell and provide sufficient airflow to evaporate the water, without recourse to any fan. This is done with a fairly open cell construction with a cell spacing of between 5 and 10 mm per cell (Daugherty, 1999). The fact that damp air is less dense than dry air (perhaps counter-intuitive, but true) aids the circulation process. However, for a more compact fuel cell, small fans can be used to blow the reactant and cooling air through the cell, though a large proportion of the heat will still be lost through natural convection and radiation. As was shown with the numerical example at the end of Section 4.4.4, this does not impose a large parasitic power loss on the system – only about 1% if the fan is well chosen and the air ducting is sensibly designed.

However, when the power of the fuel cell rises, and a lower proportion of the heat is lost by convection and radiation from and around the external surfaces of the cell, problems begin to arise. In practice, this simplest of all methods of cooling a fuel cell can only be used for systems of power up to about 100 W.

4.5.2 Separate reactant and cooling air

The need to separate the reactant air and the cooling air for anything but the smallest of PEM fuel cells can be shown by working through a specific example where the reactant gas and the cooling gas **are** combined.

Suppose a fuel cell of power P_e Watts is operating at 50°C. The average voltage of each cell in the stack is 0.6 V – a very typical figure. Let us suppose that cooling air enters the cell at 20°C and leaves at 50°C. (In practice the temperature change will probably not be so great, but let us take the best possible case for the present.) Let us also assume that only 40% of the heat generated by the fuel cell is removed by the air – the rest is radiated or naturally convected from the outer surfaces.

In Appendix 2, Section A2.6, it is shown that the heat generated by a fuel cell, if the water exits as a vapour, is

$$\text{Heating rate} = P_e \left(\frac{1.25}{V_c} - 1 \right) \text{ watts}$$

Just 40% of this heat is removed by air, of specific heat capacity c_p flowing at a rate of $\dot{m}$ kg s^{-1}, and subject to a temperature change ΔT. So we can say that

$$0.4 \times P_e \left(\frac{1.25}{V_c} - 1 \right) = \dot{m} c_p \Delta T$$

Substituting known values, that is, $c_p = 1004 \, \text{J kg}^{-1} \, \text{K}^{-1}$, $\Delta T = 30 \, \text{K}$, and $V_c = 0.6 \, \text{V}$, and rearranging, we obtain the following equation for the cooling air flow rate:

$$\dot{m} = 1.4 \times 10^{-5} \times P_e \, \text{kg s}^{-1}$$

In Appendix 2, Section A2.2, it is shown that the reactant air flowrate is

$$\dot{m} = 3.57 \times 10^{-7} \times \lambda \times \frac{P_e}{V_c} \, \text{kg s}^{-1}$$

If the reactant air and the cooling air are one and the same, then these two quantities are equal, and so the two equations can be equated. Cancelling P_e, substituting $V_c = 0.6 \, \text{V}$, and solving for λ we obtain

$$\lambda = \frac{14 \times 0.6}{0.357} \approx 24$$

A glance at Table 4.2 above will show that at 50°C this gives an exit air humidity of 26%. This is drier than the Sahara desert! The figures in Table 4.2 are based on the assumption that the entry air had a humidity of 70%, so the relative humidity *decreases* as the air goes through the cell, and so the proton exchange membrane will quickly dry out.

Note that if the assumptions made at the beginning of this section are made more realistic, that is, more heat having to be taken out by the air and less of a temperature change, then the situation becomes even worse. The only way to reduce λ, which should be somewhere between 3 and 6 at 50°C in order to stop the cell from drying out, is to reduce the air flowing over the electrodes and to have a separate cooling system. This point comes when more than about 25% of the heat generated has to be removed by a cooling fluid. In practice, this seems to be in the region of about 100 W. Fuel cells larger than this will generally need a separate reactant air supply and cooling system. This will mean two air blowers or pumps, but there is no alternative!

The usual way of cooling cells in the range from about 100 to 1000 W is to make extra channels in the bipolar plates through which cooling air can be blown, as shown in Figure 4.16. Alternatively, separate cooling plates can be added, through which air is blown. A real commercial 1-kW PEMFC that uses separate reactant and cooling air, is shown in Figure 4.17. The holes for the air cooling channels cannot really be seen from the angle of the photograph, but can be deduced from the roughness of the top of the stack. A similar cooling system is used for the 2-kW system shown in Figure 4.29.

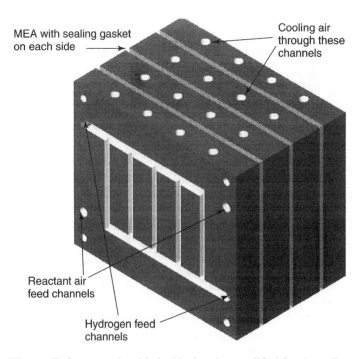

Figure 4.16   Three cells from a stack, with the bipolar plate modified for air cooling using separate reactant and cooling air.

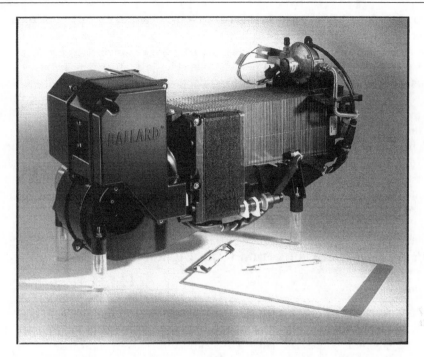

Figure 4.17 The Ballard Nexa fuel cell is an example of a commercial PEM fuel cell that uses air cooling. The blower for the cooling air can clearly be seen at the bottom left of the unit, and it blows air up through channels in the bipolar plates as per Figure 4.16. The reactant air passes through the humidifier on the front of the unit and is driven by a pump in the box on the top left of the system. (Reproduced by kind permission of Ballard Power Systems.)

Using separate cooling air in this way works for cells between about 100 W and 1 kW or so, but for larger cells this becomes impractical, and water cooling is preferred.

4.5.3 Water cooling of PEM fuel cells

The issues of when to change from air cooling to water cooling are much the same for fuel cells as they are for other engines, such as internal combustion engines. Essentially, air cooling is simpler, but it becomes harder and harder to ensure that the whole fuel cell is cooled to a similar temperature, as it gets larger. Also, the air channels make the fuel cell stack larger than it needs to be − 1 kg of water can be pumped through a much smaller channel than 1 kg of air, and the cooling effect of water is much greater.

With fuel cells, the need to water cool is perhaps greater than with a petrol engine, as the performance is more affected by variation in temperature. On the other hand, the water cooling of a petrol engine also helps in sound proofing, which is why it is sometimes used, say, on motorcycles when air cooling might otherwise suffice. On balance then, we would expect the 'changeover' from water cooling to air cooling to come at around the same sort of power levels, that is, at a few kilowatts. PEM fuel cells above 5 kW will be water cooled, those below 2 kW air cooled, with the decision for cells in between being a matter of judgement.

One factor that will certainly influence the decision of whether or not to water cool will be the question of 'what is to be done with the heat'. If it is to be just lost to the atmosphere, then the bias will be towards air cooling. On the other hand, if the heat is to be recovered, for example, in a small, domestic CHP system, then water cooling becomes much more attractive. It is far easier to 'move heat about' if the energy is held in hot water than if it is in air.

The method of water cooling a fuel cell is essentially the same as for air in Figure 4.16, except that water is pumped through the cooling channels. In practice, cooling channels are not always needed or provided at every bipolar plate. Variations on this theme are mentioned when we look more closely at different fuel cell construction methods in Section 4.6.

4.6 PEM Fuel Cell Connection – the Bipolar Plate

4.6.1 Introduction

Most PEM fuel cells are constructed along the general lines of multiple cells connected in series with bipolar plates outlined in Section 1.3, and illustrated in Figure 1.10. Internal manifolding, as in Figures 1.12 and 1.13, is almost universally used. However, there are many variations in the way the bipolar plate is constructed and the materials they are made from. This is a very important topic, because as we have seen, the MEAs for PEM fuel cells are very thin, and so the bipolar plates actually comprise almost all the volume of the fuel cell stack, and typically about 80% of the mass (Murphy et al., 1998). We have also mentioned that the platinum usage has been drastically reduced, making this have little impact on the cost. The result is that the bipolar plates are usually also a high proportion of the cost of a PEM fuel cell stack.

Another important issue is that there are different ways of making stacks, which avoid the need for a bipolar plate. These are sometimes used, though only for very small cells, and methods of doing this are also discussed here.

4.6.2 Flow field patterns on the bipolar plates

In the bipolar plate in Figure 1.14, the reactant gases are fed over the electrode in a fairly simple pattern of parallel grooves. However, this is not the only way it could be done. Figure 4.18 shows several of the almost limitless possibilities. Among the designers of PEM fuel cells there is no clear consensus as to which is best.

The supposed problem of the parallel systems, such as Figure 4.18a is that it is possible for water, or some reactant impurity such as nitrogen, to build up in one of the channels. The reactant gas will then quite happily move along the other channels, and the problem will not be shifted, leaving a region of the electrode unsupplied with reactants. This leads to the more serpentine systems, such as Figure 4.18b. Here it can be guaranteed that if the reactants are moving, they are moving everywhere – a blockage will be cleared. The problem with the serpentine systems is that the path length and the large number of turns

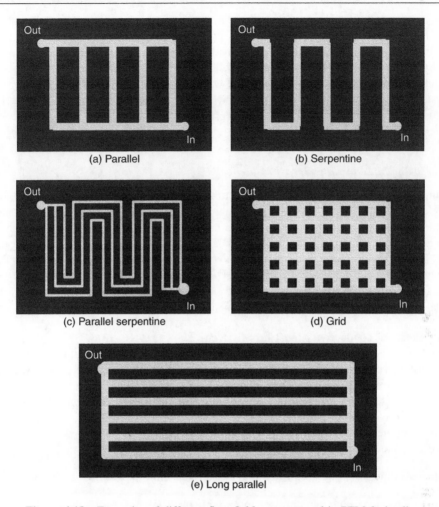

Figure 4.18   Examples of different flow field patterns used in PEM fuel cells.

mean that excessive work has to be done in pushing the gases through. The patterns such as Figure 4.18c are something of a compromise.

The pattern of Figure 4.18d could be described as 'intensely parallel'. The gases can swirl all over the face of the electrode. The idea is that any pockets of impure gases will be shifted by the swirling process of the probably unsteady flow of gas through the system. However, it would still be possible for water droplets to form, and not be shifted.

The grooves in the flow field are usually a little less than 1 mm in width and height. In order for water droplets not to form and stick in the channels, the system should be arranged so that the pressure drop along each channel is greater than the surface tension holding a water drop in place. That way, if the gas flow is stopped, there would be sufficient pressure to move the water droplet and get the gas moving again.

Recently some attempts have been made to analyse the different flow field patterns, for example Bewer et al. (2001). However, no definitive conclusions can be drawn from research published to date. Perhaps an indication of the way things may go is the pattern used by Ballard in their latest generation of PEMFC. This pattern uses bipolar plates that are rectangular in form, not square, with the width several times the height. The flow field consists of long straight runs, as in Figure 4.18d.[5] This has the advantage of being long, and so having a reasonable 'water shifting' pressure drop, and also of being straight with no inefficient bends and turns.

4.6.3 Making bipolar plates for PEM fuel cells

The bipolar plate has to collect and conduct the current from the anode of one cell to the cathode of the next, while evenly distributing the fuel gas over the surface of the anode, and the oxygen/air over the surface of the cathode. In addition to this, it often has to carry a cooling fluid through the stack and keep all these reactant gases and cooling fluids apart. If this was not enough, it also has to contain the reactant gases within the cell, so the edges of the cell must be of sufficient size to allow space for sealing.

Ruge and Büchi (2001) give a very good summary of the material requirements of the bipolar plate if it is to fulfill these requirements. These are as follows:

- The electrical conductivity must be $>10\,S\,cm^{-1}$.
- The heat conductivity must exceed $20\,W\,m^{-1}K^{-1}$ for normal integrated cooling fluids or must exceed $100\,W\,m^{-1}K^{-1}$ if heat is to be removed only from the edge of the plate.
- The gas permeability must be less that $10^{-7}\,mBar\,L\,s^{-1}\,cm^{-2}$.
- It must be corrosion resistant in contact with acid electrolyte, oxygen, hydrogen, heat, and humidity.
- It must be reasonably stiff; flexural strength $>25\,Mpa$.
- The cost should be as low as possible.

The material must also be able to be manufactured within the following requirements:

- The bipolar plate must be slim, for minimum stack volume.
- It must be light, for minimum stack weight.
- The production cycle time should be reasonably short.

The **methods for forming** these bipolar plates, and the **materials** they are made from, vary considerably. As with humidification methods considered in the previous section, there is no single method or material that is clearly the best. It would be good if we could consider the **materials** separately from the **manufacturing methods**. Unfortunately, this is not possible, as many of the candidate materials are formed during the manufacturing process, and so the two aspects of the bipolar plate – how it is made and what it is made from – must be considered as one.

Before describing some of the most promising and widely used processes, we should note one feature of bipolar plate manufacture that is common to many – that the plate is

---

[5] Information derived from pictures presented, but not published, at the Grove Fuel Cell Symposium, London, 2001.

often made in two halves. This makes the manufacture of the channels for the cooling air or water that pass right through the middle of the cell much easier. Look back to Figure 4.16 – the air cooling channels are formed by cutting grooves in the back of a half-plate, *not* by drilling holes through a whole plate.

We will now consider a range of the different materials and processes used.

One of the most established is the **machining of the graphite sheet**. Graphite is electrically conductive and reasonably easy to machine. It also has a low density, less than that for any metal that might be considered suitable. Fuel cell stacks made in this way have achieved a competitive power density. However, they have three major disadvantages:

- The machining of the graphite may be done automatically, but the cutting still takes a long time on an expensive machine.
- Graphite is brittle, and so the resulting plate needs careful handling, and assembly is difficult.
- Finally, graphite is quite porous, and so the plates need to be a few millimetres thick to keep the reactant gases apart. This means that although the material is of low density, the final bipolar plate is not particularly light.

A process that is certainly cheap is the **injection moulding of graphite filled polymer** resins. Most of these materials have such poor conductivity that the scope of their applications is likely to be limited in the foreseeable future. However, Spitzer et al. (2001) have argued that epoxy moulding compounds can be used with about 75% filling of coarse graphite – higher than other polymers. The claimed conductivity is $20\,S\,cm^{-1}$ (through plane, higher in plane) and a flexural strength of $19\,MPa$. These properties are close to, or exceed, the requirements given at the beginning of this section, and there is no reason to suppose the permeability or heat conduction should be a problem. It could be that these materials may be developed to give sufficient performance.

An alternative to injection moulding that lends itself well to bipolar plate construction is **compression moulding**. This has been described by Ruge and Büchi (2001). The mould has a top and bottom part. Graphite and thermoplastic polymer granules are mixed and spread over the lower part. The mould is then closed and pressure is applied. The mould is then heated to the glass transition temperature, at which point the materials mix and flow to fill the mould. The mould must then be cooled to solidify the mixture, before releasing the compound. This process allows a much greater proportion of carbon to be used in the polymer/carbon mixture than is possible with injection moulding, and so adequate conductivity can be achieved. However, the process is not nearly as fast as injection moulding, and a production cycle time of nearly 15 min is needed – and this is for half a plate. Nevertheless, the moulding machine is not complex, and the cycle time is less than machining, and so this material and process is one that shows promise.

Another method that is reasonably well established, though not widely used, is the use of **carbon–carbon composites**. At least two different forms of this construction have been described. The common feature is that the complex shape on the face of the plate is made by moulding a carbon composite. If this is done reasonably quickly and with good conductivity, the resulting material will not be strong enough or sufficiently impermeable to gases or sufficiently flat for a good seal. These problems are solved by applying a layer of solid carbon to the back of the plate. As with all the processes described so far, two such parts are put 'back-to-back' to make a whole bipolar plate. One such process derived from the procedure used for phosphoric acid fuel cells has been described by Murphy et al. (1998). To mould the complex shape, ordinary carbon powder is used in the composite, and then this is made more conductive by graphitisation – a process achieved by heating the part to over 2500°C. The moulding process may be cheap, but the heating process is not. Furthermore, it must be very precisely controlled, otherwise a high proportion of the resulting plates will be warped or the size will be wrong. Perhaps more promising is another process described by Besman et al. (2000). Here the face of the plate, with holes and channels 0.78 mm deep is slurry moulded using 400-μm carbon fibres and a phenol resin. It is cured at 150°C for several minutes, giving a highly conductive (200 to 300 S cm^{-1}) but rather porous material. The back is then made hermetic (non-porous) by depositing a solid carbon layer, using chemical vapour infiltration – a standard technique easily adapted to mass production. Two pieces are then put back to back. The density is less than 1 kg L^{-1} and the flexural strength is about 175 MPa. Permeability appears to be good with no measured passage of H$_2$ after 1 h at 206 kPa pressure.

All the above processes use carbon or carbon/polymer composite materials. However, **metals** can also be used to make bipolar plates. These have the advantage of being very good conductors of heat and electricity can be machined easily (e.g. by stamping), and are not porous, and so very thin pieces will serve to keep the reactant gases apart. Their major disadvantage is that they have higher density and are prone to corrosion. The inside of a PEM fuel cell has quite a corrosive atmosphere – there is water vapour, oxygen, and warmth. As well as this, there is sometimes a problem of acid leaching out of the MEAs – excess sulphonic acid from the sulphonation process mentioned in Section 4.2. To avoid corrosion, at least one of the advantages of metals must be lost. For example, stainless steel can be used, but this is quite expensive and difficult to work with. Alternatively, a special coating may be applied, but the time and complexity of this operation negates the short manufacturing processing time. The result is that the issues are fairly balanced, with some important developers preferring metal, such as Siemens (Ise et al., 2001) and Intelligent Energy, but the majority, including Ballard Power Systems, use graphite-based materials.

Metal plates can be shaped from solid pieces in just the same way as graphite or carbon composites. The greater impermeability means that they can be made thinner. However, with metals some more radical approaches can be taken. For example, **perforated or foamed metal** can be used. A particularly successful attempt at this approach has been described by Murphy et al. (1998), and the broad thrust of their method is described here. The flow field is made from metal that is 'foamed' so that it has a sponge-like structure, with many small voids within it, and the voids take up more than 50% of the bulk of the material. The material can then be sawn into thin sheets. The result could be compared to a slice of bread, which is only about 1- or 2 mm thick. An alternative way of making a functionally similar material is to take a solid sheet and introduce the voids by slitting and stretching. The metal needs to be chosen so that it will remain strong and light in this form, and titanium was chosen in the study in question. Titanium is not a particularly good conductor as metals go, but has a conductivity that is 30 times higher than graphite. Titanium can also be made sufficiently corrosion-resistant by coating with a titanium nitride protective finish. This would give an electrically conductive, precious-metal-free coating that can be applied inexpensively on a large scale.

The bipolar plate is then formed from two such pieces of foamed metal, with a thin layer of solid metal between, as in Figure 4.19. The voids form the path for the gas to diffuse through. Because the foam has been sliced, like a loaf of bread, much of the surface consists of holes, and it is through these that the gas reaches the surface of the electrodes. The edges are sealed with plastic gaskets that have holes to feed the reactant gases into the edges of the metal foam. This is also shown in Figure 4.19.

The mechanism for cooling the cell stack can be provided using essentially the same technology. This time one sheet of the metal foam is placed between two sheets of the solid (but thin) metal sheets. Water is passed through the foamed metal, carrying away the heat. The advantage of this type of approach is that it uses readily available materials – metal foam sheet is made for other applications – to make fuel cell components that are thin, light, highly conductive, but good at separating gases. Furthermore, the only manufacturing process involved is cutting and moulding of the plastic edge seals. There are many ways in which this basic idea can be adapted. One has been described by Goo and Kim (2001), which involves using foamed nickel, which is then stamped to form a flow field of one of the patterns seen in Figure 4.18 – foam channels about 0.8 mm deep can easily be made on both sides, or one side can be left flat. The foam is then filled with a silicone-based filler in order to prevent gas leakage. A conductive and non-porous material is thus formed. The foam does not provide the flow field, as is done with the Murphy (1998) method, as Goo and Kim found that this tended to become clogged with water.

A wide range of materials and manufacturing methods for PEMFC bipolar plates has been outlined. None fully satisfy the requirements outlined in the beginning of this section. In the years ahead it could be that one of these technologies, or another as yet unknown, could emerge as clearly superior. We are nowhere near that position yet, and developers

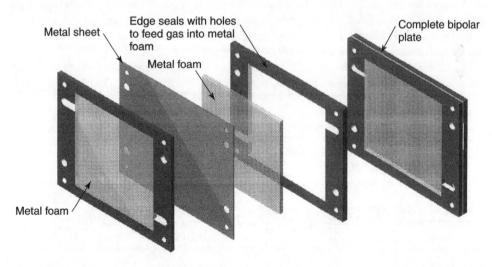

Figure 4.19 Diagram showing bipolar plate construction using metal foam, after Murphy et al. (1998).

are using all the materials and processes described above, and many variations on the same ideas incorporating proprietary improvements.

4.6.4 Other topologies

Fuel cell stack construction using bipolar plates gives very good electrical connection between one cell and the next. However, as we have seen, they are expensive and difficult to manufacture. In cases in which a fuel cell is operated at fairly low current densities, it is sometimes helpful to compromise the electrical resistance of the cell interconnects for simpler and cheaper manufacture. The flexibility and ease of handling of the MEAs used in PEM fuel cells allows many different types of construction.

One way of doing this is shown in Figure 4.20. This shows a system of three cells. The main body of the unit, shown light grey, would normally be made of plastic. A key point is that there is only one chamber containing air and only one containing hydrogen. The cells are connected in series by connecting the edge of one cathode to the edge of the next anode. This is done using metal passing through the reactant gas separator. For even less chance of leaks this could be done externally, though this would increase the current path.

The potential leaks are now much reduced, as the only key seals are those around the edges of the MEAs, but if they are carefully set into the separator as in Figure 4.20, which is not hard to do, then this will not be any problem. This type of design also reduces the problem of even humidification of the cells, since there is fairly free circulation of reactant gases in the cell. However, it is not a compact design and is only suitable for low power systems.

Another way of making small PEM fuel cells worthy of note is the use of a cylindrical design, in which the electrode is fully exposed to the air. Basically, the simple flat structure of Figure 4.20 is rolled around a hydrogen cylinder, with some extra space between the anodes and the cylinder to allow circulation of hydrogen gas. The fuel cell and hydrogen store are thus integrated. The air supply is provided very simply by natural air circulation around the outside of the fuel cell. The current collector over the top of the air cathode will be fairly rugged, perforated stainless steel being a possible material.

In addition to the problem of higher electrical resistance, a potential problem with these types of fuel cell construction methods is that of cell 'imbalance'. The individual cells

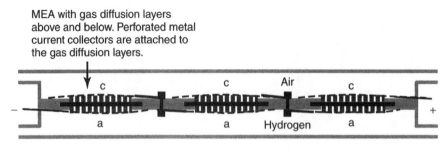

Figure 4.20 A simple method of connecting fuel cells in series, giving far simpler reactant gas supply arrangements.

are quite some way apart and they are much more physically separate than in the normal 'bipolar plate type' stack. If one cell becomes rather warmer than the others, this would lead to more rapid water evaporation, which could lead to higher internal resistance, and thus even higher temperatures, leading to more evaporation, higher resistance, and so on in a 'vicious circle' where one cell could end up dried out and with very high resistance. This would scupper the whole fuel cell, as it is a series circuit, and so they will all carry the same current. One way of reducing the likelihood of this happening is to design the system in such a way that they operate at below their optimum temperature, so that a small local rise in temperature does not lead to excessive water loss, but this is obviously not ideal. However, this is another reason why this type of construction is only practical in fairly small systems.

Another fuel cell stack construction method, which is quite close to the 'metal bipolar plates' described in the previous section, has been developed and demonstrated by Intelligent Energy (formerly Advanced Power Sources Ltd) of Loughborough in the United Kingdom. The construction is as shown in Figure 4.21.

Every single cell is constructed from a stainless steel base, MEA, and then a porous metal current collector is put on top of the cathode. This current collector is made, using patented and proprietary techniques, from sintered stainless steel powder of carefully graded size. The result is a material that is metallic, corrosion-resistant, porous, strong, conductive, and water retaining. A fuel cell stack is made by placing these rugged and

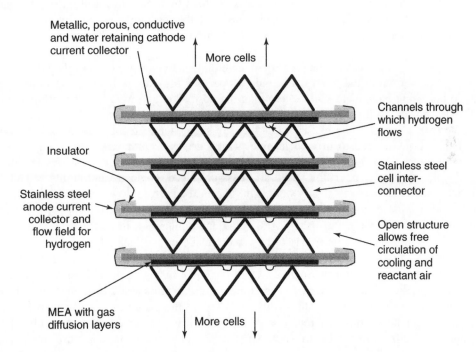

Figure 4.21 PEM fuel cell structure along the lines of that demonstrated by Advanced Power Sources Ltd.

self-contained cells one on top of the other, with a simple piece of folded stainless steel connecting the anode of one cell to the cathode of the next. Hydrogen is piped, using thin plastic tubing, to each anode. The open structure of the cell allows for free circulation of air, though this might be fan-assisted. There is good thermal (as well as electrical) connection between each cell, so the problem of cell imbalance outlined above is unlikely.

Consideration of alternative types of cell construction methods is described in the literature, and a good place to start is Heinzel et al. (1998).

4.7 Operating Pressure

4.7.1 Outline of the problem

Although small PEM fuel cells are operated at normal air pressure, larger fuel cells, of 10 kW or more, are sometimes operated at higher pressures. The advantages and disadvantages of operating at higher pressure are complex, and the arguments are not at all clear-cut, there is much to be said on both sides.

The basic issues around operating at higher pressure are the same as for other engines, such as diesel and petrol internal combustion engines, only with these machines the term used is 'supercharging' or 'turbocharging'. Indeed, the technology for achieving the higher pressures is essentially the same. Further on, in Chapter 9, we give a full description of the various types of compressors and turbines that can be used. For a full understanding of this section, the reader is encouraged to look at that chapter now; alternatively, the equations derived there can just be accepted at this stage for a quicker appreciation of the issues.

The purpose of increasing the pressure in an engine is to increase the specific power, to get more power out of the same size engine. Hopefully, the extra cost, size, and weight of the compressing equipment will be less than the cost, size, and weight of simply getting the extra power by making the engine bigger. It is a fact that most diesel engines are operated at above atmospheric pressure – they are supercharged using a turbocharger. The hot exhaust gas is used to drive a turbine, which drives a compressor that compresses the inlet air to the engine. The energy used to drive the compressor is thus essentially 'free', and the turbocharger units used are mass-produced, compact, and highly reliable. In this case the advantages clearly outweigh the disadvantages. However, with fuel cells the advantage/disadvantage balance is much closer. Above all, this is because there is little energy in the exit gas of the PEMFC,[6] and any compressor has to be driven largely or wholly using the precious electrical power produced by the fuel cell.

The simplest type of pressurised PEM fuel cell is that in which the hydrogen gas comes from a high-pressure cylinder. Such a system is shown in Figure 4.22. Only the air has to be compressed. The hydrogen gas is coming from a pressurised container, and thus its compression 'comes free'. The hydrogen is fed to the anode in a way called *dead-ended*; there is no venting or circulation of the gas – it is just consumed by the cell. The compressor for the air must be driven by an electric motor, which of course uses up some

---

[6] Note that this is NOT true for hot fuel cells such as the SOFC considered in Chapter 7.

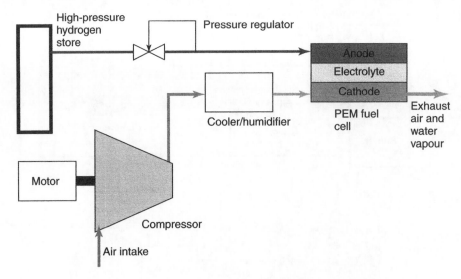

Figure 4.22   Simple motor driven air compressor and PEM fuel cell.

of the valuable electricity generated by the fuel cell. In a worked example in Chapter 9, it is shown that the typical power consumption will be about 20% of the fuel cell power for a 100-kW system. Other systems described in the literature (Barbir, 1999) report even higher proportions of compressor power consumption. It is also shown in Chapter 9 that the compressed air would need to be cooled before entry into a PEM cell – on internal combustion engines such coolers are also used, and are called *intercoolers*.

When the hydrogen fuel is derived from other hydrocarbons, such as methane, the situation is much more complex. Here the concept of 'balance of plant (BOP)', mentioned in Chapter 1, comes into play, and the fuel cell becomes apparently 'lost' in a mass of other equipment. The methods of reforming the fuel are dealt with in some detail in Chapter 8, but an introductory diagram of one possible method is given in Figure 4.23. From the point of view of 'whether to pressurise or not', the key point is that there is a burner from which hot gas is given out. This gas can be used to drive a turbine, which can drive the compressor. A motor will still often be needed to start the system, but once in operation it might not be needed at all. Indeed, it is possible to run such a system so that there is an excess of turbine power, and the motor becomes a generator, boosting the electrical power output.

4.7.2 Simple quantitative cost/benefit analysis of higher operating pressures

Running a fuel cell at higher pressure will increase the power, but it also involves the expenditure of power, and there is the cost, weight, and space taken up by the compression equipment. To consider the pros and cons of adding the extra apparatus, we have to consider more *quantitatively* the costs and the benefits of running at higher pressure, which we do in this section.

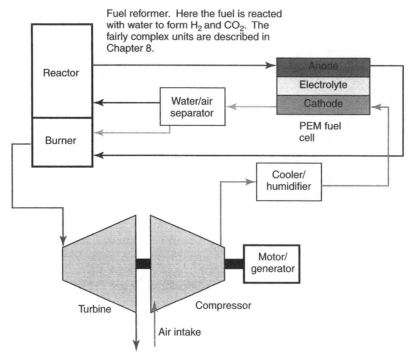

Figure 4.23 One possible method of using a turbo-compressor unit to harness the energy of the exhaust gas from the fuel reformation system to drive the compressor. Fuel reformation is covered in Chapter 8 and turbo-compressors in Chapter 9.

The increase in power resulting from operating a PEM fuel cell at higher pressure is mainly the result of the reduction in the cathode activation overvoltage, as discussed in Section 3.4. The increased pressure raises the exchange current density, which has the apparent effect of lifting the open circuit voltage (OCV), as shown in Figure 3.4. The OCV is also raised, as described by the Nernst equation, and as discussed in Section 2.5, specifically Section 2.5.4. As well as these benefits, there is also sometimes a reduction in the mass transport losses, with the effect that the voltage begins to fall off at a higher current. The effect of raising the pressure on cell voltage can be seen from the graph of voltage against current shown in Figure 4.24. In simple terms, for most values of current, the voltage is raised by a fixed value.

What is not shown in this graph is that this voltage 'boost', ΔV, is proportional to the *logarithm* of the pressure rise. This is both an experimental and a theoretical observation. In Section 2.5.4, we saw that the rise in OCV due to the change in Gibbs free energy was given by the equation

$$\Delta V = \frac{RT}{4F} \ln \left(\frac{P_2}{P_1} \right)$$

We also saw in equation 3.4 in Chapter 3 that the activation overvoltage is related to the exchange current by a logarithmic function. So, we can say that if the pressure is

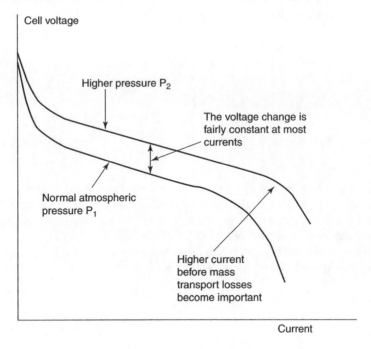

Figure 4.24   The effect of increasing pressure on the voltage current graph of a typical fuel cell.

increased from P_1 to P_2, then there is an increase or gain in voltage

$$\Delta V_{\text{gain}} = C \ln \left(\frac{P_2}{P_1} \right) \text{ volts} \qquad [4.7]$$

where C is a constant, whose value depends on how the exchange current density i_o is affected by pressure, and also on R, T, and F. Unfortunately, it is not clear what value should be used for C. In Parsons Inc. (2000)[7] figures are reported ranging from 0.03 to 0.06 V. It is certainly known that the benefit depends on the extent of the cell humidification, with more benefit if the cell is not humidified (Büchi and Srinivasan, 1997). Higher values, of up to 0.10 V, can be inferred from some results. However, for each system design, the value for the particular cell in question will have to be found.

For an initial attempt at a cost benefit analysis of pressurisation, the simple system shown in Figure 4.22 is a good place to start. In power terms the cost is straightforward – it is the power needed to drive the compressor. The benefit is the greater electrical power from the fuel cell. As we have seen, for each cell in the stack the increase in voltage ΔV is given by equation 4.7. To consider the power gain, we suppose a current of I Amps and a stack of n cells. The increase in power is then given by

$$\text{Power gain} = C \ln \left(\frac{P_2}{P_1} \right) In \text{ watts} \qquad [4.8]$$

---

[7] Derived from data on page 3–10 of this handbook.

However, as we have seen, this increase in power is not without cost. In our simple system in Figure 4.22, the power cost is that needed to drive the compressor. We show further on, in Chapter 9, Section 9.4, that an equation can be written for the power consumed by the compressor in terms of the compressor efficiency η_c, the entry temperature of the air T_1, and the pressure ratio P_2/P_1. This power is power that is lost, and so, using equation 9.7, we can say that

$$\text{Compressor power} = \text{power lost} = c_p \frac{T_1}{\eta_c} \left(\left(\frac{P_2}{P_1} \right)^{\frac{\gamma-1}{\gamma}} - 1 \right) \dot{m} \text{ watts}$$

However, this power loss is just the power delivered to the *rotor* of the compressor. This power comes from an electric motor, which has an efficiency of less than 1, and there are also losses in the connecting shaft and the bearings of the compressor rotor. If we express the combined efficiency of the motor and drive system as η_m, then the electrical power used will be greater than the compressor power by a factor of $1/\eta_m$, and so there will be a loss of *electrical* power given by the equation

$$\text{Power lost} = c_p \frac{T_1}{\eta_m \eta_c} \left(\left(\frac{P_2}{P_1} \right)^{\frac{\gamma-1}{\gamma}} - 1 \right) \dot{m} \text{ watts} \qquad [4.9]$$

In this equation $\dot{m}$ is the flow rate of the air, in kg s^{-1}. It is shown in Appendix 2, equation A2.4, that this is connected to the fuel cell electrical power output, the average cell voltage, and the air stoichiometry, by the equation

$$\dot{m} = 3.57 \times 10^{-7} \times \lambda \times \frac{P_e}{V_c} \text{ kg s}^{-1}$$

Substituting this, and electrical power $P_e = nIV_c$, and values of c_p and γ for air into equation 4.9 gives

$$\text{Power lost} = \boxed{3.58 \times 10^{-4} \times \frac{T_1}{\eta_m \eta_c} \left(\left(\frac{P_2}{P_1} \right)^{0.286} - 1 \right) \lambda} \text{ In watts} \qquad [4.10]$$

This term here in equation 4.10 is equivalent to a voltage – compare it with equation 4.8. The term is multiplied by current, and the number of cells, to derive the value of power. It is thus a *loss* equivalent to the voltage *gain* of equation 4.7. It represents the voltage of each cell that is 'used up' for driving the compressor. We can therefore write an equation that gives us the voltage loss resulting from an increase in pressure from P_1 to P_2 as

$$\Delta V_{\text{loss}} = 3.58 \times 10^{-4} \times \frac{T_1}{\eta_m \eta_c} \left(\left(\frac{P_2}{P_1} \right)^{0.286} - 1 \right) \lambda \text{ volts} \qquad [4.11]$$

We are now in a position to quantitatively consider whether a pressure increase will improve the net performance of the fuel cell system. We have a simple equation for the increase in voltage, equation 4.7. We also have a straightforward algebraic equation for the voltage loss, or rather voltage 'used up', by providing the increased pressure, equation 4.11. What we are now able to do is plot graphs of

$$\text{Net } \Delta V = \Delta V_{\text{gain}} - \Delta V_{\text{loss}} \quad \text{for different values of} \quad \frac{P_2}{P_1}.$$

However, before we can do this we need to estimate suitable values for the various constants in equations 4.7 and 4.11. In Figure 4.25 this is done for two sets of values, one designated 'optimistic', the other 'realistic'. The values for both these models are given below.

- The voltage gain constant C has already been discussed. 0.06 V would be a realistic value, though it might be as high as 0.10 V, which is used as the 'optimistic' value.
- The inlet gas temperature T_1 is taken as 15°C (288 K) for both cases.
- The efficiency of the electric drive for the compressor, η_m, is taken as 0.9 for the 'realistic' model, and 0.95 for the 'optimistic'.
- The efficiency of the compressor η_c is taken as 0.75 for the 'optimistic' model, and 0.70 for the 'realistic'.
- The lowest reasonable value for the stoichiometry, λ, is 1.75, and so this is used in the 'optimistic' model. A more realistic value is 2.0.

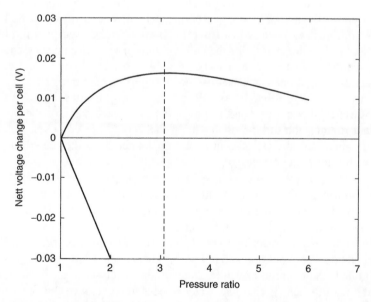

Figure 4.25 Graphs of net voltage change resulting from operating at higher pressure, for two different PEM fuel cell models.

These values are used in equations 4.7 and 4.11 to plot graphs of the net voltage change, as shown in Figure 4.25. It can be seen that for the optimistic model, there is a net gain of about 17 mV per cell when the pressure is boosted by a ratio of about 3, but the gains diminish at higher pressures. However, for the more 'realistic' model, there is *always* a net loss as a result of the higher pressure. The power gained is always exceeded by the power needed to drive the compressor. We can see why the practice of operating at above atmospheric pressure is by no means universal even with larger PEMFCs. With smaller fuel cells, less than about 5 kW, it is rare, as the compressors are likely to be less efficient and more expensive, as is explained in Section 7 of Chapter 9.

The system we have considered is very simple, yet we have been able to build a model against which to quantitatively judge the issue of the optimum pressure. However, this is only an introduction to the topic of fuel cell system modelling. There are other issues that must be considered, other factors that might make higher operational pressures more or less attractive. These we consider in the next section.

4.7.3 Other factors affecting choice of pressure

Although it is the simplest to quantify, the voltage boost is not the only benefit from operating at higher pressure. Similarly, loss of power to the compressor is not the only loss.

One of the most important gains can be shown by reference to Figure 4.23. In this system there is a fuel reforming system. Such a chemical plant usually benefits from higher pressure. In this particular case, as we shall see in Chapter 8, the kinetics of the reaction indicate that a lower pressure would be preferred. However, this is more than compensated for by the reduction in size that is achieved by higher pressure. So, as well as the fuel cell itself, the fuel reformation system also gains from higher pressure.

In the system of Figure 4.23 there is a burner, which is needed to provide heat for the fuel reformation process. The exhaust from this burner can be used by a turbine to drive the compressor – thus the energy for the compression process becomes essentially 'free'. This is not really true, because, as can be seen by a careful study of the diagram, the energy comes from hydrogen that has passed through the fuel cell, but was not turned into electricity. Typically, about 10 to 20% of the hydrogen will be needed for the fuel reformer. However, this burning and heating process is necessary for the fuel reformer, so we are not running the turbine off anything that would not otherwise be burnt. This greatly affects the analysis, as, of course, we no longer have the electrical power loss associated with running a compressor.

Another factor of great importance in favour of pressurisation is the problem of reactant humidification. Humidification is a great deal easier if the inlet air is hot and needs cooling, because there is plenty of energy available to evaporate the water. A temperature rise is always associated with compression of gases. In Chapter 9, where we consider compressors in much more detail, we show in a worked example that a typical temperature rise would be about 150°C. Air at such a temperature could be very easily humidified, and the process would also cool the air closer to a suitable temperature. However, the main benefit is that less water is needed to achieve the same humidity at higher pressures, and at higher temperatures the difference is particularly great. It can be seen from Table 4.1 that above 80°C the saturated vapour pressure rises very markedly, which means that the partial pressure of the dry air *falls* very greatly. This means that the humidity ratio increases even more. However, if the pressure is high, then the partial pressure of the air will not be so greatly affected. The reader is encouraged to look back to Section 4.4.5. Here the humidification of a 10 kW cell operating at 90°C and at about 2 bar was used as an example. It was shown that at 2 bar the humidification was not a great problem, but that if the pressure was reduced to 1.2 bar it became very difficult, and still the exit gases were too dry. In essence, **it is very difficult to arrange proper PEMFC humidification at temperatures above 80°C unless the system is pressurised to about 2 bar or more.**

With larger fuel cells, the flow paths for the reactant gases will be quite long, and if size has been minimised, they will be narrow. Thus, a certain pressure will be needed to get them through the cell in any case. If this pressure is about 0.3 bar, the designer might then be faced with the problem of choosing an available compressor or blower, and might find that they are not available at that pressure, but only at higher pressures. In which case the choice will have been forced by the very practical issue of product availability – air blowers and compressors in the pressure range of around 0.2/0.3 bar are not readily available at the smaller flow rates required by fuel cells.

On the negative side there are the issues of size, weight, and cost. However, it must be borne in mind that some sort of air blower for the reactant air would be needed anyway, so it is the *extra* size, weight, and cost of higher pressure compressors compared to lower pressure blowers that is the issue. The practical issue of product availability means that this difference will often be quite small for fuel cells of power in the region of tens of kilowatts.

With systems such as that in Figure 4.22, the compressor will often be a screw compressor, as described in Chapter 9. A problem with this type of compressor is that they can be very noisy, certainly compared to the silence of the rest of the system. This is an important problem, since it negates one of the major advantages of fuel cells over other electricity generators. The centrifugal compressors (also described in Chapter 9) associated with the turbo units of Figure 4.23 are somewhat quieter, but nevertheless increased noise will often be an important disadvantage of a pressurised system.

In the discussion thus far we have assumed that air is being used to supply the cathode with oxygen. There are some systems, notably in space applications, in which pure oxygen from pressurised cylinders is used. In these cases, the choice of operating pressure will be made on completely different grounds. The issues will be the balance between increased performance and increased weight due to the mechanical strength needed to operate at the higher pressures. The optimum pressure will probably be much higher than that for air systems.

In the last section we constructed a simple model to try to find an optimum operating pressure. Real models would have to take into account the issues we have been discussing in this section, most notably the effect of pressure on the fuel reformer and the effect of the turbine. Such models would be highly non-linear and discontinuous – pumps and blowers are not readily available at all pressure ratios and all flow rates. There will be a huge number of variables and many different ways in which the system could be run. For example, the system could be run at higher fuel stoichiometry, giving more heat out of the burner and more power at the turbine, which could be converted into electricity in the motor/generator unit. Would this give more total electrical power output than a set-up optimised for the fuel cell itself? A careful model would have to be constructed to find out. This can be done using commercially available mathematical modelling computer programs.

In practice, most small systems (<1 kW) operate at approximately ambient air pressure. It is the larger systems (>5 kW) that usually operate at higher pressure.

4.8 Reactant Composition

4.8.1 Carbon monoxide poisoning

Up to now we have generally assumed that our PEM fuel cells have been running on pure hydrogen gas as the fuel and air as the oxidant. In small systems, this will frequently be the case. However, in larger systems the hydrogen fuel will frequently come from some kind of fuel reforming system. These are discussed in much more detail in Chapter 8, and have been mentioned briefly back in Section 4.7.1, and in Figure 4.23. These fuel reformation systems nearly always involve a reaction producing carbon monoxide, such as the reaction between methane and steam

$$CH_4 + H_2O \rightarrow 3H_2 + CO$$

Some of the high-temperature fuel cells described in Chapter 7 can use this carbon monoxide as a fuel. However, fuel cells using platinum as a catalyst most certainly cannot. Even very small amounts of carbon monoxide have a very great effect on the anode. If a reformed hydrocarbon is to be used as a fuel, the carbon monoxide must be 'shifted' to carbon dioxide using more steam

$$CO + H_2O \rightarrow H_2 + CO_2$$

This reaction is called the *water gas shift reaction*. It does not easily go to completion, and there will nearly always be some carbon monoxide left in the reformed gas stream. A 'state-of-the-art' system will still have CO levels on the order of 0.25 to 0.5% ($= 2500-5000$ ppm) (Cross, 1999).

The effect of the carbon monoxide is to occupy platinum catalyst sites – the compound has an affinity for platinum and it covers the catalyst, preventing the hydrogen fuel from reaching it. Experience suggests that a concentration of carbon monoxide as low as 10 ppm has an unacceptable effect on the performance of a PEM fuel cell. This means that the CO levels in the fuel gas stream need to be brought down by a factor of 500 or more.

The methods used for removal of carbon monoxide from reformed fuel gas streams is discussed in some detail in Section 8.4.9, in Chapter 8. The extra processing needed adds considerably to the cost and size of a PEMFC system.

In some cases the requirement to remove carbon monoxide can be made somewhat less rigorous by the addition of small quantities of oxygen or air to the fuel stream (Stumper et al., 1998). This reacts with the carbon monoxide at the catalyst sites, thus removing it. Reported results show, for example, that adding 2% oxygen to a hydrogen gas stream containing 100-ppm carbon monoxide eliminates the poisoning effect of the carbon monoxide. However, any oxygen not reacting with CO will certainly react with hydrogen, and thus waste fuel. Also, the method can only be used for CO concentrations of 10s or 100s of ppm, not the 1000s of ppm concentration from typical fuel reformers. In addition, the system to feed the precisely controlled amounts of air or oxygen will be fairly complex, as the flow rate will need to carefully follow the rate of hydrogen use.

Typically, a gas clean-up unit will be used on it's own, or a less effective (and lower cost) gas clean-up unit combined with some air feed to the fuel gas.

Another important point to note is that as the molecule length of the hydrocarbon fuel to be reformed becomes longer, the problem becomes worse. With methane, CH_4, the initial reaction produced three hydrogen molecules for each CO molecule to be dealt with. If we go to a fuel such as C_8H_{18}, then the initial reaction is

$$C_8H_{18} + 8H_2O \rightarrow 17H_2 + 8CO$$

The ratio of H_2 to CO is now about 2:1 – significantly more CO to be dealt with.

4.8.2 Methanol and other liquid fuels

An ideal fuel for any fuel cell would be a liquid fuel already in regular use, such as petrol. Unfortunately such fuels simply do not react at a sufficient rate to warrant consideration – with the exception of methanol. The use of methanol can either be 'direct', where it is used as a reactant at the anode, or 'indirect', where it is reformed into hydrogen first in a separate unit. Both these uses show great promise. The former is considered in depth in Chapter 6. The use of methanol as a hydrogen carrier, releasing the hydrogen in a reformer unit, is considered in some detail in the 'fueling fuel cells' chapter, Chapter 8.

4.8.3 Using pure oxygen in place of air

It will be very rare for a designer to have a real choice between the use of air or oxygen in a PEM fuel cell system. Oxygen is used in air-independent systems such as submarines and spacecraft; otherwise air is used. However, the use of oxygen does markedly improve the performance of a PEM fuel cell. This results from three effects:

- The 'no loss' open circuit voltage rises because of the increase in the partial pressure of oxygen, as predicted by the Nernst equation, as noted in Section 2.5
- The activation overvoltage reduces because of better use of catalyst sites, as noted in Section 3.4.3.
- The limiting current increases, thus reducing the mass transport or concentration overvoltage losses. This is because of the absence of nitrogen gas, which is a major contributor to this type of loss at high current densities (see Section 3.7).

Published results (Prater, 1990) suggest that the effect of a change from air to oxygen will increase the power of a PEM fuel cell by about 30%. However, this will vary depending on the design of the cell. For example, a stack designed with poor reactant air flow will benefit more from a switch to oxygen.

Before we leave the topic of PEM fuel cells, three example systems will be presented, covering a power range from 12 W to about 250 kW.

4.9 Example Systems

It will be apparent that there are many parameters of PEM fuel cells that can be changed. These changes, such as operating temperature and pressure, air stoichiometry, reactant humidity, water retaining properties of the gas diffusion membrane, and so on, affect the performance of the cell in fairly predictable ways. This naturally lends itself very well to the construction of computer models. This can be done for things as specific as the water and heat movement; alternatively, the performance of a single cell under different conditions can be modelled (Thirumalai and White, 1997 and Wohn et al., 1998). Given the performance of a single cell, the performance of a whole stack can also be modelled (Lee and Lalk, 1998). An infinite variety of designs can thus be tried. However, by way of example, three PEM fuel cell systems are presented here to illustrate the issues we have been discussing in this chapter.

4.9.1 Small 12-W system

The first is a small 12-W fuel cell designed for use in remote conditions, such as when camping, in boats, for military applications and for remote communications, and is shown in Figure 4.26. The electrodes are in the form of a disc or ring, as in Figure 4.27. The

Figure 4.26 Twelve-watt fuel cell with no need for ventilation fans or other moving parts, as developed by DCH Technology. The fuel cell is the small cylinder in the middle of the picture. It is intended for uses such as camping, and is shown connected to a CD player, radio, and lamp. The connection to these devices is via a DC/DC converter (see Chapter 10).

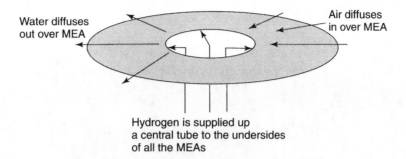

Water diffuses
out over MEA

Air diffuses
in over MEA

Hydrogen is supplied up
a central tube to the undersides
of all the MEAs

Figure 4.27 Principle of the 'disc type' PEM fuel cell.

hydrogen feed is up a central tube. Strategically placed gaps in this tube feed the fuel to the anodes on the underside of each MEA. The top of each MEA has a thick gas diffusion layer that allows oxygen to diffuse in and water to diffuse out to the edge, where it evaporates. The entire periphery is exposed to the air, so there is sufficient air circulation without any need for fans or blowers. Underneath the anode there is a gas diffusion layer and an enclosure to prevent escape of hydrogen. This will typically be about 1.5 mm deep and made of stainless steel.

The general construction method has features in common with the cell already described and shown in Figure 4.21, except that the gas diffusion layer is a more open material, and thicker, and extends up to the base of the next cell, thus also serving as the cell inter-connect. The radial symmetry of the design simplifies manufacture. Such cells have been described by Daugherty et al. (1999) and Bossel (1999), and that shown in Figure 4.26 is made along these lines.

The hydrogen supply is 'dead-ended', there is no circulation or venting – though there may be a system of periodic manual purging to release impurities from time to time.

The air circulation is basically self-regulating. When more power is drawn, more heat and more water are produced, both of which will decrease the density of the air, and thus increase circulation. On the other hand, because the air access to the cathode is limited, with water having to diffuse to the edge, the MEA does not dry out when the current is low. Although it does have the ability to respond to changes in power demand, the cell is well suited to continuous operation, and can very effectively be used in parallel with a rechargeable battery, which could provide higher power, though obviously only for short periods.

This then is an attempt to produce a 'pure' fuel cell system, with the absolute minimum of extra devices. The efficiency, as with all fuel cells, varies with current, but at rated power it is liable to be around 48% (refer lower heating value (LHV)), that is, the voltage of each cell will be about 0.6 V.

A way of indicating the various energy flows and power losses in a system is the 'Sankey diagram'. They are useful in all energy conversion systems, including fuel cells. This very 'pure' fuel cell is a good place to introduce them, as it gives almost the simplest possible diagram, which is shown in Figure 4.28. The supply is 25 W of hydrogen power, which is converted into 12 W of electrical power. The losses are simply 13 W of

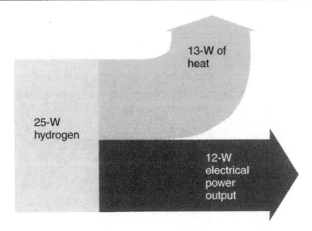

Figure 4.28   Sankey diagram for the very simple 'disc type' 12-W fuel cell.

heat – there are no other losses. The losses are swept up or down out of the left to right line of the diagram.

4.9.2 Medium 2-kW system

The second fuel cell to be discussed here is a 2.0-kW demonstrator designed and built by the Paul Scherrer Institute in Switzerland. A photograph is shown in Figure 4.29, a Sankey diagram in Figure 4.31, but we start with the system diagram in Figure 4.30. The fuel cell stack is made using the standard bipolar plate method. The bipolar plates are compression moulded graphite/polymer mixture (as in Section 4.6.3), with separate air cooling, as in Figure 4.16. The need to separate the cooling air and reactant air was explained in Section 4.5.2. To increase the maximum power of the fuel cell the reactant air is humidified, using the exit air as illustrated in Figure 4.15. The hydrogen gas is not simply 'dead-ended', but rather is circulated using a pump. This is to help with the even humidification of the larger fuel cell stack. The circulation of the hydrogen is also a way of clearing impurities out of the cell.

In this case, the Sankey diagram is rather more interesting. The final output power is just 1.64 kW. Heat loss amounting to 1.8 kW is shown, together with 360 W of electrical loss being swept downwards. This 360 W is needed to drive the three blowers and pumps and to supply an electronic controller.

Although the reactant airflow will be less than the cooling airflow, it is against a pressure of about 0.1 bar, 10 kPa above the atmospheric pressure. This is due to the length and narrowness of the reactant air path through the stack and humidifier. This is not a high pressure, but, as we shall see in Chapter 9, it makes a big difference to the blower that can be used and the power consumed. The electric power consumed by the reactant air blower is 200 W, whereas for the cooling air only 70 W is needed. The hydrogen gas flow rate is of course much less than that for the reactant air, and the pump only requires 50 W. All these pumps need controlling, and so the system has an electronic controller, which consumes 40 W. The electrical power lost amounts to 18% of

Figure 4.29 Photograph of a 2-kW fuel cell system as demonstrated by the Paul Scherrer Institute. The reactant air is pumped by pump A, through the humidifier B, to the stack C. Coolant air is blown up through the stack by blower D. The hydrogen fuel is circulated using the membrane pump E. (Reproduced by kind permission of PSI.)

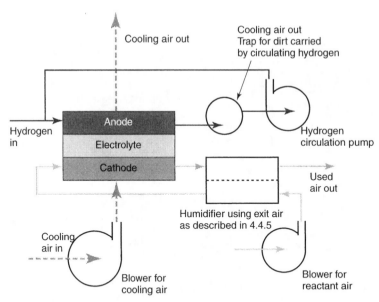

Figure 4.30 System diagram of the 2.0-kW fuel cell shown in Figure 4.29.

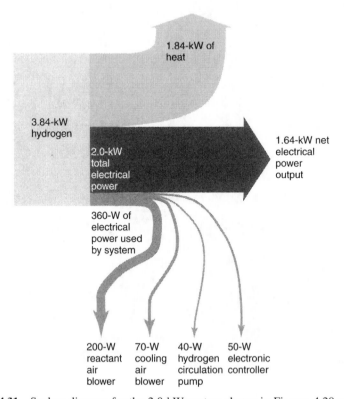

Figure 4.31 Sankey diagram for the 2.0-kW system shown in Figures 4.29 and 4.30.

Figure 4.32 Fuel cell engine that fits in the place normally taken by the diesel engine in a bus. The maximum output power of the motor is 205 kW. The power of the fuel cell is about 260 kW, with the difference driving the various pumps and compressors needed by the system. The unit is about 2.5 m wide, 1.6 m deep, and 1.33 m high. (Reproduced by kind permission of Ballard Power Systems.)

the electrical power generated by the fuel cell. The net output power is 1.64 kW, making an efficiency of 42.5% (refer LHV).

The photograph of this system in Figure 4.29 allows these various parts of the system to be seen. It is a very clearly laid out system, which illustrates many points of good PEM fuel cell system design practice.

4.9.3 205-kW fuel cell engine

Among the largest PEM fuel cell systems demonstrated is the fuel cell engine for buses produced by Ballard Power Systems, and shown in Figure 4.32. Buses are particularly suitable for fuel cell operation. Because they all refuel at one place – the bus depot – they can cope with novel fuels. Also, they operate in city centres, where pollution is often a major issue.

The fuel cell in this case has a power of about 260 kW, and the final motor drive is built into the system – hence, the manufacturers call this a 'fuel cell engine'. The power of the electric motor is 205 kW. Its dimensions are about 2.5 m wide, 1.6 m deep, and 1.33 m high. The stacks are made by using 'mainstream' methods with graphite-based bipolar plates. The system has two more or less identical stacks, as can be seen from Figure 4.32. Each half produces about 130 kW and consists of about 750 cells in series. They are water and glycol cooled. At maximum power the system is pressurised to 3 bar. The compression system and other ancillaries consume about 60 kW of power and are mounted between the two main stacks.

This fairly complex high-power system uses compressors of the type discussed in Chapter 9 and various electrical subsystems discussed in Chapter 10. So we will use this system as a 'stepping stone' to these later chapters. We will look in detail at an earlier version of this system in Chapter 11 near the end of the book, where we will use it as a case study. Readers who are only interested in PEMFCs could now move on to Chapter 8, where the very important question of 'where does the hydrogen come from?' is addressed. In Chapters 5, 6, and 7 we consider other fuel cell types in more detail.

References

Barbir, F. (1999) "Fuel cell powered utility vehicles." *Proceedings of the European Fuel Cell Forum Portable Fuel Cells Conference*, Lucerne, pp. 113–121.

Besman T.M., Klett J.W., Henry J.J., and Lara-Curzio E. (2000) "Carbon/carbon composite bipolar plate for proton exchange membrane fuel cells", *Journal of the Electrochemical Society*, **147**(11), 4083–4086.

Bevers D., Wagner N., and VonBradke M. (1998) "Innovative production procedure for low cost PEFC electrodes and electrode/membrane structures", *International Journal of Hydrogen Energy*, **23**(1), 57–63.

Bewer T., Beckmann T., Dohle H., Mergel J., and Stolten D. (2001) "Evaluation and optimisation of flow distribution in PEM and DMFC fuel cells." *Proceedings of the First European PEFC forum* (EFCF), pp. 321–330.

Bossel U.G. (1999) "Portable fuel cell charger with integrated hydrogen generator." *Proceedings of the European Fuel Cell Forum Portable Fuel Cells Conference*, Lucerne, pp. 79–84.

Büchi F.N. and Srinivasan S. (1997) "Operating proton exchange membrane fuel cells without external humidification of the reactant gases. Fundamental aspects", *Journal of the Electrochemical Society*, **144**(8), 2767–2772.

Cross J.C. (1999) "Hydrocarbon reforming for fuel cell application." *Proceedings of the European Fuel Cell Forum Portable Fuel Cells Conference*, Lucerne, pp. 205–213.

Daly B.B. (1979) *Woods Practical Guide to Fan Engineering*, Woods of Colchester Ltd, Colchester, UK, p. 339.

Daugherty M., Haberman D., Stetson N., Ibrahim S., Lokken D., Dunn D., Cherniak M., and Salter C. (1999) "Modular PEM fuel cell for outdoor applications." *Proceedings of the European Fuel Cell Forum Portable Fuel Cells Conference*, Lucerne, pp. 69–78.

Floyd D.E. (2001) "A simplified air humidifier using a porous metal membrane." *Proceedings of the First European PEFC Forum* (EFCF), pp. 417–423.

Giorgi L., Antolini E., Pozio A., and Passalacqua E. (1998) "Influence of the PTFE content in the diffusion layer of low-Pt loading electrodes for polymer electrolyte fuel cells", *Electrochimica Acta*, **43**(24), 3675–3680.

Goo Y. and Kim J. (2001) "New concept of bipolar plate and humidity control in PEMFC." *Proceedings of the First European PEFC Forum* (EFCF), pp. 211–220.

Heinzel A., Nolte R., Ledjeff-Hey K., and Zedda M. (1998) "Membrane fuel cells – concepts and system design", *Electrochimica Acta*, **43**(24), 3817–3820.

Ise M., Schmidt H., and Waidhas M. (2001) "Materials and construction principles for PEM fuel cells." *Proceedings of the First European PEFC Forum* (EFCF), pp. 285–295.

Kiefer J., Brack H-P., Huslage J., Büchi F.N., Tsakada A., Geiger F., and Schere G.G. (1999) "Radiation grafting: a versatile membrane preparation tool for fuel cell applications." *Proceedings of the European Fuel Cell Forum Portable Fuel Cells Conference*, Lucerne, pp. 227–235.

Koppel T. (1999) *Powering the Future – The Ballard Fuel Cell and the Race to Change the World*, John Wiley & Sons, Toronto.

Lee J.H. and Lalk T.R. (1998a) "Modelling fuel cell stack systems", *Journal of Power Sources*, **73**(2), 229–241.

Lee S.J. (1998b) "Effects of Nafion impregnation on performance of PEMFC electrodes", *Electrochimica Acta*, **43**(24), 3693–3701.

Lee W-K., Ho C-H., Van Zee J.W., and Murthy M., (1999) "The effects of compression and gas diffusion layers on the performance of a PEM fuel cell", *Journal of Power Sources*, **84**, 45–51.

Murphy O.J., Cisar A., and Clarke E. (1998) "Low cost light weight high power density PEM fuel cell stack", *Electrochimica Acta*, **43**(24), 3829–3840.

Parsons Inc., EG&G Services (2000) *Fuel Cells: A Handbook*, 5th ed., US Department of Energy, Springfield, pp. 3–10.

Prater K. (1990) "The renaissance of the solid polymer fuel cell", *Journal of Power Sources*, **29**(1&2), 239–250.

Ralph T.R., Hards G.A., Keating J.E., Campbell S.A., Wilkinson D.P., Davis H., StPierre J., and Johnson M.C. (1997) "Low cost electrodes for proton exchange membrane fuel cells", *Journal of the Electrochemical Society*, **144**(11), 3845–3857.

Ren X. and Gottesfeld S. (2001) "Electro-osmotic drag of water in a poly(perfluorosulphonic acid) membrane", *Journal of the Electrochemical Society*, **148**(1), A87–A93.

Rozière J. and Jones D. (2001) "Recent progress in membranes for medium temperature fuel cells." *Proceedings of the First European PEFC Forum* (EFCF), pp. 145–150.

Ruge M. and Büchi F.N. (2001) "Bipolar elements for PE fuel cell stacks based on the mould to size process of carbon polymer mixtures." *Proceedings of the First European PEFC Forum* (EFCF), pp. 299–308.

Spitzer M., Reitmajer G., Willis P.D., Buchmann H., and Glauch D. (2001) "Efficient manufacture of bipolar plates with epoxy molding components." *Proceedings of the First European PEFC Forum* (EFCF), pp. 269–276.

Stumper J., Campbell S.A., Wilkinson D.P., Johnson M.C., and Davis M. (1998) "In-situ methods for the determination of current distributions in PEM fuel cells", *Electrochimica Acta*, **43**(24), 3773–3783.

Thirumalai D. and White R.E. (1997) "Mathematical modelling of proton-exchange-membrane fuel-cell stacks", *Journal of the Electrochemical Society*, **144**(5), 1717–1723.

Warshay M., Prokopius P.R., Le M., and Voecks G. (1997) "NASA fuel cell upgrade program for the Space Shuttle Orbiter." *Proceedings of the Intersociety Energy Conversion Engineering Conference*, Vol. **1**, pp. 228–231.

Warshay M. and Prokopius P.R. (1990) "The fuel cell in space: yesterday, today and tomorrow", *Journal of Power Sources*, **29**(1&2), 193–200.

Watanabe M., Uchida H., Seki Y., Emon M., and Stonehart P. (1996) "Self humidifying polymer electrolyte membranes for fuel cells", *Journal of the Electrochemical Society*, **143**(12), 3847–3852.

Wilson M.S. and Gottesfeld S. (1992) "High performance catalysed membranes of ultra-low Pt loadings for polymer electrolyte fuel cells" *Journal of the Electrochemical Society*, **139**(2), L28–L30.

Wohn M., Bolwin K., Schnurnberger W., Fisher M., Neubrand W., and Eigernberger G. (1998) 'Dynamic modelling and simulation of a polymer membrane fuel cell including mass transport limitations", *International Journal of Hydrogen Energy*, **23**(3), 213–218.

Wood D.L., Yi J.S., and Nguyen T.V. (1998) "Effect of direct liquid water injection and interdigitated flow field on the performance of proton exchange membrane fuel cells", *Electrochimica Acta*, **43**(24), 3795–3809.

Zawodzinski T.A., Derouin C., Radzinski S., Sherman R.J., Smith V.T., Springer T.E., and Gottesfeld S. (1993) 'Water uptake by and transport through Nafion 117 membranes", *Journal of the Electrochemical Society*, **140**(4), 1041–1047.

5

Alkaline Electrolyte Fuel Cells

5.1 Historical Background and Overview

5.1.1 Basic principles

The basic chemistry of the alkaline electrolyte fuel cell (AFC) was explained in Chapter 1, Figure 1.4. The reaction at the anode is

$$2H_2 + 4OH^- \rightarrow 4H_2O + 4e^- \qquad [5.1]$$

The electrons released in this reaction pass round the external circuit, reaching the cathode, where they react, forming new OH^- ions.

$$O_2 + 4e^- + 2H_2O \rightarrow 4OH^- \qquad [5.2]$$

The electrolyte obviously needs to be an alkaline solution. Sodium hydroxide and potassium hydroxide solution, being of lowest cost, highly soluble, and not excessively corrosive, are the prime candidates. It turns out that potassium carbonate is much more soluble than sodium carbonate, and as we shall see in Section 5.6, this is an important advantage. Since the cost difference is small, potassium hydroxide solution is always used as the electrolyte. However, as we shall see, this is about the only common factor among different AFCs. Other variables such as pressure, temperature, and electrode structure vary greatly between designs. For example, the Apollo fuel cell operated at 260°C, whereas the temperature of the Orbiter alkaline fuel cell is about 90°C. The different arrangements for dealing with the electrolyte, the different types of electrodes used, the different catalysts, the choice of pressure and temperature, and so on are discussed in the sections that follow.

5.1.2 Historical importance

The alkaline fuel cell has been described since at least 1902,[1] but it was only in the 1940s abd 1950s that they were proved to be viable power units by F.T. Bacon at Cambridge.

---

[1] J.H. Reid, US patent no. 736 016 017 (1902).

Fuel Cell Systems Explained, Second Edition James Larminie and Andrew Dicks
© 2003 John Wiley & Sons, Ltd   ISBN: 0-470-84857-X

Although the proton exchange membrane (PEM) fuel cell was chosen for the first NASA manned spacecraft, it was the alkaline fuel cell that took man to the moon with the Apollo missions. The success of the alkaline fuel cell in this application, and the demonstration of high power working fuel cells by Bacon, led to a good deal of experiment and development of alkaline fuel cells during the 1960s and early 1970s. Demonstration alkaline fuel cells were used to drive agricultural tractors (Ihrig, 1960), power cars (Kordesh, 1971), provide power to offshore navigation equipment and boats, drive fork lift trucks, and so on. Descriptions and photographs of these systems can be found in McDougall (1976) and Kordesh and Simader (1996), as well as the papers already referred to.

Although many of these systems worked reasonably well as demonstrations, other difficulties, such as cost, reliability, ease of use, ruggedness, and safety were not so easily solved. When attempts were made to solve these wider engineering problems, it was found that, at that time, fuel cells could not compete with rival-energy-conversion technologies, and research and development was scaled down. The success of PEM fuel cell developments in recent years has furthered the decline in interest in alkaline fuel cells, and now very few companies or research groups are working in this field. The space program remained the one shining star in the alkaline fuel cell world, with the Apollo system being improved and developed for the space shuttle Orbiter vehicles. However, the fact that the new fuel cells for the space shuttle Orbiter vehicles will be proton exchange membrane fuel cells (PEMFCs) (Warshay et al., 1996) only serves to further underline their decline in importance. Nevertheless, because of their success with the space programme, alkaline fuel cells played a hugely important role in keeping fuel cell technology development going through the latter half of the twentieth century. Also, it could well be that the problems that we will be discussing in this chapter can be solved.

5.1.3 Main advantages

Despite their decline in relative importance, and the lack of high-profile research interest, the outlook for the AFC is not so bleak. A number of companies in the United States, Canada, and Europe are all producing cells, and quietly getting on with building fuel cell businesses. The reason is that the AFC does have a number of clear and fundamental advantages over other types of fuel cells.

The first important advantage of the AFC is that the activation overvoltage at the cathode (discussed in Chapter 3) is generally less than that with an acid electrolyte. This is the most important voltage loss in low-temperature fuel cells. It is not clearly understood why the reduction of oxygen proceeds more rapidly in an alkaline system, but it has long been observed that it does. This allows AFCs to have operating voltages as high as 0.875 V per cell, considerably higher than that with, for example, PEMFCs.

Another advantage of great importance relates to the system cost. There are very few standard chemicals that are cheaper than potassium hydroxide. It really is a very low-cost material; the electrolyte cost of the AFC is thus far less than any other type – and it always will be. Also, the electrodes, particularly the cathode, can be made from non-precious metals, and no particularly exotic materials are needed. The electrodes are thus considerably cheaper than other types of fuel cells, and there is no reason to suppose this will change in the near future either.

Figure 5.1 1.5-kW fuel cell from the Apollo spacecraft. Two of these units were used, each weighing 113 kg. These fuel cells provided the electrical power, and much of the potable water, for the craft that took man to the moon. (Photograph by kind permission of International Fuel Cells.)

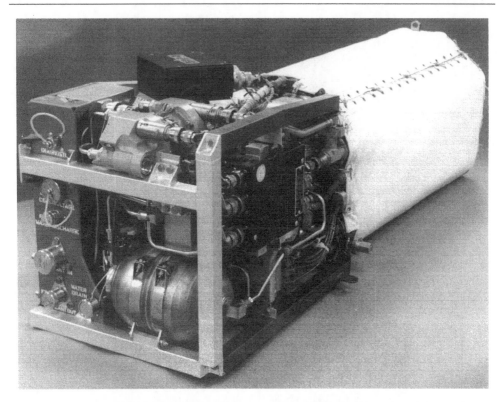

Figure 5.2 A 12-kW fuel cell used on the Space Shuttle Orbiter. This unit weighs 120 kg and measures approximately $36 \times 38 \times 114$ cm. The nearer part is the control system, pumps, and so on, and the stack of 32 cells in series under the white cover. Each cell operates at about 0.875 V. (Photograph by kind permission of International Fuel Cells.)

The final advantages are not quite so telling. We shall see in Section 5.5 that AFCs do not usually have bipolar plates. We have noticed in the previous chapter that these are a very significant contributor to the cost of the PEMFC, for example. This then has the benefit of reducing the cost, but it must be said that it is also an important cause of the lower power density of the AFC, in comparison with the PEMFC.

Finally, it should be noted that while AFCs are usually fairly complex systems, they are generally somewhat less complex than the PEMFC. For example, the water management problem is considerably more easily solved.

5.2 Types of Alkaline Electrolyte Fuel Cell

5.2.1 Mobile electrolyte

The basic structure of the mobile electrolyte fuel cell is shown below in Figure 5.3. The KOH solution is pumped around the fuel cell. Hydrogen is supplied to the anode,

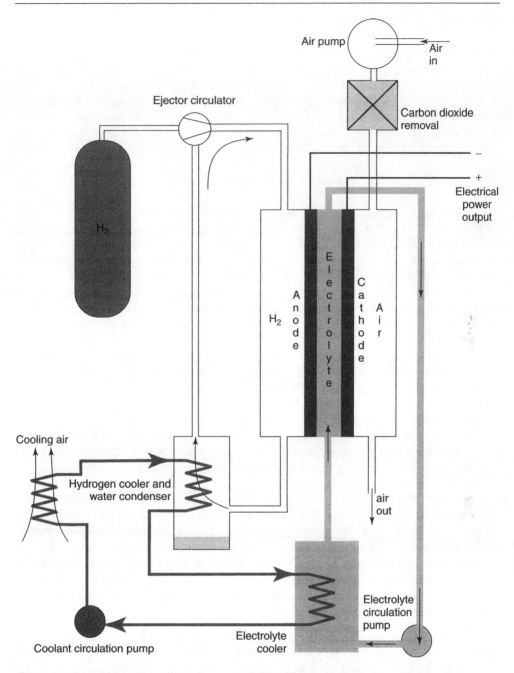

Figure 5.3 Diagram of an alkaline fuel cell with mobile electrolyte. The electrolyte also serves as the fuel cell coolant. Most terrestrial systems are of this type.

but must be circulated, as it is at the anode where the water is produced. The hydrogen will evaporate the water, which is then condensed out at the cooling unit through which the hydrogen is circulated. The hydrogen comes from a compressed gas cylinder, and the circulation is achieved using an ejector circulator, as described in Chapter 9, Section 9.10 – though this may be done by a pump in other systems.

The majority of alkaline fuel cells are of this type. The main advantage of having the mobile electrolyte is that it permits the electrolyte to be removed and replaced from time to time. This is necessary because, as well as the desired fuel cell reactions of equations 5.1 and 5.2, the carbon dioxide in the air will react with the potassium hydroxide electrolyte

$$2KOH + CO_2 \rightarrow K_2CO_3 + H_2O \qquad [5.3]$$

The potassium hydroxide is thus gradually changed to potassium carbonate. The effect of this is that the concentration of OH^- ions reduces as they are replaced with carbonate CO_3^{2-} ions, which greatly affects the performance of the cell. This major difficulty is discussed further in Section 5.6, but one way of at least reducing it is to remove the CO_2 from the air as much as possible of, and this is done using a CO_2 scrubber in the cathode air supply system. However, it is impossible to remove **all** the carbon dioxide, so the electrolyte will inevitably deteriorate and require replacing at some point. This mobile system allows that to be done reasonably easily, and of course potassium hydroxide solution is of very low cost.

The disadvantages of the mobile electrolyte centre around the extra equipment needed. A pump is needed, and the fluid to be pumped is corrosive. The extra pipework means more possibilities for leaks, and the surface tension of KOH solution makes for a fluid that is prone to find its way through the smallest of gaps. Also, it becomes harder to design a system that will work in any orientation.

There is a further very important problem that cannot be deduced from Figure 5.3, since it only shows one cell. Because the electrolyte is pumped through all the cells in a stack, they are effectively joined together. Ionic conduction between the cells within a stack can seriously affect the stack performance. This is mitigated by making the circulation system give the longest and narrowest possible current path between the KOH solution in each cell – otherwise the electrolyte of each and every cell will be connected together, and there will be an internal 'short-circuit'.[2] The problem is also reduced in some systems by connecting the cells in series **and** parallel, to reduce the internal voltages. So, for example, a 24-cell stack may be connected as two systems of 12 cells in parallel. This is mentioned again when cell interconnection is explained in Section 5.5.

To summarise then, the main advantages of the mobile electrolyte-type alkaline fuel cell are as follows:

- The circulating electrolyte can serve as a cooling system for the fuel cell.
- The electrolyte is continuously stirred and mixed. It will be seen from the anode and cathode equations given at the beginning of the chapter that water is consumed at the cathode and produced (twice as fast) at the anode. This can result in the electrolyte

---

[2] This internal current can be measured by finding the hydrogen consumption at open circuit. For example, in the cells used by Kordesh (1971), the internal current was found to be about $1.5\,mA\,cm^{-2}$.

becoming too concentrated at the cathode – so concentrated that it solidifies. Stirring reduces this problem.

- Having the electrolyte circulate means that if the product water transfers to the electrolyte, rather than evaporating at the anode, then the electrolyte can be passed through a system for restoring the concentration (i.e. an evaporator).
- It is comparatively straightforward to pump out all the electrolyte and replace it with a fresh solution, if it has become too dilute by reaction with carbon dioxide, as in equation 5.3.

This mobile electrolyte system was used by Bacon in his historic alkaline fuel cells of the 1950s and in the Apollo mission fuel cells. It is almost universally used in terrestrial systems, but the Shuttle Orbiter vehicles use a static electrolyte, as described in the next section.

5.2.2 Static electrolyte alkaline fuel cells

An alternative to a 'free' electrolyte, which circulates as in Figure 5.3, is for each cell in the stack to have its own, separate electrolyte that is held in a matrix material between the electrodes. Such a system is shown in Figure 5.4. The KOH solution is held in a matrix material, which is usually asbestos. This material has excellent porosity, strength, and corrosion resistance, although, of course, its safety problems would be a difficulty for a fuel cell system designed for use by members of the public.

The system of Figure 5.4 uses pure oxygen at the cathode, and this is almost obligatory for a matrix-held electrolyte, as it is very hard to renew if it becomes carbonated as in equation 5.3. The hydrogen is circulated, as with the previous system, in order to remove the product water. In the spacecraft systems, this product water is used for drinking, cooking, and cabin humidification. However, a cooling system will also be needed, and so cooling water, or other fluid, is needed. In the Apollo system it was a glycol/water mixture, as is used in car engines. In the Orbiter systems the cooling fluid is a fluorinated hydrocarbon dielectric liquid (Warshay and Prokopius, 1990).

This matrix-held electrolyte system is essentially like the PEM fuel cell – the electrolyte is, to all intents and purposes, solid and can be in any orientation. A major advantage is, of course, that the electrolyte does not need to be pumped around or 'dealt with' in any way. There is also no problem of the internal 'short circuit' that can be the result of a pumped electrolyte. However, there is the problem of managing the product water, the evaporation of water, and the fact that water is used at the cathode. Essentially, the water problem is similar to that for PEM fuel cells, though 'inverted', in that water is produced at the anode and removed from the cathode. (In the PEMFC, water is produced at the cathode and removed from the anode by electro-osmotic drag, as explained in Section 4.4.) The fuel cell must be designed so that the water content of the cathode region is kept sufficiently high by diffusion from the anode. Generally though, the problem of water management is much less severe than with the PEMFC. For one thing, the saturated vapour pressure of KOH solution does not rise so quickly with temperature as it does with pure water, as we will see in Section 5.3 below. This means that the rate of evaporation is much less.

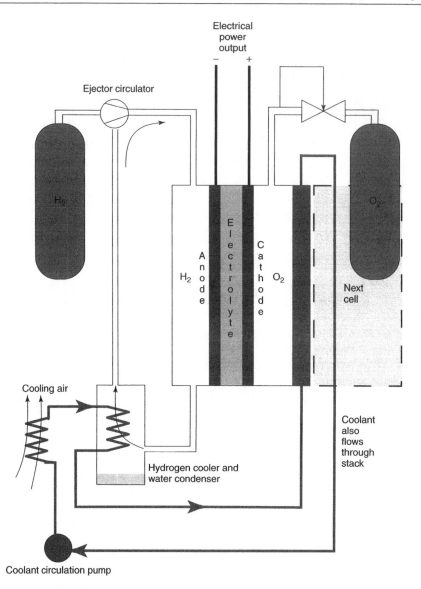

Figure 5.4 An alkaline electrolyte fuel cell with static electrolyte held in a matrix. This system uses pure oxygen instead of air.

In space applications, the advantages of greater mechanical simplicity mean that this approach is now used. However, for terrestrial applications, where the problem of carbon dioxide contamination of the electrolyte is bound to occur, renewal of the electrolyte must be possible. For this matrix-type cell, this would require a complete fuel cell rebuild. Also, the use of asbestos is a severe problem, as it is hazardous to health, and in some countries its use is banned. A new material needs to be found, but research in this area is unlikely to be undertaken while the likelihood of its eventual use is so low.

5.2.3 Dissolved fuel alkaline fuel cells

This type of fuel cell is unlikely to be used for high power generation, but is included as it is the simplest type of fuel cell to manufacture, and it shows how the alkaline electrolyte can be the basis of very simple fuel cells. It has been used in a number of successful fuel cell demonstrators, and older books on fuel cells (Williams, 1966) give this type of cell a great deal of coverage.

The principle is shown in Figure 5.5. The electrolyte is KOH solution, with a fuel, such as hydrazine, or ammonia, mixed with it. The fuel anode is along the lines discussed in Section 5.4.4, with a platinum catalyst. The fuel is also fully in contact with the cathode. This makes the 'fuel crossover' problem discussed in Section 3.5 very severe. However, in this case, it does not matter greatly, as the cathode catalyst is not platinum, and so the rate of reaction of the fuel on the cathode is very low. There is thus only one seal that could leak, a very low-pressure joint around the cathode. The cell is refuelled simply by adding more fuel to the electrolyte. A possible variation is to put a membrane across the electrolyte region and only provide fuel below the membrane – a suitably chosen membrane will stop the fuel from reaching the air cathode.

An ideal fuel for this type of fuel cell is hydrazine, H_2NNH_2, as it dissociates into hydrogen and nitrogen on a fuel cell electrode. The hydrogen then reacts normally. (Other reaction routes with similar results are possible, as described by McDougall (1976), p. 71.) Low cost, compact, simple, and easily refuelled cells can be made this way. This chemistry was the basis of several impressive fuel cell demonstrators in the 1960s. Unfortunately, however, hydrazine is toxic, carcinogenic, and explosive! It could possibly be used as a fuel in certain very well-regulated circumstances, but it is not suitable for widespread use, and so cells using it can only be used for demonstration purposes.

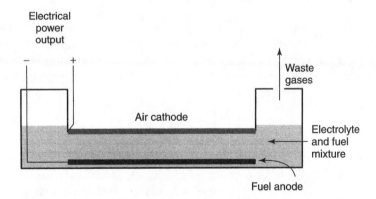

Figure 5.5 The diagram shows the principle of the dissolved fuel fuel cell, arguably the simplest of all types. This one has a selective catalyst on the cathode that does not react with the fuel; in other cases, a membrane within the electrolyte can keep the fuel away from the air cathode, but this adds to cost and complexity.

The dissolved fuel principle can, in theory, be used with acid electrolyte fuel cells. However, practical considerations mean that only alkali electrodes are viable. It is very difficult, for example, to make an active catalyst in a low-temperature acid electrolyte fuel cell that does not use precious metals and that will not therefore oxidise the fuel.

An interesting type of dissolved fuel fuel cell that has received renewed interest in recent times is based on sodium borohydride (sodium tetrahydridoborate). This is being considered as a hydrogen carrier (see Chapter 8), but its use as a direct fuel is actually slightly better thermodynamically. The principle has been known and demonstrated for many years (Indig and Snyder, 1962 and Williams, 1966), but interest has recently been revived (Amendola et al., 1999). The fuel ($NaBH_4$) is dissolved in the electrolyte, and the fuel anode reaction is

$$NaBH_4 + 8OH^- \rightarrow NaBO_2 + 6H_2O + 8e^- \qquad [5.4]$$

The impressive fact to note is that we get **eight** electrons for just one molecule of fuel. Things get even more interesting if we look at the Gibbs free energy changes and thus the reversible voltage. The air cathode reaction is exactly the same as with hydrogen, as in equation 5.2. The overall reaction is thus

$$NaBH_4 + 2O_2 \rightarrow NaBO_2 + 2H_2O$$

For this reaction,

$$\Delta \bar{g}_f = (-920.7 - (2 \times 237.2)) + 123.9 = -1271.2 \, kJ \, mol^{-1}$$

So, from equation 2.2,

$$E = \frac{-\Delta \bar{g}_f}{zF} = \frac{1271.2 \times 10^3}{8 \times 96485} = 1.64 \, V$$

This voltage is noticeably higher than that obtained from hydrogen, and at eight electrons per molecule it indicates a fuel of remarkable potency. Unfortunately, the voltages actually obtained are not so different from a hydrogen fuel cell, because the catalysts that promote the direct borohydride oxidation of equation 5.4 also promote the hydrolysis reaction:

$$NaBH_4 + 2H_2O \rightarrow NaBO_2 + 4H_2 \qquad [5.5]$$

This hydrolysis reaction is the main cause for this technology being abandoned in the 1960s. The electrodes in use at that time were not able to properly use the hydrogen, and so the hydrogen loss made the system very inefficient. However, since these four hydrogen molecules are formed right on the electrode, with modern porous electrodes, it is possible to arrange things so that they will immediately oxidise, giving us eight electrons – two for each hydrogen molecule (see equation 5.1). This is possible provided the hydrolysis reaction is not proceeding too quickly. So, we still get eight electrons per molecule of borohydride, but the voltage is the same as for a hydrogen fuel cell. However, if the concentration of the borohydride in the electrolyte is low and the electrolyte concentration is high, the hydrolysis reaction is significantly slowed down.

All these reactions, the oxidation of borohydride (5.4), the hydrolysis of borohydride (5.5), and the oxidation of hydrogen (5.1), all proceed quite readily on platinum catalysts. The result is that this type of fuel cell performs very well, even at room temperature. The electrodes can be of very low cost, as the platinum loading can be low, and it is only needed on the fuel anode. The cell is also extremely simple to make, as the electrolyte and the fuel are mixed. An example demonstration cell is shown in Figure 5.6, which costs not more than a few dollars to make, yet just a single cell is sufficient to power a radio. The energy density of the fuel is far in excess of any other candidate fuel cell fuel.

In principle then, this type of fuel cell looks very attractive. The problem is the cost of the fuel – sodium borohydride is very expensive. This point is given a little more consideration in Chapter 8, when its use as a hydrogen carrier is considered. However, this could change, and so this type of fuel cell is certainly one to watch out for in the future.

Figure 5.6 This simple, very low-cost cell is of the 'dissolved fuel' type. When using sodium borohydride as the fuel, it operates at about 0.75 V. A single cell, with DC/DC converter, can drive light loads such as a radio.

5.3 Operating Pressure and Temperature

Historically, the major alkaline electrolyte fuel cells have operated at well above ambient pressure and temperature. The pressures and temperatures, together with information about the electrode catalyst, is given for a selection of important alkaline fuel cells in Table 5.1.

The advantages of higher pressure have been seen in Chapter 2, where it was shown in Section 2.5.4 that the open circuit voltage (OCV) of a fuel cell is raised when the pressure increases according to the formula

$$\Delta V = \frac{RT}{4F} \ln \left(\frac{P_2}{P_1} \right)$$

The demonstration cell of F.T. Bacon is operated at around 45 bar, and 200°C. However, even these high pressures would only give a 'lift' to the voltage of about 0.04 V if this 'Nernstian' effect was the sole benefit. We also saw, in Chapter 3, Section 3.4, that a rise in pressure (and/or temperature) increases the exchange current density, which reduces the activation overvoltage on the cathode. The result is that the benefit of increased pressure is much more than the equation above would predict. As a consequence, the very high pressure gave the 'Bacon cell' a performance level that even today would be considered remarkable: 400 mA cm^{-2} at 0.85 V or 1 A cm^{-2} at 0.8 V.

The choice of operating pressure and temperature, the KOH concentration, and the catalyst used are all linked together. A good example is the movement from the Bacon cell to the system used in the Apollo spacecraft. As we have seen, the performance of the 'Bacon cell' is very good, but it has a heavy engineering design, a very high-pressure system. To obtain a low enough mass for space applications, the pressure had to be reduced. However, to maintain the performance at an acceptable level, the temperature had to be increased. It was then necessary to increase the concentration of the KOH to 75%, otherwise the electrolyte would have boiled. Increasing the KOH concentration considerably lowers the vapour pressure, as can be seen from Figure 5.7. At ambient temperature, this concentration of KOH is solid, and so heaters had to be provided to start the system. In the Orbiter system, the electrolyte concentration was reduced back to 32% and the temperature

Table 5.1 Operating parameters for certain alkaline electrolyte fuel cells. [Data from Warshay and Prokopius, 1990 and Strasser, 1990.] The pressure figures are approximate, since there are usually small differences between each reactant gas

Fuel cell	Pressure bar	Temp. °C	KOH (% conc.)	Anode catalyst	Cathode catalyst
Bacon	45	200	30	Ni	NiO
Apollo	3.4	230	75	Ni	NiO
Orbiter	4.1	93	35	Pt/Pd	Au/Pt
Siemens	2.2	80		Ni	Ag

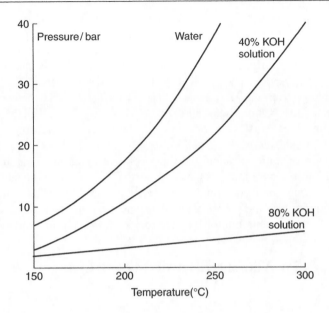

Figure 5.7 Graphs showing the change in vapour pressure with temperature for different concentrations of KOH solution.

to 93°C. To maintain performance, an electro-active catalyst is needed, and $20\,mg\,cm^{-2}$ of gold/platinum alloy is used on the cathode, and $10\,mg\,cm^{-2}$ Pt on the anode.

Most alkaline electrolyte fuel cells operate using reactant gases from high pressure or from cryogenic storage systems. In each case the gas is supplied at fairly high pressure, and so the 'costs' involved with operating at high pressure are those connected with the extra mass involved with designing a high-pressure system. In addition to the problem of containing the gases, and preventing leaks, there is also the problem of internal stresses if the reactants are at different pressures, so the two pressures must be accurately controlled.

The problem of leaks from high-pressure systems is obviously a concern. Apart from the waste of gas, there is also the possibility of the build-up of explosive mixtures of hydrogen and oxygen, especially when the cell is intended for use in confined spaces such as submarines. One solution to this problem is to provide an outer envelope for the fuel cell stack that is filled with nitrogen, and to have this nitrogen at a higher pressure than any of the reactant gases. As an example, Strasser (1990) describes the Siemens system in which the hydrogen is supplied at 2.3 bar, the oxygen at 2.1 bar, and the surrounding nitrogen gas is at 2.7 bar. Any leak would result in the flow of nitrogen into the cells, which would reduce the performance, but would prevent an outflow of reactant gas.

In AFCs, there is often a difference in the pressure of the reactant gases and/or of the electrolyte. The pressure differences in the Siemens AFC, where the hydrogen pressure is slightly higher, have already been noted. In the Orbiter fuel cell, the hydrogen gas pressure is kept at 0.35 bar *below* the oxygen pressure. In the Apollo system the gases were at the same pressure, but both were about 0.7 bar above the pressure of the electrolyte. No general rules can be given about this. The small pressure differences will be required for a variety of reasons – electrode diffusion and product water management being two examples. The needs will strongly depend on the details of the system.

We saw in Chapter 2 that raising the temperature actually reduces the 'no loss' OCV of a fuel cell. However, in practice, this change is far exceeded by the reduction in the activation overvoltage, especially on the cathode. As a result, increasing the temperature increases the voltage from an AFC. Hirschenhofer et al. (1995) has shown, from a wide review of results, that below about 60°C there is a very large benefit in raising the temperature, as much as $4\,mV\,°C^{-1}$ for each cell. At this rate, increasing the temperature from 30 to 60°C would increase the cell voltage by about 0.12 V, a big increase in the context of fuel cells operating at about 0.6 V per cell. At higher temperatures, there is still a noticeable benefit, but only in the region of $0.5\,mV\,°C^{-1}$. We could conclude that about 60°C would be a minimum operating temperature for an AFC. Above this, the choice would depend strongly on the power of the cell (and thus any heat losses), the pressure, and the effect of the concentration of the electrolyte on the rate of evaporation of water.

5.4   Electrodes for Alkaline Electrolyte Fuel Cells

5.4.1 Introduction

We have already pointed out that alkaline fuel cells can be operated at a wide range of temperatures and pressures. It is also the case that their range of applications is quite restricted. The result of this is that there is no standard type of electrode for the AFC, and different approaches are taken depending on performance requirements, cost limits, operating temperature, and pressure. Different catalysts can also be used, but this does not necessarily affect the electrode structure. For example, platinum catalyst can be used with any of the main electrode structures described here.

5.4.2 Sintered nickel powder

When F.T. Bacon designed his historically important fuel cells in the 1940s and 1950s, he wanted to use simple, low-cost materials and avoided precious metal catalysts. He thus opted for nickel electrodes. These were made porous by fabricating them from powdered nickel, which was then sintered to make a rigid structure. To enable a good three-phase contact between the reactant gas, the liquid electrolyte, and the solid electrode, the nickel electrode was made in two layers using two sizes of nickel powders. This gave a wetted fine-pore structure to the liquid side and more open pores to the gas side (Appleby, 1990). This structure gave very good results, though careful control of the differential pressure between the gas and the electrolyte was needed to ensure that the liquid gas boundary sat at the right point, as wet proofing materials such as polytetrafluoroethylene (PTFE) were not available at that time.

This electrode structure was also used in the Apollo mission fuel cells. Such structures may or may not be combined with catalysts. In the Apollo and Bacon cells, the anode was formed from the straightforward nickel powder as described above, whereas the cathode was partially lithiated and oxidised.[3]

---

[3] The addition of lithium salts to the surface of the cathode reduced oxidation of the nickel.

5.4.3 Raney metals

An alternative method of achieving a very active and porous form of a metal, which has been used for alkaline fuel cells from the 1960s to the present, is the use of Raney metals. These are prepared by mixing the active metal (e.g. nickel) with an inactive metal, usually aluminium. The mixing is done in such a way that distinct regions of aluminium and the host metal are maintained – it is not a true alloy. The mixture is then treated with a strong alkali that dissolves out the aluminium. This leaves a porous material, with very high surface area. This process does not require the sintering of the nickel powder method, yet does give scope for changing the pore size, since this can be varied by altering the degree of mixing of the two metals.

Raney nickel electrodes prepared in this way were used in many of the fuel cell demonstrations mentioned in the introduction to this chapter. Often Raney nickel was used for the anode and silver for the cathode. This combination was also used for the electrodes of the Siemens alkaline fuel cell used in submarines in the early 1990s (Strasser, 1990). They have also been used more recently, in a ground up form, in the rolled electrodes to be described in Section 5.4.4 (Gulzow, 1996).

5.4.4 Rolled electrodes

Modern electrodes tend to use carbon-supported catalysts, mixed with PTFE, which are then rolled out onto a material such as nickel mesh. The PTFE acts as a binder, and its hydrophobic properties also stop the electrode from flooding and provide for controlled permeation of the electrode by the liquid electrolyte. A thin layer of PTFE is put over the surface of the electrode to further control the porosity and to prevent the electrolyte passing through the electrode, without the need to pressurise the reactant gases, as has to be done with the porous metal electrodes. Carbon fibre is sometimes added to the mix to increase strength, conductivity, and roughness. Such an electrode is shown in Figure 5.8.

The manufacturing process can be done using modified paper-making machines, at quite low cost. Such electrodes are not just used in fuel cells but are also used in metal/air batteries, for which the cathode reaction is much the same as for an alkali fuel cell. For example, the same electrode can be used as the cathode in a zinc air battery (e.g. for hearing aids), an aluminium/air battery (e.g. for telecommunications reserve power), and an alkaline electrolyte fuel cell. The carbon-supported catalyst is of the same structure as that shown in Figure 4.6 in the previous chapter. However, the catalyst will not always be platinum. For example, manganese can be used for the cathode in metal air batteries and fuel cells.

With a non-platinum catalyst, such electrodes are readily available at a cost from about \$0.01 per cm^2, or around \$10 per ft^2, which is very low compared to other fuel cell materials. Adding a platinum catalyst increases the cost, depending on the loading, but it might only be by a factor of about 3, which still gives, in fuel cell terms, a very low-cost electrode. However, there are problems.

One problem is that the electrode is covered with a layer of PTFE, and so the surface is non-conductive, and thus a bipolar plate cannot be used for cell interconnection. Instead, the cells have to be edge-connected. This is not so bad, as with the nickel mesh running right through the electrode, its surface conductivity is quite good. A more serious problem

Figure 5.8 Photograph shows the structure of a rolled electrode. The catalyst is mixed with a PTFE binder and is rolled onto a nickel mesh. The thin layer of PTFE on the gas side is shown partially rolled back.

is the effect of carbon dioxide on the electrode performance. This is in addition to the effect that carbon dioxide has on the electrolyte, which we noted in equation 5.3, though it might be connected in that there could be an effect of contamination by the carbonate. In reviewing the literature in this area, in Hirschenhofer et al. (1995) it is noted that the lifetime of such carbon-supported catalyst electrodes was from 1600 to 3400 h at $65°C$ and $100\,mA\,cm^{-2}$ when using air containing CO_2. However, if the carbon dioxide is removed from the air, the lifetime of the electrodes increases to at least 4000 h under similar conditions. This is not a particularly high current density, and the lifetime is less at higher currents. Lower temperatures also shorten life, presumably because the solubility of the carbonate decreases. Sixteen hundred hours is only about 66 days, and is not suitable for any but the smallest number of applications. (Note, however, that it is quite adequate for the operating time of a battery, for which this type of electrode is generally used.)

When using such carbon-supported electrodes, it is clear that the carbon dioxide must be removed from the air. Another approach to the problem is the use of a different type of rolled electrode, which does not use carbon-supported catalyst. Gulzow (1996) describes an anode based on granules of Raney nickel mixed with PTFE. This is rolled onto a metal net in much the same way as the PTFE/carbon-supported catalyst. A cathode can be prepared in much the same way, only using silver instead of nickel. It is claimed that these electrodes are not damaged by CO_2.

5.5 Cell Interconnections

The great majority of current AFC systems use electrodes of the type shown in Figure 5.8, which have a PTFE layer over the 'gas side'. PTFE is an insulator, and so it is difficult to make an electrical connection to the face of the electrode. The result is that the 'bipolar plate–type cell interconnect' usually used in fuel cells, and shown in Figures 1.9 and 1.10, cannot normally be used in the AFC. This is also encouraged by the fact that the electrode is built onto a metal mesh, and so has a higher than normal conductivity across the plane, making edge connections less of a problem.

The connections are thus normally made to the edge of the electrode. Simple wires connect the positive of one cell to the negative of another. This gives a certain flexibility: it is not necessary to connect the positive of one cell to the negative of the adjacent cell, as must occur with bipolar plates. Instead, complex series/parallel connections can be made, as in Figure 5.9.

It is often the practice in AFCs to have the cathodes of adjacent cells facing each other, so that one air channel feeds two cathodes. The same applies to the anodes, that is, two are together. Figure 5.9 illustrates this, with the orientation of the cells alternating along the stack, $-++--++-$, and so on. The complex connection patterns are sometimes used in order to reduce the problem of internal currents within the electrolyte of the fuel cell stack. These arise because the ionically conductive electrolyte is in contact with all the cells within a stack. Having cells in parallel reduces the voltage while increasing the current. Also, the alternating pattern can sometimes helpfully balance electric fields within the electrolyte.

5.6 Problems and Development

We have already noted that activity in the area of alkaline fuel cells is currently rather low. The main problem with AFCs for terrestrial applications is the problem of carbon

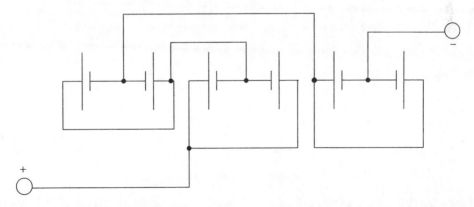

Figure 5.9 The cell in an alkaline fuel cell stack is often connected in a complex series/parallel circuit to reduce internal currents within the electrolyte. Notice that the $+ -$ orientation of the cells alternates along the stack.

dioxide reactions with the alkaline electrolyte. This occurs with the carbon dioxide in the air and would happen even more strongly if hydrogen derived from hydrocarbons (such as methane) was used as the fuel.

The problem is that the carbon dioxide reacts with the hydroxide ion, forming carbonate, as in equation 5.3. The effects of this are as follows:

- The OH^- concentration is reduced, thus reducing the rate of reaction at the anode.
- The viscosity is increased, reducing the diffusion rates, thus lowering the limiting currents and increasing the mass transport losses, as discussed in Chapter 3, Section 3.7.
- The carbonate salt is less soluble, and so will eventually precipitate out, blocking the pores and causing other damage to the electrodes.
- Oxygen solubility is reduced, increasing the activation losses at the cathode.
- The electrolyte conductivity is reduced, increasing the ohmic losses.
- The electrodes performance may be degraded, as noted above in Section 5.4.4.

Most of the notable achievements of AFCs have been the use of pure hydrogen and oxygen supplies for air-independent power sources. However, the scope for these is limited.

For an alkaline electrolyte fuel cell to work well over a long period, it is essential to remove the carbon dioxide from the air. This can be done, but of course increases costs, complexity, mass, and size. One way that shows particular promise has been described by Ahuja and Green (1998), though it can only be used when hydrogen is stored cryogenically. Basically, it takes advantage of the fact that heat exchangers are needed to warm the hydrogen and cool the fuel cell. These could be combined with a system to freeze the carbon dioxide out of the incoming air, which can then be reheated.

Another possibility, which is actually what Bacon had in mind when developing his AFC designs in the mid 20th century, is to incorporate the cells into a regenerative system. Electricity from renewable sources is used to electrolyse water when the energy is available, and the fuel cell turns it back into electricity as needed. Of course, any fuel cell could be used in such a system, but here the AFC's disadvantages would be largely removed, since both hydrogen and oxygen would be generated. Thus their advantages of low cost, simplicity, lack of exotic materials, good cathode performance, and wide range of operating temperatures and pressures might bring them to the fore again.

References

Amendola S.C., Onnerud P., Kelly M.T., Petillo P.J., Shar-Goldman S.L., and Binder M. (1999) "A novel high power density borohydride-air cell", *Journal of Power Sources*, **84**, 130–133.
Ahuja V. and Green R. (1998) "Carbon dioxide removal from air for alkaline fuel cells operating with liquid nitrogen – a synergistic advantage", *International Journal of Hydrogen Energy*, **23**(20), 131–137.
Appleby A.J. (1990) "From Sir William Grove to today, fuel cells and the future", *Journal of Power Sources*, **29**, 3–11.
Gulzow E. (1996) "Alkaline fuel cells: a critical view", *Journal of Power Sources*, **61**, 99–104.
Hirschenhofer J.H., Stauffer D.B., and Engleman R.R. (1995) *Fuel Cells: A Handbook*, revision 3, Business/Technology Books, Orinda, CA, Section 6, pp. 10–15.
Ihrig H.K. (1960), 11th Annual Earthmoving Industry Conference, SAE Paper No. S-253.

Indig M.E. and Snyder R.N. (1962) "Sodium borohydride, and interesting anodic fuel", *Journal of the Electrochemical Society*, **109**, 1104–1106.

Kordesh K.V. (1971) "Hydrogen-air/lead battery hybrid system for vehicle propulsion", *Journal of the Electrochemical Society*, **118**(5), 812–817.

Kordesh K.V. and Simader G. (1996) *Fuel Cells and Their Applications*, VCH Verlagsgesellschaft, Weinheim.

McDougall A. (1976) *Fuel Cells*, Macmillan, London.

Strasser K. (1990) "The design of alkaline fuel cells," *Journal of Power Sources*, **29**, 149–166.

Warshay M. and Prokopius P.R. (1990) "The fuel cell in space: yesterday, today and tomorrow", *Journal of Power Sources*, **29**, 193–200.

Warshay M., Prokopius P.R., Le M., and Voecks G. (1996) "NASA fuel cell upgrade program for the space shuttle orbiter." *Proceedings of the Intersociety Energy Conversion Engineering Conference*, Vol. 1, pp. 1717–1723.

Williams K.R. (ed.) (1966) *An Introduction to Fuel Cells*, Elsevier, Amsterdam, pp. 110–112.

6

Direct Methanol Fuel Cells

6.1 Introduction

There is a problem as to the most logical place to put this chapter. There is a good case for putting it later in the book because the major advantages of the direct methanol fuel cell (DMFC) cannot be appreciated unless the problems of supplying hydrogen to portable fuel cells are grasped. If the readers look ahead to Chapter 8, where this question is addressed, they will see a very long chapter, indicating the difficulty of the problem and the many issues that need to be considered. It would be useful at this point to look at Table 8.20 right at the end of the chapter. Here, it will be seen that the usual methods of storing hydrogen are remarkably inefficient. Hydrogen itself has a high energy density, but it is very difficult to contain. Nevertheless, the DMFC is a fuel cell type, and so it should be described here in this earlier part of the book, where different fuel cell technologies are explained.

Table 8.20 features methanol as a carrier of hydrogen – indeed, it is arguably the best. But that is **not** what this chapter is concerned with. This chapter is concerned with the use of methanol **directly** as a fuel. We have already mentioned this possibility at various points in Chapters 2 and 3. If methanol can be used as a fuel, then all the problems of storing or making hydrogen are swept aside. Methanol is a readily available and low-cost liquid fuel that has an energy density not very different from gasoline. If it could be used directly in fuel cells, the resulting weight of any portable fuel cell system would be vastly reduced. Table 6.1 shows how methanol compares with the main hydrogen storage technologies.

The net energy density of methanol is higher than any of the other options in Table 6.1, and much higher than the first two options involving hydrogen storage. This is the main advantage of the direct methanol system. The fact that the system is simpler to use and very quick to refill are also important advantages. The question of safety is rather more balanced, since there are almost as many questions about the safety of methanol as there are for hydrogen. This will be discussed in Section 6.5.

Fuel Cell Systems Explained, Second Edition James Larminie and Andrew Dicks
© 2003 John Wiley & Sons, Ltd   ISBN: 0-470-84857-X

Table 6.1 Energy density comparison for methanol and the most important hydrogen storage technologies. Estimated mass of the reformer is included in the case of indirect methanol, where it is chemically reacted to produce H_2 in the ways outlined in Chapter 8

Storage method	Energy density of fuel	Storage efficiency (%)	Net energy density
H_2 at 300 bar pressure in composite cylinders	$119.9\,MJ\,kg^{-1}$ $33.3\,kWh\,kg^{-1}$	0.6	$0.72\,MJ\,kg^{-1}$ $0.20\,kWh\,kg^{-1}$
H_2 in metal hydride cylinders	$119.9\,MJ\,kg^{-1}$ $33.3\,kWh\,kg^{-1}$	0.65	$0.78\,MJ\,kg^{-1}$ $0.22\,kWh\,kg^{-1}$
H_2 from methanol – indirect methanol	$119.9\,MJ\,kg^{-1}$ $33.3\,kWh\,kg^{-1}$	6.9	$8.27\,MJ\,kg^{-1}$ $2.3\,kWh\,kg^{-1}$
Methanol in strong plastic tanks for direct use as fuel	$19.9\,MJ\,kg^{-1}$ $5.54\,kWh\,kg^{-1}$	95	$18.9\,MJ\,kg^{-1}$ $5.26\,kWh\,kg^{-1}$

Section 6.5 will also consider the problems of storing and producing methanol, and how it compares with other alcohols such as ethanol.

The most pressing problem associated with the DMFC is that the fuel anode reactions proceed so much more slowly than with hydrogen. The oxidation of hydrogen occurs readily – the oxidation of methanol is a much more complex reaction, and proceeds much more slowly. This results in a fuel cell that has a far lower power for a given size. The anode reactions of the DMFC will be explained in Section 6.2

The second major problem is that of fuel crossover. This was discussed briefly in Section 3.5. It is particularly acute in the DMFC because the electrolyte used is usually a proton exchange membrane (PEM), as described in Chapter 4. These readily absorb methanol, which mixes well with water, and so quickly reaches the cathode. This shows itself as a reduced open circuit voltage but affects the performance of the fuel cell at all currents. DMFC electrolytes and this fuel crossover problem are further discussed in Section 6.3.

The result of both these problems is that the performance of the DMFC is markedly worse than other types, such as a hydrogen-fuelled proton exchange membrane fuel cell (PEMFC). Figure 6.1 shows the performance of a state-of-the-art DMFC in 2002, compared with the V/I graph for a hydrogen-fuelled PEMFC. The shape of the graph is broadly similar, but the voltages and current densities are considerably lower.

If the problems are solved, then the DMFC could be used in all mobile fuel cell applications, including such high-power applications as motor vehicles. However, the rate of development is such that this sort of application is a long way off. The first applications of the DMFC will almost certainly be in cases where a power of only a few watts is sufficient, but a high energy density is required. Good examples are third-generation (3G) mobile phones, or a high specification personal digital assistant (PDA), or digital movie cameras. The DMFC should be able to provide a better alternative to the rechargeable lithium-ion battery, which is the battery type almost universally used in such electronic equipment. This Li-ion battery provides a very clear target for DMFCs to compete with. They can certainly be recharged more quickly – simply pour some

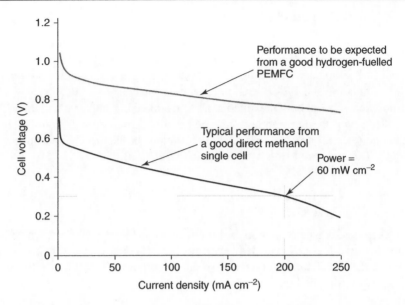

Figure 6.1   Voltage/current density graph for 2003 state-of-the-art direct methanol FC operating under ambient conditions.

methanol in. However, as yet they are not competitive in terms of reliability, energy density, and cost. What needs to be done to bring this about is discussed in Section 6.4, where we consider the major applications of the DMFC.

6.2 Anode Reaction and Catalysts

6.2.1 Overall DMFC reaction

The overall reaction in the DMFC is represented by the equation

$$CH_3OH + 1\tfrac{1}{2}O_2 \rightarrow 2H_2O + CO_2 \qquad [6.1]$$

We have already noted in Section 2.2 that the change in molar Gibbs energy for this reaction is $-698.2\,kJ\,mol^{-1}$. As we shall see below, six electrons are transferred for each molecule of methanol, and so, from equation 2.2, the reversible 'no-loss' cell voltage is

$$E = \frac{-\Delta \overline{g}_f}{zF} = \frac{698.2}{6} = 1.21\,V$$

As can be seen from Figure 6.1, the practical voltages obtained are considerably less than this, and the losses are greater than for other types of fuel cell. One of the main features that sets apart the DMFC is that there is considerable voltage loss at the fuel anode, as well as the cathode losses that are a feature of all cells. In this section, we will look at the anode reaction of the DMFC in more detail.

6.2.2 Anode reactions in the alkaline DMFC

The anode and cathode reactions depend on the electrolyte used, and methanol can, in principle, be used in a fuel cell with any of the standard electrolytes. However, if we are to retain the simplicity of a liquid fuel system, then it is only the low-temperature electrolytes – alkaline and PEM – that are serious contenders.

Methanol fuel works in both the alkaline FC considered in Chapter 5 and the PEMFC of Chapter 4. However, the reaction between the product carbon dioxide and the electrolyte of the alkaline (electrolyte) fuel cell (AFC) is a major problem, which would appear to be insoluble. The anode reaction of the alkaline DMFC is

$$CH_3OH + 6OH^- \rightarrow CO_2 + 5H_2O + 6e^- \qquad [6.2]$$

Here we see the six electrons produced for each molecule of methanol, which is why it is such an attractive fuel cell fuel. However, for this reaction to occur the reactants must interface with the alkaline electrolyte, and so the carbon dioxide will inevitably react with the hydroxide, forming carbonate:

$$2OH^- + CO_2 \rightarrow CO_3^{2-} + H_2O \qquad [6.3]$$

The electrolyte thus inevitably and steadily loses its alkalinity, and has a very limited life. In alkaline DMFCs, it is not possible to see **any** product carbon dioxide, as it all reacts according to equation 6.3. For this reason the alkaline DMFC is not practical, even though it could have many advantages in terms of cost and performance in comparison with PEM-based systems.

6.2.3 Anode reactions in the PEM direct methanol FC

Current DMFC research and development is centred on and around PEM electrolytes. In this case, the overall anode reaction is

$$CH_3OH + H_2O \rightarrow 6H^+ + 6e^- + CO_2 \qquad [6.4]$$

The H^+ ions move through the electrolyte and the electrons move round the external circuit. Note that water is required at the anode, though it is produced more rapidly at the cathode via the reaction

$$1\tfrac{1}{2}O_2 + 6H^+ + 6e^- \rightarrow 3H_2O$$

The result is the same as for the alkaline DMFC – six electrons per molecule of methanol fuel. However, the reaction of 6.4 above does not proceed simply, but takes place in stages, which can take a variety of routes. The oxidation reactions of methanol have been extensively studied, and they are described in much more detail in the literature

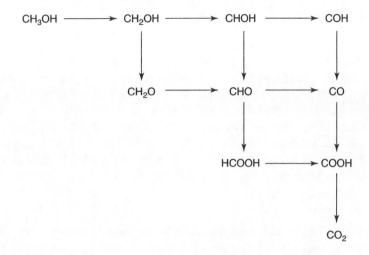

Figure 6.2 Stages in the oxidation of methanol at a DMFC anode.

(Hamnett, 1997). The chart in Figure 6.2 is an attempt to explain the stages and the different possibilities.[1]

At the top left is methanol; at the bottom right is carbon dioxide. The right moving steps involve 'hydrogen stripping' – the removal of a hydrogen atom and the generation of a proton (H^+) and electron (e^-) pair. The downward moving steps also involve the removal of a hydrogen atom and the generation of a proton-electron pair, but these downward steps also involve the addition or destruction of an OH group.

Any route through the compounds of Figure 6.2 from top left to bottom right is possible, and all have the same result – three right steps, three down steps, producing carbon dioxide and six proton–electron pairs as in equation 6.4. However, the compounds along the lower left edge (the hypotenuse) are the only stable compounds, and so this might be considered a 'preferred' route. We can divide it neatly into three steps. First the methanol is oxidised to methanal (formaldehyde), by taking one 'right' and one 'down' step in Figure 6.2.

$$CH_3OH \rightarrow CH_2O + 2H^+ + 2e^- \qquad [6.5]$$

The methanal then reacts to form methanoic (formic) acid via another 'right' and another 'down' step:

$$CH_2O + H_2O \rightarrow HCOOH + 2H^+ + 2e^- \qquad [6.6]$$

Finally, via another 'one step right, one step down' in Figure 6.2, the formic acid is oxidised to carbon dioxide:

$$HCOOH \rightarrow CO_2 + 2H^+ + 2e^- \qquad [6.7]$$

---

[1] This chart is taken from Carrette et al. (2001).

The sum of the reactions 6.5 to 6.7 is the same as 6.4. Notice also from Figure 6.2 that the formation of carbon monoxide is possible. This impacts the choice of catalyst – an issue discussed in Section 6.2.5 below.

We should note in passing that it would be possible to use either of the stable intermediate compounds (formaldehyde or formic acid) as fuels instead of methanol. Their energy density would be considerably less, as only 4 or 2 electrons would be produced for each molecule of fuel. Some interest has been shown recently in using formic acid in a 'direct formic acid' fuel cell (Rice et al., 2002), the particular advantage being that the problem of fuel crossover is greatly reduced. Whether the disadvantages of greatly reduced energy density, increased cost, lowered overall efficiency, increased toxicity, and reduced availability are outweighed by the increase in performance is somewhat doubtful.

6.2.4 Anode fuel feed

The reactions described by equations 6.4 and 6.6 both show how water is needed with the methanol to enable the oxidation reaction. This has important implications for the method of feeding the fuel to the anode. Pure methanol cannot be used, but a mixture with water must be provided. For the energy density of the fuel to be maintained, the original fuel supply must by pure methanol, and so there must be water stored in the fuel cell system, and the methanol should be added to this water, as in Figure 6.3.

This might appear to add considerably to the complexity of the system, which indeed it does. However, having a dilute solution of methanol in contact with that anode mitigates other problems. Firstly, the reduced methanol concentration ameliorates the problem of fuel crossover to be considered in Section 6.3 below. Secondly, the fact that water contacts the cell means that the PEM remains very well hydrated, which, as we know from Chapter 4, is an issue of critical importance.

The concentration of the methanol has to be around 1 molar ($= \sim 3\%$ by weight), according to most published work (Shukla et al., 1998, Scott et al., 1999b, and Dohle et al., 2002). This is to prevent the crossover of methanol to the cathode. It will usually be necessary to have a methanol sensor in the fuel feed system to ensure that correct concentration is maintained, though the concentration might be deduced by a circuit monitoring the cell's output. (See Barton et al., 1998, for a description of a methanol sensor design.)

The net DMFC reaction given by equation 6.1 indicates that water is produced within the fuel cell. This water will evaporate when the air passes over the cathode – indeed, it is probable that the rate of water evaporation will exceed the rate of water production. (This issue is discussed at length in the PEMFC chapter, Section 4.4.) In this case, the water circuit for the anode will need replenishing. This would normally be done by including a condenser/water separator in the exit air flow, collecting some of the evaporated water and supplying it to the anode water circuit as needed. This may seem complex and will not always be needed, but it should be noted that the water management system is still less complex than for most hydrogen PEM fuel cells, and that the PEM is more or less perfectly hydrated everywhere.

In Figure 6.3, the carbon dioxide is shown simply bubbling off from the anode and being vented from the system. Often this cannot be so simple, especially where the fuel cell can be tipped over and used at any angle. It is known that the efficient removal of

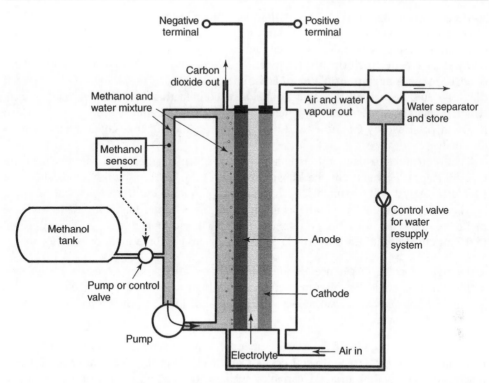

Figure 6.3 The main components of a DMFC. Not all the components will always be present. However, larger systems may have **additional components**, such as heat exchangers in the fuel system for cooling, air pumps, and methanol condensers in the carbon dioxide outlet pipe. See text, especially Section 6.2.4. Note that the electrode connections are shown at the edge for simplicity; normally the current will be taken off the whole face of the electrode, as in Figure 1.8.

carbon dioxide is one area where DMFCs are far from optimised. Shukla et al. (2002) have shown that adding air to the methanol/water mixture at the anode notably improves cell performance. This is not what would be expected, as it should cause a mixed potential and reduced voltage in the way that the methanol fuel at the air cathode does. It is supposed that 'the air bubbles promote higher liquid saturation in the anode diffusion layer and the faster removal of carbon dioxide'. Except in cells running at quite low temperatures, the carbon dioxide gas will contain significant evaporated methanol, which represents an inefficient loss of fuel, and in larger systems the methanol will be recovered by condensing it out of the exit carbon dioxide. Considerable thought is given to the practicalities of this, and other solutions to the problem, in Scott, 1999a.

6.2.5 Anode catalysts

Because none of the methanol oxidation reactions proceed as readily as the oxidation of hydrogen, there are considerable activation overvoltages[2] at the fuel anode, as well as

---

[2] See Section 3.4 and Figure 3.12.

at the cathode in the DMFC. This is the main cause for the lower performance. Much work has been done to develop suitable catalysts for the anode of the DMFC. Initially platinum was used, as with the hydrogen fuel cell, but it has long been known that bimetal catalysts give better performance (Hamnett and Kennedy, 1988). It is now more or less standard to use a mixture of platinum and ruthenium in equal proportions. Other bimetal catalysts have been tried, but this 50:50 Pt/Ru combination seems to be hard to beat (Hogarth and Hards, 1996 and Hamnett, 1997). One modification worthy of note, which has been described by Otomo et al. (2001), is the addition of tungstophosphoric acid to the catalyst.

The fact that the oxidation of methanol involves a step-wise reaction, with two different types of steps (the 'right' and the 'down' of Figure 6.2), gives an intuitive indication that a bimetal catalyst will be suitable, each catalyst promoting the different types of reaction.

The catalyst loading tends to be as much as ten times higher than with hydrogen PEM fuel cells, $2 \, mg \, cm^{-2}$ or more (Havranek et al., 2001 and Dohle et al., 2002), compared to $0.2 \, mg \, cm^{-2}$ for standard hydrogen PEMFCs. There are three main reasons for this:

- the higher loading that is needed to reduce the activation losses to reasonable levels,
- the DMFC is competing in a market where the higher costs are sustainable,
- a more active anode catalyst reduces the problem of fuel crossover to be considered in Section 6.3 that follows.

The catalyst is usually of the carbon-supported type, exactly as in the normal PEMFC of Chapter 4, except that the metal crystals are made from the Pt/Ru mixture. There is much scope for optimisation of variables such as the method of depositing the anode catalyst and the amount of ionomer to be added to the catalyst layer – see for example, Hajbolouri et al., 2001. (The purpose of adding some of the electrolyte ionomer to the electrode is explained in Section 4.3; see especially Figure 4.8. The amount of ionomer will be different in a hydrogen PEMFC, as the electrode and diffusion layer will have to behave quite differently – flooding by the methanol/water mixture is required, and the gas (product carbon dioxide) needs to be expelled as quickly as possible. All this is very different from the anode of a hydrogen fuel cell, where the gas has to be drawn in, and flooding by (product) water is highly undesirable.)

6.3 Electrolyte and Fuel Crossover

6.3.1 How fuel crossover occurs

We have already noted that the PEM type of electrode is currently viewed as the only near-term possibility for DMFCs. The use of alkaline electrolytes results in the inevitably fatal problem of carbonate formation. The PEM electrolytes also have a major problem, but steps can be taken to reduce it, and it does not inevitably destroy the cell.

The problem for PEM electrolytes in connection with using methanol fuel is that of fuel crossover. This was considered in general in Section 3.5, as it occurs to some extent in all fuel cells. However, in the DMFC with a PEM electrolyte it is particularly severe. The reason is that methanol mixes very readily with water, and so spreads into the water that is such an essential part of the structure of the PEM electrolyte, which has been described in detail in Section 4.2. The methanol will thus reach the air cathode. This has a platinum catalyst, and although it will not oxidize the fuel as effectively as the Pt/Ru catalyst on the anode, it will do so fairly readily. The reaction of the fuel at the cathode is not only a waste of fuel – it will also reduce the cell voltage, for the reasons explained in Section 3.4. This phenomenon is sometimes called a *mixed potential*.

The loss of methanol is often transposed into a 'crossover current' – the current equivalent to that which would be produced by the methanol, had it reacted properly on the fuel anode. This current, i_c, can be used with the useful output current i to give an important 'figure of merit' for a DMFC, which is the fuel utilisation coefficient η_f. This gives the ratio of the fuel that is usefully and properly reacted on the anode to the total fuel supplied, the difference being accounted for by some fuel crossing-over and being lost at the cathode.

$$\eta_f = \frac{i}{i + i_c} \qquad [6.8]$$

Using the techniques described below, it is possible to bring this figure up to as high as 0.85 or even 0.90 (Gottesfeld, 2002), though 0.80 (or 80%) would probably be a more realistic figure to expect.

6.3.2 Standard techniques for reducing fuel crossover

There are four principle ways that DMFC designers use to reduce fuel crossover, and there are other ideas that are more at the experimental stage. The four key established methods are the following:

1. The anode catalyst is made as active as possible, within the bounds of reasonable cost. This results in the methanol reacting properly at the anode and not being available to diffuse through the electrolyte and on to the cathode.
2. The fuel feed to the anode is controlled, as mentioned in Section 6.2.4 above, so that at times of low current there is no excess of methanol. Clearly, the lower the methanol concentration at the anode, the lower it will be in the electrolyte, and hence at the cathode. See Figure 6.4. The effect of methanol concentration on DMFC fuel cell performance has been extensively studied, for example, in Scott et al., 1999b and Dohle et al., 2002. Mathematical models have also been developed, as in, for example, Dohle et al., 2000. The conclusion is that the concentration should always be about 1 M, though a more accurate optimum will need to be found for every type of cell under all conditions.
3. Thicker electrolytes than what is normal for PEMFCs are used. Clearly this will reduce fuel crossover, though it will also increase the cell resistance. There is a compromise to be found, and in the DMFC that optimum is with a somewhat thicker electrolyte than for hydrogen fuel cells. The thickness membrane normally used by DMFC developer is between 0.15 and 0.20 mm,[3] whereas for a hydrogen PEMFC it would be between 0.05 and 0.10 mm.[4]

---

[3] e.g. Du Pont's Nafion 117 at 0.18 mm.

[4] e.g. Du Pont's Nafion 112 at 0.05 mm.

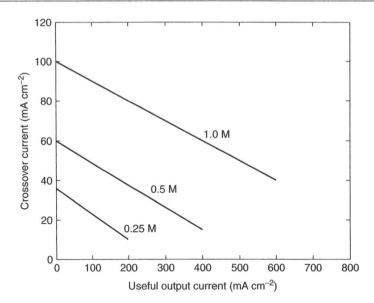

Figure 6.4 Graph showing how the crossover of methanol to the cathode changes with fuel concentration at the anode and with load current. This is a simplified presentation of results reported in Ren, Zelanay et al., 2000.

4. In addition to the thickness of the PEM, its *composition* also has an effect. It has been shown by Ren, Springer and Gottesfeld (2000) that the diffusion and water uptake for 1100 EW Nafion is about half that for 1200 EW Nafion.

It should also be noted that fuel crossover reduces as the current from the cell increases. This is linked to points 1 and 2 above – the fuel reacts promptly at the anode and is not made available to crossover. In Ren, Zelanay et al., 2000, it is shown how the crossover equivalent current falls with methanol concentration and with increasing current. These results are summarised in Figure 6.4. This means that it is helpful if a DMFC is designed to run more or less continuously at quite near its maximum power. To do this, it will usually need to be run in parallel with a supercapacitor or a small rechargeable battery. Such arrangements are very attractive in the case of portable electronics equipment, which will probably be the first application of DMFCs (e.g. studies of such hybrid systems, see Jarvis et al., 1999 and Park et al., 2002).

6.3.3 Fuel crossover techniques in development

In addition to the (almost) universally applied techniques outlined above, there are other ideas being tried that are more experimental or at a very early stage of development. Among these are the following:

1. The use of selective (non-platinum) catalysts on the air cathode. These will stop the fuel reacting on the cathode and so eliminate the voltage drop due to the 'mixed-potential'.

However, there are problems with this approach. The first is that all catalysts that do not promote the fuel oxidation tend only to very slowly promote the reaction of oxygen with the H^+ ions. Thus, the activation losses on the cathode are made even worse than normal, and there is no increase in performance. Another problem is that although the mixed-potential problem may be solved, the fuel is still crossing over, and while it may not be reacting on the cathode, it will probably just evaporate instead. Thus, it will still be wasted. So, although it may be possible in the future to find selective cathode catalysts that ameliorate the fuel crossover problem, this approach does not offer a complete solution.

2. The use of a layer in the electrolyte that is porous to protons but less so to methanol. If such a material could be found, then this would obviously be a solution to the problem. Some of the ideas being tried in this area include treating the surface of the Nafion™ membrane and also coating it with a very thin layer of palladium, applied by sputtering (Choi et al., 2001 and Yoon et al., 2002). Another idea is to use a membrane made from two layers, one of which is much less permeable to methanol, as reported by Jung et al. (2002). Related to this is the possibility of including additives to the PEM, which will discourage methanol crossover, while at the same time reducing the proton conductivity. The use of caesium ions has been proposed by Trivoli (1998).

3. The development of more conductive proton exchange membranes – these would allow thicker membranes to be used, thus reducing the fuel crossover. This approach is being actively pursued by several research groups all over the world. Some are taking the approach of using different polymers than the standard Nafion-type materials (Gieger et al., 2001), while others are experimenting with additives to the normal sulphonated PTFE materials (Dimitrova et al., 2001).

An interesting example (and special case) of the use of selective catalysts mentioned above is the possibility of abandoning any attempt to prevent crossover and of feeding *all* the reactants (methanol, air, water) as a mixture over *both* electrodes. Experiments with very small cells made in this way have been reported by Barton et al. (2001).

6.4 Cathode Reactions and Catalysts

The cathode reaction in the DMFC is the same as that for the hydrogen fuel cell with acid electrolyte.

$$1\tfrac{1}{2}O_2 + 6H^+ + 6e^- \rightarrow 3H_2O \qquad\qquad [6.9]$$

Since the reaction is the same, the same catalyst is used. There is no advantage in using the more expensive Pt/Ru bimetal catalyst used on the anode here as well. Essentially then there is very little difference between the operation of the cathode on the PEM electrolyte direct methanol FC and that of the PEM electrolyte hydrogen FC. As can be seen from Figure 6.3, the general arrangement is similar, except that the cathode air supply will not need humidifying – the anode arrangements ensure a well-hydrated electrolyte. However, the water vapour from the exit air will often need to be processed to extract some of the water in order to keep the anode feed system topped-up.

As has been mentioned earlier, some experiments are being done with non-platinum cathode catalysts. The idea of using these cathode catalysts is that they will not oxidise the methanol crossing to the cathode, and hence solve the mixed-potential problem. Research on this is still at a fairly early stage, and such catalysts do not yet have sufficient oxygen reduction performance to be used.

6.5 Methanol Production, Storage, and Safety

6.5.1 Methanol production

The potential of the DMFC, and also the indirect use of methanol given consideration in Chapter 8, relies on the fact that it is a fuel that is produced in bulk at reasonable cost. Methanol is currently produced at the rate of well in excess of 20 000 000 tonnes per year.[5] It has a wide range of uses, but a high proportion (about 40%) is used to make formaldehyde, and about 20% is used in the manufacture of the fuel additive MTBE. Large amounts are used in cleaners, for example, the windshield wash for cars often contains methanol, and an astonishing 250 million US gallons[6] ($= \sim$750 000 tonnes) are used in this way each year. Only a very small proportion, about 2%, of methanol is currently used directly as a fuel.

Methanol can be quite efficiently produced from almost any hydrocarbon fuel. Natural gas is very suitable, as is the well-head gas that is often just burnt off if an oil well is at all remote. The first stage is the simple and well-established process of reacting with steam – a process covered in some depth in Chapter 8 that follows. This produces a mixture of hydrogen, carbon monoxide, and carbon dioxide, the proportions depending on the fuel feedstock, temperature, and pressure. These gases then react to form methanol using either of the reactions

$$2H_2 + CO \rightarrow CH_3OH$$

or

$$3H_2 + CO_2 \rightarrow CH_3OH + H_2O$$

Both these reactions result in less moles of substance, and so are helped by high pressure. The reaction proceeds fairly readily, over a suitable catalyst at high pressure, but at about 50 bar, these high pressures are not excessive. Note that if these reactions were to occur in the hydrogen generation systems described in Chapter 8, they would be highly undesirable – fortunately, the need for high pressures and a suitable catalyst means that the reactions do not occur unless the conditions are right.

Because methanol is required in large quantities for industrial purposes, considerable effort has been put into making the production process as efficient as possible. State-of-the-art plants are currently estimated to use about 29 kJ per kilogram of product – this is the lower heating value (LHV) of the feedstock fuel and the energy used to operate the

[5] Unless otherwise stated, the data in this section is taken from the Kirk–Othmer Encyclopedia of Chemical Technology.
[6] Figure provided by the American Methanol Association.

process. The LHV of methanol is $19.93\,kJ\,kg^{-1}$, so this corresponds to an efficiency of about 70%.

Although the great majority of methanol is currently produced using natural gas and other fossil fuels, it can also be produced from renewable biomass. Hamelinck and Faaij (2002) give a very full description and analysis of the cost involved in such processes. They show that in a few years such processes should yield methanol with a process efficiency of about 60% and at a cost per Joule similar to current refined diesel and gasoline prices, though at the moment these are higher by a factor of about 3.

The price of methanol is somewhat variable, owing to the fluctuations in the supply and demand of such industrial feedstock. Bearing in mind the efficiency of the production process, and the fact that it can be made from such a wide range of hydrocarbons, including some, such as well-head gas, that are quite low in cost, the price should not be greatly different from that of gasoline. In early 2003, the price was $200 per tonne in bulk – making the cost of the methanol needed to run a 2-W fuel cell, at 20% efficiency, for a whole year ($= 16\,kg$ of methanol) about $3. This is negligible compared to the packaging and distribution costs of the fuel.

We may conclude that methanol is indeed a low cost and readily available fuel for use in small fuel cells.

6.5.2 Methanol safety

The use of methanol in consumer products, such as power supplies for portable electronics equipment, raises potential safety problems – as does the use of hydrogen.

One safety issue is that of the flammability and burning properties of methanol. It is, of course, a highly flammable product. A possible problem adding to the danger is that methanol burns with an invisible flame. It is used as the fuel for the Indy Car series races in the United States, and it is said by some that this is because it is safer – the invisible flame makes it possible to see what is happening in a burning vehicle, and so rescue can be more easily achieved. Table 8.1 in Chapter 8 gives basic physical data about methanol and other fuels, and it can be seen that methanol also has a higher auto-ignition temperature than gasoline, and a lower flammability limit. Others say this is nonsense, and the invisible flame is a hazard, because a fire can start and not be noticed and that the only reason methanol is used is that it allows higher performance in internal combustion engines than gasoline. In a thorough study of all the issues relating to pollution, production costs, and flammability, Adamson and Pearson (2000) concluded that hydrogen and methanol were both about equal from a safety point of view, and both safer than gasoline. However, this was only from a fire safety point of view and did not take into account the question of toxicity.

Methanol is a poison, made worse by the fact that it can easily mix with any water-based fluid, such as water supplies or almost any drink. Furthermore, it does not have a taste that makes it immediately repellent, which means it can be drunk. This 'drinkability' problem makes it considerably more dangerous in everyday use than other fuels – such as gasoline – which are also poisonous, and in wide circulation. However, the safety arguments are fairly complex, as methanol is naturally present in the human body, and in small quantities it is perfectly safe.

Methanol is produced in the body by the action of the digestive system on a wide range of products, particularly natural products such as fruit and also on some man-made additives such as the sweeteners used in 'diet' drinks. It is interesting to note that the body decomposes the methanol, in the liver, to carbon dioxide in exactly the same step-wise fashion as the fuel cell. The steps are methanol to formaldehyde, formaldehyde to formic acid, and formic acid to carbon dioxide. The problem is not the methanol *per se*, but the formic acid formed in the breakdown process. This acidifies the blood and causes fatal problems if allowed to reach excess levels.

The main ways of treating patients who have taken methanol is to slow down the initial stages of the process – to keep the methanol as methanol – so that the formic acid concentrations do not become too high.

An important corollary to this is that methanol is not a cumulative poison in the way of some others such as lead. It does not accumulate in the body. Quite the reverse, the body can deal perfectly well with frequent small exposures – indeed this occurs in nature. It should also be pointed out that methanol is not carcinogenic and has none of the insidious mutagenic properties of some other chemicals. Another important point about methanol safety is that it is much less harmful to the ecology than most other fuels. It quickly breaks down to carbon dioxide in the soil and when exposed to sunlight. It is used as a windshield wash in cars, where all of it is released to the environment, and this poses no hazards.

Table 6.2 gives some information about the levels of methanol in the human body in various situations. Note that there is a normal 'background' level of methanol present, resulting from the digestion of food and drinks as mentioned earlier. A problem to note is that **methanol vapour is particularly dangerous** – presumably this is because entry through the lungs gives faster access to the blood, and hence to the liver, than the digestive system. This is a very strong argument against designing DMFCs that use methanol vapour rather than liquid, which has been proposed by some. Note that 200 ppm is the threshold limit value (TLV) for methanol. The immediately dangerous vapour level of 2.5% corresponds to 25 000 ppm.

As mentioned in the previous section, methanol is already widely used, and there is experience in dealing with exposure to it. Data is available on accidents with methanol

Table 6.2 Effects of different methanol exposures on the human body. The figures are for an average 70 kg body and are mostly taken from data provided by the Methanex Corporation

Exposure/dose	Body methanol
Background in a 70 kg body	~35 mg
Hand in liquid methanol for 2 min	170 mg
Drink 0.8 L of aspartame sweetened 'diet' drink	Background + ~40 mg
Drink 0.2 mL of methanol	170 mg
Drink over 25 mL of methanol	~20 000 mg FATAL
Inhalation of 40 ppm vapour for 8 h	170 mg
Inhalation of 200 ppm for 8 h	~850 mg DANGEROUS
Inhalation of 2.5% vapour for 1 s	VERY DANGEROUS

Table 6.3 Information about dangerous incidents with methanol gathered by the American Association of Poison Control Centres. Table 6.3a gives data on the outcome of methanol exposure and Table 6.3b gives the motivating factor in the case of the fatal exposures

a) Outcome	1995	1996	1997
No treatment	699	736	782
Minor treatment	752	814	780
Evasive treatment	31	45	28
Death	11	10	14
Total	1493	1605	1604

b) Motivation	1995	1996	1997
Suicide	4	2	6
Intentional self-harm	3	4	6
Unknown	1	3	2
Other	3	1	0
Total	11	10	14

from which information can be drawn to aid designers to make the compound safe. Some basic information about reported incidents in connection with methanol in the United States are shown in Table 6.3. A particular point to note is that a very high proportion of the high-level exposures that led to death were deliberate suicide or self-harm actions. Systems must be designed so that even deliberate and premeditated drinking of methanol is very difficult – stopping accidental consumption is not enough.

Provided we are aware of the dangers, we can conclude that it should be possible to design methanol fuel cell systems that are safe. It will be important that designers of small systems make arrangements so that the fuel cartridges do not permit the liquid to be easily drunk – for example, by making them hard to open, or hold open, unless they are connected to the fuel cell system. An approach proposed by some is the inclusion of additives to the methanol so that, for example, it tastes so repulsive that it is impossible to drink by accident and difficult to do so deliberately. The fuel cell community would resist this, as the additives would almost certainly harm a fuel cell. Good mechanical design is surely the way, certainly in the case of DMFCs, where the fuel containers will be small (though still holding potentially fatal quantities). Safe locking devices that open the fuel containers only when attached to the fuel cell can be designed with reasonable ease.

6.5.3 Methanol compared to ethanol

It is sometimes suggested that ethanol be used in place of methanol. It is a similar hydrogen carrier and has the major advantage that it is not nearly so poisonous – indeed it is deliberately produced for human consumption in many forms in huge quantities.[7] Industrial ethanol is also produced in large quantities, of the same order of magnitude as methanol, though a little less than 20 000 000 tonnes per year. Most of this ($\sim$75%) is used as fuel, which is an interesting contrast with methanol. One of the main users of ethanol as fuel is Brazil.

However, ethanol can only be made from a much narrower range of feedstock than methanol. Crucially, it cannot be made nearly so directly from natural gas, though it can be made from ethylene, which can be made from a fairly wide range of hydrocarbons. The result is that it is considerably more expensive.

The proportion of hydrogen in ethanol (C_2H_5OH) is 13%, marginally higher than the 12.5% in methanol (CH_3OH). However, the reformation of the compound to hydrogen is considerably more difficult, requiring higher-temperature plants and more complex carbon monoxide clear-up equipment. So, even in the 'indirect' cell, methanol is better.

In a direct fuel cell, methanol is vastly better than ethanol. The initial oxidation step to remove two hydrogen atoms and generate two electrons proceeds equally readily for both alcohols. However, any further oxidation appears to be impossible[8] in the case of ethanol fuel cells – though it does occur in the human body, which completely oxidises the ethanol to carbon dioxide. The result is that ethanol is a fuel that gives just *two* electrons per molecule, as against the *six* from methanol, which is also lighter. Its practical energy density is thus worse by a factor of about three. This completely rules out its use in direct fuel cells.

6.5.4 Methanol storage

The main problems of methanol storage relate to its ability to mix with water and the need to make it difficult to accidentally – or even deliberately – consume it.

Because methanol mixes so readily with water, it will actually draw water out of the atmosphere, and if left unsealed a sample of methanol will steadily become a mixture of methanol and water. Such a mixture is quite corrosive, and tanks made from materials such as ordinary steel must not be used for methanol. Stainless steel can be used, as can glass. Methanol is a good solvent, and so great care is needed when using plastic bottles, as not all are suitable. Methanol can also attack rubber, so gasket materials must be chosen with care.

Large-scale storage of methanol requires very special measures to be taken in terms of venting and fire safety, but these are beyond the scope of this book.

We have already made the point that any methanol fuel containers would have to be designed so that it is very difficult to remove the methanol other than by connecting it into a DMFC system such as that of Figure 6.3.

---

[7] In the fairly dilute form usually consumed, ethanol does little harm, and the other compounds in alcoholic drinks may even have health benefits. However, if pure ethanol were drunk in anything other than small quantities, it would be very harmful. If between 250 and 500 mL of pure industrial ethanol were drunk, it might be fatal – this is at least 10 times the figure required for methanol.

[8] See Section 2.5.5 of Chapter 2, where blood alcohol level measurement is shown.

6.6 Direct Methanol Fuel Cell Applications

Recent developments in DMFCs, including published results of demonstration systems (Gottesfeld, 2002; Dohle et al., 2002; and Ren, Zelanay et al., 2000) indicate that a power of about 60 mW per square centimeter of electrode area is feasible, but is unlikely to be exceeded in the near future. This is considerably lower than the performance of hydrogen fuel cells and considerably constrains the area of application of this type of cell.

DMFCs lend themselves well to applications where the *power* density can be low, but the *energy* density must be high. To put it another way, they are suited to applications where the average power is only a few watts, but that power must be provided for a very long time – typically for several days.

Fortunately there are many examples where this is indeed the case. Good examples are 3G 'always on' mobile telephones, PDAs that combine with communication equipment, traffic systems, remote monitoring and sensing equipment, navigation systems, and so on. The antithesis of a good DMFC system is the motorcar, where a high-power density is essential and usage times will often be quite short.

In many potential applications of fuel cells the competition is quite varied. In this case it is not – the rival technology is clearly the rechargeable battery. The best performing rechargeable battery is the Lithium-ion cell. If a DMFC can offer significant improvement on this, then serious impact on a huge market can be assured.

The key figure of merit in this case is the system energy density – how many watt-hours of energy can be stored in a given space. The specific energy, or the energy per kilogram, is also important, but not as much. The size of the portable device is usually more important then its weight. In any case, this is less of a challenge for the DMFC – if it can win on energy density, it will certainly also win on specific energy, as batteries are quite dense.

The volume of a DMFC system obviously depends not only on the energy to be stored but also on the average power and the efficiency. If we take the recently published performance of DMFCs (Gottesfeld, 2002 and Dohle et al., 2002), then we see that $60\,\mathrm{mW\,cm^{-2}}$ is a good guide figure for what should be possible in the near-term future. We can then estimate the power density as follows:

- 60 mW for each square centimeter of electrode
- the cell thickness is about 0.3 cm per cell average, a figure achieved or exceeded by most recent designs.
- so power density $= 60/0.3\,\mathrm{cm^3} = 200\,\mathrm{mW\,cm^{-3}}$

However, this figure is far in excess of what could actually be achieved. Looking at Figure 6.3, we see that there is quite a large 'balance of plant' associated with DMFCs. These controllers and such like will be miniaturised, but there will often also be components not shown in the diagram. For example, a supercapacitor or small rechargeable battery will almost certainly have to be connected in parallel with the DMFC, as there will be short bursts of much higher power demand. In addition, there will be the space needed for sealing and manifolding around the edge of the electrode. For example, in the system described by Dohle et al. (2002), the cell area is $256\,\mathrm{cm^2}$, though the active electrode area is only $144\,\mathrm{cm^2}$. This reduces the power density by a factor of 0.56, though it results from a sealing and manifolding area of a hardly excessive 2-cm band round the electrode edge.

A reasonable target figure would thus be a quarter of this maximum $(200\,\text{mW cm}^{-2})$ value, say $50\,\text{mW cm}^{-3} = 0.05\,\text{W cm}^{-3}$. This figure can be used to give the size of the cell itself.

The next parameter of importance is the efficiency of the cell – this will set how much electrical energy is obtained from the methanol. The specific energy of methanol is $5.54\,\text{Wh kg}^{-1}$ (Table 6.1). Its density is $0.792\,\text{kg L}^{-1}$, so the energy density it $4.39\,\text{kWh L}^{-1}$, so also $4.39\,\text{Wh cm}^{-3}$. However, by no means is all this energy converted into electrical energy. Following the method explained in Chapter 2, Section 2.4, we can find the fuel cell efficiency directly from its voltage. The LHV of methanol is $-638.5\,\text{kJ mol}^{-1}$. So, a cell of 100% efficiency, relative to the LHV, would have a voltage of

$$\frac{\Delta H}{zF} = \frac{-638.5}{-6 \times 96\,485} = 1.10\,\text{V}$$

To get powers up to the value of $60\,\text{mW cm}^{-2}$ a DMFC will typically operate at a cell voltage of about 0.3 V – see Figure 6.1. So the cell efficiency is

$$\eta_{\text{cell}} = \frac{0.3}{1.10} = 0.27 = 27\%$$

But we must also consider the question of fuel utilisation. Some of the fuel, as we have seen, is wasted at the cathode, so

$$\text{Overall efficiency} = \eta_f \times \eta_{\text{cell}} = 0.8 \times 0.27 = 0.22 = 22\%$$

The effective energy content of the fuel is thus 22% of $4.39\,\text{Wh cm}^{-3}$, which is almost exactly $1.0\,\text{Wh cm}^{-3}$ – a convenient and simple number to use. To confirm our reasoning, this figure of $1\,\text{Wh cm}^{-3}$ is quoted as the energy density by Smart Fuel Cell GmbH, the first company to commercialise this type of cell.

So, if P_e is the mean electrical power output of the DMFC and E is the energy to be stored in watt-hours, then the approximate volume is given by the formula

$$\text{Vol}_{\text{DMFC}} = \frac{P_e}{0.05} + E \qquad [6.10]$$

The comparative size of a rechargeable L-ion cell is found from the energy density of such cells. Currently, this is in the range 130 to $160\,\text{Wh L}^{-1}$, or 0.13 to $0.16\,\text{Wh cm}^{-3}$. The technology is quite mature, and so we are unlikely to see any major improvements in this figure. However, we will use the slightly improved figure of $0.2\,\text{Wh cm}^{-3}$. The volume of the L-ion battery is thus

$$\text{Vol}_{\text{Li}} = \frac{E}{0.2} \qquad [6.11]$$

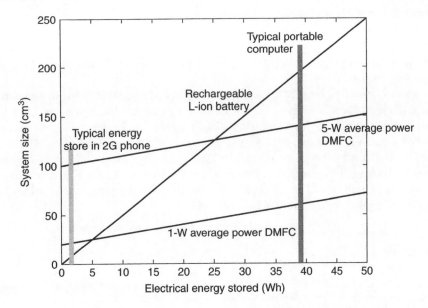

Figure 6.5 Graph showing the change in volume of an L-ion rechargeable battery, and two hypothetical DMFCs, for different values of electrical energy stored.

Figure 6.5 shows a graph of the size of two types of direct methanol systems and a good L-ion rechargeable battery. It can be seen that for devices of higher energy requirement, the DMFC becomes more and more attractive.[9] The DMFC is not better than L-ion for all applications, such as the currently used type of mobile phone. However, for many other applications, such as laptop computers, it should soon be able to provide a more compact alternative. It should be noted that the graph assumes that a power density of $0.05\,\mathrm{W\,cm^{-3}}$ can be achieved. The assumed efficiency is already available with commercially available systems, but current DMFC systems only have a power density of about $0.01\,\mathrm{W\,cm^{-3}}$. Nevertheless, very good progress is being made.[10]

It should be noted that the performance of DMFCs depends very strongly on issues such as temperature and pressure. Temperature is very important, as it has a particularly strong impact on DMFCs. A rise in temperature greatly improves the anode reactions, as well as those at the cathode. The improved anode performance reduces the problem of fuel crossover, as it reacts properly on the anode and is not available for crossover. This not only has a further improving effect on the cathode but also improves efficiency. It has been shown by Dohle et al. (2002) that raising the temperature from 22°C to 77°C increases the power available by a factor of 4. Increasing the pressure can make further improvements, but whether the power required to compress the air makes this

---

[9] Dyer (2002) shows a similar comparison and also extends the argument to include system requirements such as battery chargers – not needed for the fuel cell. He also speculates on system costs to show the advantage of fuel cells even more clearly.

[10] A good place to start for examples of current products is the website of SFC Smart Fuel Cell GmbH, Germany, www.smartfuel cell.com.

worthwhile is not clear. (See the argument in Section 4.7 of Chapter 4.) This means that the performance that will be obtained from a DMFC will vary greatly according to the application. A very small cell will not be able to operate very much above ambient temperature, and so its performance in terms of power density would probably be worse than the figures used above.

We should close this chapter by pointing out that a strong rival to the *direct* methanol fuel cell is the *indirect* methanol FC. In Chapter 8, we see how hydrogen can be comparatively easily obtained from methanol. Furthermore, in Section 8.6.3, we describe a design for a micro-reactor that is suitable for systems at the 5 to 50 W scale. These would provide hydrogen for fairly conventional PEMFCs that have much higher power densities than the DMFCs we have been describing. These micro-reactors are at a very early stage of development, and it will be fascinating to watch, over the years ahead, to see which approach eventually proves the most successful.

References

Adamson K-A and Pearson P. (2000) "Hydrogen and methanol; a comparison of safety, economics, efficiencies, and emissions", *Journal of Power Sources*, **86**, 548–555.

Barton S.C., Murach B.L., Fuller T.F., and West A.C. (1998) "A methanol sensor for portable direct methanol fuel cells", *Journal of the Electrochemical Society*, **145**, 3783–3788.

Barton S.C., Patterson T., Wang E., Fuller T.F., and West A.C. (2001) "Mixed-reactant, strip-cell direct methanol fuel cells", *Journal of Power Sources*, **96**, 329–336.

Carrette L., Friedric K.A., and Stimming U. (2001) "Fuel cells – fundamentals and applications", *Fuel Cells*, **01**, 5–39.

Choi W.C., Kim J.D., and Woo S.I. (2001) "Modification of proton conducting membrane for reducing fuel crossover in a direct methanol fuel cell", *Journal of Power Sources*, **96**, 411–414.

Dimitrova P., Friedrich K., Stimming U., and Vogt B. (2001) "Recast Nafion-based membranes for methanol fuel cells." *Proceeding of the First European PEFC Forum-Lucerne*, pp. 97–107.

Dohle H., Divisek J., and Jung R. (2000) "Process engineering of the direct methanol fuel cell", *Journal of Power Sources*, **86**, 469–477.

Dohle H., Schmitz H., Bewer T., Mergel J., and Stolten D. (2002) "Development of a compact 500 W class direct methanol fuel cell stack", *Journal of Power Sources*, **106**, 313–322.

Dyer C.K. (2002) "Fuel cells for portable applications", *Journal of Power Sources*, **106**, 31–34.

Gieger A.B., Rager T., Matejcek L., Schere G.G., and Wokaun A. (2001) "Radiation grafted membranes for application in direct methanol fuel cells." *Proceeding of the First European PEFC Forum-Lucerne*, pp. 129–134.

Gottesfeld S. (2002) "Development and demonstration of direct methanol fuel cells for consumer electronics applications." *The Fuel Cell World – Proceedings, Lucerne*, EFCF, pp. 35–41.

Hajbolouri F., Lagergren C., Hu Q-H., Lindbergh G., Kasemo B., and Ekdunge P. (2001) "Pt and PtRu model electrodes for electrochemical oxidation of methanol." *Proceeding of the First European PEFC Forum-Lucerne*, pp. 21–27.

Hamelinck C.N. and Faaij A.P. (2002) "Future prospects for production of methanol and hydrogen from biomass", *Journal of Power Sources*, **111**, 1–22.

Hamnett A. and Kennedy B.J. (1988) "Bimetallic carbon supported anodes for the direct methanol-air fuel cell", *Electrochimica Acta*, **33**, 1613–1618.

Hamnett A. (1997) "Mechanism and electrocatalysis in the direct methanol fuel cell", *Catalysis Today*, **38**, 445–457.

Havranek A., Klafki K., and Wippermann K. (2001) "The influence of the catalyst loading and the ionomer content on the performance of DMFC anodes." *Proceeding of the First European PEFC Forum-Lucerne*, pp. 221–230.

Hogarth M. and Hards G. (1996) "Direct methanol fuel cells: technological advances and further requirements", *Platinum Metals Review*, **40**, 150–159.

Jarvis L.P., Terrill B.A., and Cygan P.J. (1999) "Fuel cell/electrochemical capacitor hybrid for intermittent high power applications", *Journal of Power Sources*, **79**, 60–63.

Jung D.H., Cho S.Y., Peck D.H., Shin D.R., and Kim J.J. (2002) "Performance evaluation of a Nafion/silicon oxide hybrid membrane for direct methanol fuel cell", *Journal of Power Sources*, **106**, 173–177.

Otomo J., Ota T., Wen C-J., Eguchi K., and Takahashi H. (2001) "Heteropoly compound anode electrocatalysts for direct methanol fuel cell." *Proceeding of the First European PEFC Forum-Lucerne*, pp. 29–37.

Park K-W., Ahn H-J., and Sung Y-F. (2002) "All solid state supercapacitor using a Nafion polymer membrane and its hybridization with a direct methanol fuel cell", *Journal of Power Sources*, **96**, 329–336.

Ren X., Zelanay P., Thomas S., Davey J., and Gottesfeld S. (2000) "Recent advances in direct methanol fuel cells at Los Alamos National Laboratory", *Journal of Power Sources*, **86**, 111–116.

Ren X., Springer T.E., and Gottesfeld S. (2000) "Water and methanol uptakes in Nafion membranes and membrane effects on direct methanol performance", *Journal of the Electrochemical Society*, **147**, 92–98.

Rice C., Ha S., Masel R.J., Waszczuk P., Wieckowski A., and Barnard T. (2002) "Direct formic acid fuel cells", *Journal of Power Sources*, **111**, 83–89.

Scott K., Taama W.M., and Argyroloulos P. (1999a) "Engineering aspects of the direct methanol fuel cell system", *Journal of Power Sources*, **79**, 43–59.

Scott K., Taama W.M., Argyroloulos P., and Sundmacher K. (1999b) "The impact of mass transport and methanol crossover on the direct methanol fuel cell", *Journal of Power Sources*, **83**, 204–216.

Shukla A.K., Christensen P.A., Dickinson A.J., and Hamnett A. (1998) "A liquid feed solid polymer electrolyte direct methanol fuel cell operating at near ambient conditions", *Journal of Power Sources*, **76**, 54–59.

Shukla A.K., Jackson K., Scott K., and Murgia G. (2002) "A solid polymer electrolyte direct methanol fuel cell with a mixed reactant and air anode", *Journal of Power Sources*, **111**, 43–51.

Trivoli V. (1998) "Proton and methanol transport in poly(perfluorosulphonate) membranes containing Cs^+ and H^+ cations", *Journal of the Electrochemical Society*, **145**(11), 3798–3801.

Yoon S.R., Hwang C.H., Cho W.I., Oh I-H., Hong S-A., and Ha H.Y. (2002) "Modification of polymer electrolyte membrane for DMFC using Pd films formed by sputtering", *Journal of Power Sources*, **106**, 215–223.

7

Medium and High Temperature Fuel Cells

7.1 Introduction

In Chapter 2 we noted that the 'no loss' open circuit voltage (OCV) for a hydrogen fuel cell reduces at higher temperatures. Indeed, we noted that above about 800°C, the theoretical maximum efficiency of a fuel cell was actually less than that of a heat engine. It might, therefore, seem unwise to operate fuel cells at higher temperatures. However, these apparent problems are in many cases outweighed by the advantages of higher temperature. The main advantages are as follows:

- The electrochemical reactions proceed more quickly, which manifests in lower activation voltage losses. Also, noble metal catalysts are often not needed.
- The high temperature of the cell and the exit gases means that there is heat available from the cell at temperatures high enough to facilitate the extraction of hydrogen from other more readily available fuels, such as natural gas.
- The high-temperature exit gases and cooling fluids are a valuable source of heat for buildings, processes, and facilities near the fuel cell. In other words, such fuel cells make excellent 'combined heat and power (CHP) systems'.
- The high-temperature exit gases and cooling fluids can be used to drive turbines that can drive generators, producing further electricity. This is known as a 'bottoming cycle'. We will show in Section 7.2.4 that this combination of a fuel cell and a heat engine allows the complementary characteristics of each to be used to great advantage, providing very efficient electricity generation.

There are three types of medium- and high-temperature fuel cells that we shall be considering in this chapter. The phosphoric acid electrolyte fuel cell (PAFC) is the most well developed of the three. Many PAFC 200-kW CHP systems are installed at hospitals,

Fuel Cell Systems Explained, Second Edition James Larminie and Andrew Dicks
© 2003 John Wiley & Sons, Ltd   ISBN: 0-470-84857-X

military bases, leisure centres, offices, factories, and even prisons, all over the world. Their performance and behaviour is well understood. However, because they operate at only about 200°C, they need a noble metal catalyst, and so, like the proton exchange membrane (PEM) fuel cell, are poisoned by any carbon monoxide in the fuel gas. This means that their fuel processing systems are necessarily complex. Since the PAFC has similar characteristics to the PEM fuel cell, and is more commercially advanced than the molten carbonate (electrolyte) fuel cell (MCFC) and the solid oxide fuel cell (SOFC), this will be described first.

The molten carbonate electrolyte fuel cell (MCFC) has a history that can be traced back at least as far as the 1920s.[1] It operates at temperatures around 650°C. The main problems with this type of cell relate to the degradation of the cell components over long periods. The MCFC does, however, show great promise for use in CHP systems, and this is discussed in detail in Section 7.4.

The solid oxide fuel cell (SOFC) has also been the object of research for many years. Its development can be traced back to the Nernst 'Glower' of 1899. Since the SOFC is a solid-state device, it has many advantages from the point of view of mechanical simplicity. The SOFC is also very flexible in the way it can be made, and its possible size. It therefore has scope for a wide variety of applications. SOFCs can be made from a range of different materials, with different operating temperatures, from about 650 to 1000°C. These issues are described in Section 7.5.

However, before we consider the details of the three different types of medium- and high-temperature fuel cells, we should consider the main features that are *common* to all the cells. Three out of the four main advantages listed on the previous page related to what could be done with the waste heat. It can be used to reform fuels, provide heat, and drive engines. This means that the PAFC, MCFC, and SOFC **can never be considered simply as fuel cells, but they must always be thought of as an integral part of a complete fuel processing and heat generating system**. The wider system issues are largely the same for all three fuel cell types. The common features are considered under four headings as outlined below.

- These medium- and high-temperature fuel cells will nearly always use a fuel that will need processing. Fuel reforming is a large topic and is covered in some detail in the chapter that follows, Chapter 8. The basics of how this impinges on fuel cells, and how it is integrated into the fuel cell system, is explained in Section 7.2.1.
- The question of *fuel utilisation* is a problem with all these cells. The fuel will nearly always be a mixture of hydrogen, carbon oxides, and other gases. As the fuel passes through the stack, the hydrogen will be used, and so its concentration in the mixture will reduce. This reduces the local cell voltage. If all the hydrogen was used up, then in theory the voltage at the exit of the stack would be at zero. The problem of how much of the hydrogen fuel to use in the fuel cell is a tricky one, and this is addressed in Section 7.2.2
- The high-temperature gases leaving these cells carry large amounts of heat energy. In many cases it is desirable to use turbines to convert this heat energy into further

---

[1] Baur and his associates in Switzerland.

electrical energy. This can be done with all three of the fuel cells considered here. It is explained in Section 7.2.3 how *this combination of fuel cell and heat engine can lead to unsurpassed levels of efficiency,* with each machine compensating for the practical problems of the other.

- Heat from the stack exhaust gases can be used to preheat fuel and oxidant using suitable heat exchangers. Best use of heat within high-temperature fuel cell systems is an important aspect of system design, and chemical engineers often referred to this as 'process integration'. To ensure high electrical and thermal system efficiencies, the minimising of *exergy* loss is a key element, and 'pinch technology' is a method that system designers have in their toolbox to help with this. Such heat management aspects of system design are covered in Section 7.2.4.

We first consider these features common to all medium- and high-temperature fuel cells. Note that since these fuel cells use processed fuel and since use can be made of the exhaust heat, they have been mainly applied to stationary power generation systems. Although we shall see that the SOFC may find application in some mobile applications, the complexity of fuel processing usually rules out the application of high-temperature fuel cells for mobile use.

7.2 Common Features

7.2.1 An introduction to fuel reforming

Fuel reforming is described in detail in Chapter 8. It will suffice to say at this stage that the production of hydrogen from a hydrocarbon usually involves 'steam reforming'. In the case of methane, the steam reforming reaction may be written as

$$CH_4 + H_2O \rightarrow 3H_2 + CO \qquad\qquad [7.1]$$

However, the same reaction can be applied to any hydrocarbon C_xH_y, producing hydrogen and carbon monoxide as before.

$$C_xH_y + xH_2O \rightarrow \left(x + \frac{y}{2}\right)H_2 + xCO \qquad\qquad [7.2]$$

In most cases, and certainly with natural gas, the steam reforming reactions are *endothermic*, that is, heat needs to be supplied to drive the reaction forward to produce hydrogen. Again, for most fuels, the reforming has to be carried out at relatively high temperatures, usually above about 500°C. For the medium- and high-temperature fuel cells, heat required by the reforming reactions can be provided, at least in part, from the fuel cell itself in the form of exhaust heat. In the case of the PAFC, the heat from the fuel cell at around 200°C has to be supplemented by burning fresh fuel gas. This has the effect of reducing the efficiency of the overall system, and typically the upper limit of efficiency of the PAFC is around 40 to 45% (ref. HHV). For the MCFC and the SOFC, heat is available from the fuel cell exhaust gases at higher temperatures. If all this heat is used to promote the reforming reactions (especially when reforming is carried out inside the stack), then the efficiency of these fuel cells can be much higher (typically $>50\%$ (ref. HHV)).

The carbon monoxide produced with hydrogen in steam reforming is a potential problem. This is because it will poison any platinum catalyst, as used in PEMFCs or PAFCs. With these fuel cells, the reformate gas must be further processed by means of the 'water-gas shift' reaction (usually abbreviated to the 'shift' reaction):

$$CO + H_2O \rightarrow CO_2 + H_2 \qquad\qquad [7.3]$$

This reaction serves to reduce the CO content of the gas by converting it into CO_2. As is explained in Section 8.4.9, this process usually needs two stages, at different temperatures, to achieve CO levels low enough for PAFC systems.

A further complication is that fuels such as natural gas nearly always contain small amounts of sulphur or sulphur compounds that must be removed. Sulphur is a well-known catalyst poison and will also deactivate the electrodes of all types of fuel cells. Therefore, sulphur has to be removed before the fuel gas is passed to the reformer or stack. Desulphurisation is a well-established process that is required in many situations, not just for fuel cells. The process is also described in more detail in Chapter 8. At this stage it is sufficient to know that the fuel must be heated to about $350°C$ before entering the desulphuriser, which can be considered as a 'black box' for the moment.

The total fuel processing system, especially for a PEMFC or a PAFC, that requires very low levels of carbon monoxide is therefore quite a complex system. The reader should look at Figures 8.4, 8.5 and 11.9 to see what is meant by this. However, the molten carbonate fuel cell (MCFC) and SOFC that we will be considering later operate at temperatures of about $650°C$ or more. This is hot enough to allow the basic steam reformation reaction of equation 7.1 above to occur *within the fuel cell stack itself*. Furthermore, the steam needed for the reaction to occur is also present in the fuel cell, because the product water from the hydrogen/oxygen reaction appears at the *anode* – the fuel electrode. Thus, there is the opportunity to simplify the design of MCFC or SOFC systems. Steam can be supplied for the reforming by recirculating anode exhaust gas, for example, making the system self-sufficient in water. It is also shown in the sections dedicated to these fuel cells that both of them can use carbon monoxide as a fuel – it reacts with oxygen from the air to produce electric current, just like the hydrogen.

This means that the MCFC and the SOFC systems have the potential to be much simpler than the PAFC systems. The fuel will still need to be desulphurised, but the fuel processing system is much less complex.

7.2.2 Fuel utilisation

The question of fuel utilisation arises whenever the hydrogen for a fuel cell is supplied as part of a mixture or becomes part of a mixture owing to internal reforming. As the gas mixture passes through the cell, the hydrogen is utilised by the cell, the other components simply pass through unconverted, and so the hydrogen concentration falls. If most of the hydrogen gets used, then the fuel gas ends up containing mostly carbon dioxide and steam. The partial pressure of the hydrogen will then be very low. This will considerably lower the cell voltage.

Back in Chapter 2, Section 2.5, we considered the effects of pressure and gas concentration on the 'no loss' open circuit voltage of a fuel cell. The very important Nernst equation was presented in many forms. One form, which is useful in this case, relates the open circuit voltage and the partial pressures of hydrogen, oxygen, and steam. This was equation 2.8.

$$E = E^0 + \frac{RT}{2F} \ln \left[\frac{P_{H_2} . P_{O_2}^{\frac{1}{2}}}{P_{H_2O}} \right] \qquad [7.4]$$

If we isolate the hydrogen term, and say that the hydrogen partial pressure changes from P_1 to P_2, then the change in voltage is

$$\Delta V = \frac{RT}{2F} \ln \left(\frac{P_2}{P_1} \right)$$

In this case, the partial pressure of the hydrogen is falling as it is used up, so P_2 is always less than P_1, and thus ΔV will always be negative. An issue with higher-temperature fuel cells is the RT term in this equation, which means that the voltage drop will be greater for higher-temperature fuel cells. There will also be a voltage drop due to oxygen utilisation from the air, since the partial pressure of the oxygen reduces as it passes through the cell. This is less of a problem – one reason being the 1/2 term applied to the oxygen partial pressure in equation 7.4.

These issues are shown graphically in Figure 7.1 that shows four graphs of voltage against temperature. The first line shows the voltage of a standard hydrogen fuel cell operating at 1.0 bar, with pure hydrogen and oxygen supplies. The second is for a cell using air at the cathode, and a mixture of 4 parts hydrogen, 1 part carbon dioxide at the anode – simulating the gas mixture that would be supplied by reformed methane. The next two lines show the 'Nernst Exit Voltage' for 80 and 90% fuel utilisation, with 50% air utilisation in both cases. Here, both the air and fuel are flowing in the same direction (co-flow). We can see that the drop in voltage is significant, and that this is a serious problem. If one part of a cell has a lower voltage, then this will result in reduced efficiency. Alternatively, a reduced current might be the result. As expected, the voltage drop increases with temperature and fuel utilisation.

Sometimes, the total voltage drop through a high-temperature cell can be reduced by feeding the air and fuel through the cell in opposite directions (counter-flow). This means that the part of the cell with the exiting fuel at least has the highest oxygen partial pressure. The result of this is that the Nernstian voltage drop due to fuel and air utilisation is spread more evenly through the cell. However, fuel and airflow directions are also varied to take account of temperature control, which is sometimes helped with the co-flow arrangements.

A factor that has been ignored in Figure 7.1 is that in the molten carbonate and solid oxide fuel cells, as we shall see, the product steam ends up at the anode – essentially the hydrogen is replaced by steam. So, if the partial pressure of the hydrogen decreases, the steam partial pressure will *increase*. Studying equation 7.4, we see that a rise in the partial

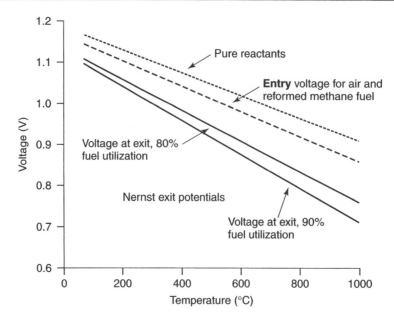

Figure 7.1 Theoretical open circuit voltage for a hydrogen fuel cell under different circumstances. The two curves for the voltage at the exit show how the voltage depends on fuel utilisation and temperature. In both cases the oxygen utilisation is 50%.

pressure of steam will cause a fall in the open circuit voltage. However, the effect of this is difficult to model – some of the steam may be used in internal fuel reforming, for example. The situation is liable, in practice, to be worse than Figure 7.1 would indicate.

The important conclusion is that in the case of a reformed fuel containing carbon dioxide or when internal reforming is applied, all the hydrogen can never be consumed in the fuel cell itself. Some of the hydrogen must pass straight through the cell, to be used later to provide energy to process the fuel or to be burnt to increase the heat energy available for heat engines, as we will be considering in Section 7.2.3. The optimum fuel utilisation for any system can only be found by extensive modelling using computer simulations of the entire system. As we have said before, the fuel cell cannot be considered in isolation.

7.2.3 Bottoming cycles

This rather curious term refers to the use of the 'waste' heat from a fuel cell exhaust gases to drive some kind of heat engine. There are two ways in which this is commonly done:

1. The heat is used in a boiler to raise steam, to drive a steam turbine. A diagram of such a system is shown in Figure 7.3.
2. The whole system, including the fuel cell, is pressurised, and the exit gases from the fuel cell power a gas turbine (GT). Such a system is outlined in Section 7.5 on SOFCs and shown in Figures 7.31 and 7.32.

A third possibility is to combine both a steam *and* gas turbine with a fuel cell into a triple cycle system. This would be a possibility for larger systems in the future, but at the time of writing none have yet been built.

These combined cycle systems offer an elegant way of producing very efficient systems of generating electricity. To understand *why* they are potentially so efficient there are two extreme positions that we must recognise and refute. The fuel cell enthusiast might say that the fuel cell acts as a 'reactor', burning the fuel and producing heat to drive the turbine. The electricity is a 'free bonus'. This is obviously a very efficient system! The opposition might say that the fuel cell is a waste. If the fuel was burnt ordinarily, the temperature would be much higher, and so the heat engine would be more efficient. Get rid of the fuel cell!

Both these positions are wrong! To show this we need to use a little thermodynamics theory.[2] In Chapter 2 we saw that the energy accessed in a fuel cell is the Gibbs free energy. We noted, in equation 2.4, that the efficiency limit of a fuel cell was

$$= \frac{\Delta \bar{g}_f}{\Delta \bar{h}_f}$$

We can make this equation a little less 'cluttered' if we use the plain forms of Gibbs free energy and enthalpy, rather than the molar specific forms. This is justified as long as we do it for both energies, and it gives

$$\text{Maximum efficiency} = \frac{\Delta G}{\Delta H}$$

However, we also noted that in the case of hydrogen the absolute value of the Gibbs free energy *decreases* with temperature. It is the same for carbon monoxide, and so we can say that for all practical fuels this is so. It is impractical to operate a fuel cell at below ambient temperature, and so the highest possible efficiency limit will occur at ambient temperature, T_A. At this temperature the Gibbs free energy is maximum and is equal to ΔG_{T_A}. So, in practice, we can refine our formula for the maximum efficiency to

$$\text{Maximum efficiency} = \frac{\Delta G_{T_A}}{\Delta H} \qquad [7.5]$$

However, at ambient temperature the losses (detailed in Chapter 3) are large, and so the practical efficiencies are well below this limit. To reduce the voltage drops, we need to raise the temperature, but this reduces ΔG, and hence the efficiency.

The reason for the fall in ΔG as the temperature rises can be seen from the fundamental thermodynamic relationship that connects the Gibbs free energy, enthalpy, and entropy:

$$\Delta G = \Delta H - T \Delta S \qquad [7.6]$$

---

[2] The approach given here is along the lines used by A.J. Appleby in Appleby & Foulkes (1993) and Blomen et al. (1993) among others. It does involve some approximations, but what it lacks in complete rigour it more than makes up for in clarity. For an alternative approach based on the T/S diagrams loved (or hated) by students of thermodynamics, the reader should consult Gardner (1997).

The enthalpy of reaction, ΔH, is more or less constant for the hydrogen reaction, as is the entropy change ΔS.[3] The Gibbs free energy becomes less negative (i.e. less energy is released) with temperature because ΔS is negative.

Now, the fraction of the enthalpy of reaction that is not converted into electricity in the fuel cell is converted into heat. This means that as the temperature of the fuel cell rises, the energy $T\Delta S$ that is converted into heat increases.

Suppose a fuel cell is operating *reversibly*, at a temperature T_F that is considerably higher than the ambient temperature T_A. Under these reversible conditions,

- ΔG_{T_F} is the electrical energy that would be supplied by the fuel cell, and
- $T_F\Delta S$ is the heat that would be produced.

Now, this heat energy could be converted into work, and hence electrical energy, using a heat engine and generator. However, even in a reversible system this conversion would be subject to the 'Carnot limit'. The upper temperature is T_F, the temperature of the fuel cell, as this is the temperature at which the gases leave the cell. The lower temperature is the ambient temperature T_A. So, the extra electrical energy that could be available in a reversible system is

$$\frac{T_F - T_A}{T_F} T_F\Delta S = (T_F - T_A)\Delta S$$

This energy would be added to the electrical energy supplied directly by the fuel cell, so the total electrical energy available from the dual cycle system would be given by the equation

$$\text{Maximum electrical energy} = \Delta G_{T_F} + (T_F - T_A)\Delta S$$

$$= \Delta G_{T_F} + T_F\Delta S - T_A\Delta S \qquad [7.7]$$

However, this equation can be simplified, because if we rewrite equation 7.6 specifically for temperature T_F and T_A, we get the following two equations:

$$T_F\Delta S = \Delta H - \Delta G_{T_F}$$

and

$$T_A\Delta S = \Delta H - \Delta G_{T_A}$$

If we substitute these two equations into equation 7.7 we get

$$\text{Maximum electrical energy} = \Delta G_{T_F} + \Delta H - \Delta G_{T_F} - \Delta H + \Delta G_{T_A}$$

$$= \Delta G_{T_A}$$

So, dividing this by the enthalpy change to give the fraction of the energy converted into electricity, we have, **exactly as before**

$$\text{Maximum efficiency} = \frac{\Delta G_{T_A}}{\Delta H}$$

---

[3] This can be confirmed by looking at Table A1.3 in Appendix 1, where a range of values is given.

This is a remarkable result! The equation is the same as equation 7.5. If we take a fuel cell and operate it at the elevated temperature of T_F, and we add a heat engine bottoming cycle, we end up with the same efficiency limit as for an ambient temperature fuel cell. Figure 7.2 is a sketch graph of the efficiency limits of a fuel cell on its own, a heat engine on its own, and a combined cycle fuel cell and heat engine. It is hoped that this diagram makes the point – the dual cycle system of fuel cell and heat engine is particularly advantageous.

The elegance of this arrangement becomes even more clear if we consider the limiting case of a heat engine. The Carnot limit is well known, that is to say

$$\text{Maximum efficiency} = \frac{T_2 - T_1}{T_2} \tag{7.8}$$

However, does this mean we can keep increasing T_2, getting nearer to 100% efficiency? No! There is clearly a temperature limit. One of these is certainly the combustion temperature of the fuel being used – it is clear that this cannot reasonably be exceeded. Now, at spontaneous combustion $\Delta G = 0$, and so, using equation 7.6, to a good approximation

$$T_c = \frac{\Delta H}{\Delta S}$$

So, we can substitute this equation for T_2 in the Carnot limit equation given above. It is obvious too that the sink temperature T_1 cannot be lower than the ambient temperature

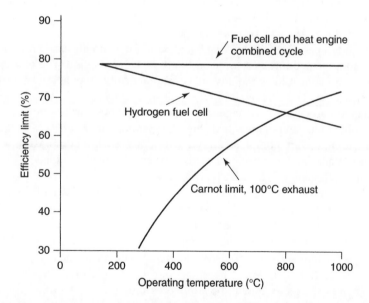

Figure 7.2 Efficiency limits for a heat engine, a hydrogen fuel cell, and a fuel cell/turbine combined cycle. For the heat engine and the combined cycle, the lower temperature is 100°C. The fuel cell efficiencies are referred to as the higher heating value.

T_A, so we can also substitute T_A for T_1. The Carnot limit equation thus becomes

$$\text{Maximum efficiency} = \frac{\dfrac{\Delta H}{\Delta S} - T_A}{\dfrac{\Delta H}{\Delta S}} = \frac{\Delta H - T_A \Delta S}{\Delta H} \qquad [7.9]$$

However, we know that
$$\Delta G_{T_A} = \Delta H - T_A \Delta S \qquad [7.10]$$

If we substitute equation 7.10 into 7.9, we see that the maximum efficiency of a fuel burning heat engine is

$$\text{Maximum efficiency} = \frac{\Delta G_{T_A}}{\Delta H}$$

This is exactly the same formula again! So we can fully appreciate the potential of the fuel cell/heat engine combined cycle. The heat engine cannot work close to its efficiency limit because of materials and practical limitations. Its efficiency limit is at the combustion temperature of the fuel, but engines cannot operate at temperatures of several thousand degrees Celsius – they would melt! Similarly, though fuel cells can operate at ambient temperatures, their efficiencies are far below the theoretical limit, because of all the problems we discussed in Chapter 3. However, a fuel cell operating at around 800/900/1000°C can approach the theoretical maximum efficiency. At these temperatures heat engines are also at their best – they do not require exotic materials and are not too expensive to produce. As A.J. Appleby has put it:

'Thus, a high-temperature fuel cell combined with, for example, a steam cycle condensing close to room temperature is a 'perfect' thermodynamic engine. The two components of this perfect engine also have the advantage of practically attainable technologies. The thermodynamic losses (i.e. irreversibilities) in a high-temperature fuel cell are low, and a thermal engine can easily be designed to operate at typical heat source temperatures equal to the operating temperature of a high-temperature fuel cell. Thus, the fuel cell and the thermal engine are complementary devices, and such a combination would be a practical 'ideal black box' (or because of its low environmental impact, a 'green box') energy system.' (Blomen et al., 1993, p. 168).

A possible combined cycle system based on an SOFC is shown in Figure 7.3. The steam turbine in the centre of the diagram drives two compressors (one for the air, the other for the fuel gas) and an alternator. The air compressor is needed to drive the air through the preheater, the fuel cell, the afterburner, the boiler, and out to the final heating system. The gas compressor needs to drive the fuel through the same components, and the desulphuriser as well, and so would be at a somewhat higher pressure. They would not normally compress the gases much above air pressure, and the fuel cell would operate at about 1.3 bar – only a little above ambient pressure. The natural gas is internally reformed in the fuel cell, but not all the hydrogen is consumed. The remainder is burnt in the afterburner, which raises the temperature of both gas streams to about 800°C. This

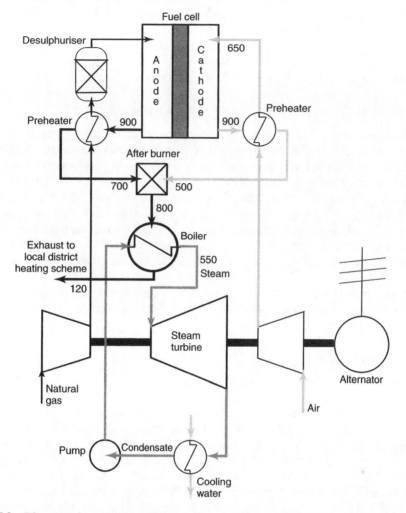

Figure 7.3 Diagram of a possible SOFC and steam turbine combined cycle. The figures indicate likely approximate temperatures of the fluids, in Celsius, at the different stages of the process.

hot gas is used in a heat exchanger–type boiler to raise steam at about 550°C, which drives a turbine that drives the alternator.

The compressors, and the associated work they do driving the gas through the system, represents the major irreversibility in this system. Fry et al. (1997) have published a design of a 36-MW system along the lines of that in Figure 7.3, and using realistic figures for turbine and fuel cell performance (as in 1997) and irreversibilities throughout the system, they came up with the following figures:

Natural gas usage	—	1.19 kg s^{-1}
Fuel cell output	—	26.9 MW
Alternator output	—	8.34 MW

This corresponds to an overall system efficiency of 60%, which is noticeably better than the 55% obtainable from combined cycle gas/steam turbine systems. The combination of fuel cell and gas turbine, where the fuel cell operates at well above ambient pressure, will be presented in the section on SOFCs. Such systems offer the possibility of even higher efficiencies.

7.2.4 The use of heat exchangers – exergy and pinch technology

It will become evident that as more fuel cell systems are described there are challenges in the way in which the various components of the system are integrated. This applies to all systems, but is particularly important in the medium- and high-temperature systems in which several of the balance of plant (BOP) items are operating at high temperatures. Examples are the desulphuriser, reformer reactor, shift reactors, heat exchangers, recycle compressors, and ejectors. In some of these components heat may be generated or consumed. The challenge for the system designer is to arrange the various components in a way that minimises heat losses to the external environment, and at the same time to ensure that heat is utilised to the best extent. In this way the energy that is fed to the fuel cell system in the fuel is converted in the most effective way to electricity and useful heat. The optimisation of such system design is known in the chemical engineering discipline as *process integration*, and designers will refer to carrying out *exergy analysis* and *pinch technology*.

Heat exchangers

In any fuel cell system there are process streams that need to be heated (e.g. fuel preheat for the reformer, and the reformer itself, and in raising and superheating steam) and streams that need to be cooled (e.g. the fuel cell stack and the outlet of the shift reactor(s) in a PEM system). The challenge of the system designer is to use the heat available from one stream to heat another in the most efficient way (i.e. avoiding unnecessary heat losses). Heat transfer from one process stream to another is carried out in a *heat exchanger*. The gas (or liquid) to be heated passes through pipework that is heated by the gas (or liquid) to be cooled. A commonly used symbol for a heat exchanger is shown in Figure 7.4. When the exit fluids are used to heat incoming fluids, the heat exchanger is often called a *recuperator*.

There are several common types of heat exchangers, examples being shell and tube, plate/fin, and printed circuit exchangers. Descriptions of these are outside the scope of this book, and several good references are available. The choice of the heat exchanger is governed by the temperature range of operation, the fluids used (e.g. liquid or gas phase), the fluid throughput, and the cost. The cost of the exchanger is, of course, governed by the material costs and also by the heat-transfer area. In Chapter 8 we shall see how heat exchangers may be used in practical fuel processing systems. Some are shown being used in Figures 7.3 and 8.4, which are good examples.

Exergy

Exergy may be defined as the quality of heat. More precisely, it is the work or the ability of energy to be used for work or converted to work. *Potential energy* as classically

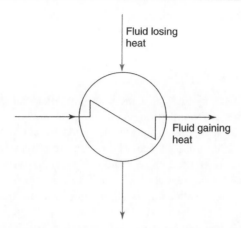

Figure 7.4 A commonly used heat exchanger symbol. The fluid to be heated passes through the 'zigzag' element.

defined is therefore *exergy*, as is the Gibbs free energy (with the sign changed) for the combustion of fuel. Energy is conserved in all processes (first law of thermodynamics), whereas *exergy* is only conserved in processes that are reversible. Real processes are of course irreversible, so that exergy is always partly consumed to give enthalpy.

The exergy of a system is measured by its displacement from thermodynamic equilibrium. Since $\partial G = V \partial P - S \partial T + \sum \mu_i \partial n_i$, where V and P are volume and pressure, respectively, and μ_i is the chemical potential of the ith chemical component with a number of molecules, n, it follows that the exergy change of a system going between an initial state and a reference state (subscript o) is given by

$$E = S(T - T_\mathrm{o}) - V(P - P_\mathrm{o}) - \sum n_i(\mu_i - \mu_\mathrm{o}) \qquad [7.11]$$

The symbol E for exergy used here should not be confused with the same symbol used for energy in classical thermodynamics, nor for fuel cell voltage! It is plain that the higher the temperature T, the greater the exergy. This link with temperature and the ability to do work can also be approached less generally via consideration of the Carnot limit.

If we consider the case in which both the SOFC and PEM systems have the same power output and efficiency, then the *heat*, which is the enthalpy content, of the exhaust streams produced by both systems will be the same. However, the heat produced by the SOFC will be at a higher temperature and therefore has a higher *exergy* than that from a PEM fuel cell. We can, therefore, say that the heat from an SOFC is more valuable. The heat that is liberated in a PEM fuel cell is around 80°C and has limited use both within the system and for external applications. For the latter, it may be limited to air space heating, or possibly integration with an absorption cooling system. The heat that is produced by an SOFC on the other hand is exhausted at temperatures of around 1000°C and is clearly more valuable, as it can also be used, as we have seen, in a bottoming cycle.

The importance of the concept of exergy is that with high-temperature fuel cell systems, the system configuration should be designed in such a way that exergy loss is minimised.

If the high-grade heat that emerges from the fuel cell stack is used inefficiently within the system, then the heat that emerges from the system for practical use will be degraded.

Pinch analysis and system design

Pinch analysis or pinch technology is a methodology that can be applied to fuel cell systems for deciding the optimum arrangement of heat exchangers and other units so as to minimise exergy loss. It was originally designed by chemical engineers as a tool for defining energy-saving options, particularly through improved heat exchanger network design. Pinch analysis has been used in many fuel cell system designs, and good examples are given in the reference by Blomen et al. (1993). The concept is straightforward enough, but for complex systems, sophisticated computer models are required. The procedure is broadly as follows:

In any fuel cell plant, there will be process streams that need heating (cold streams) and those that need cooling (hot streams) *irrespective of where heat exchangers are located*. The first stage in system design is, therefore, to establish the basic chemical processing requirements and to produce a system configuration showing all the process streams, thereby defining the hot and cold streams. By carrying out heat and mass balances, the engineer can calculate the enthalpy content of each process stream. Knowing the required temperatures of each process stream, heating and cooling curves can be produced. Examples of these for an MCFC system are shown in Figure 7.5. The individual cooling and heating curves are then summed together to make two composite curves, one showing the total heating required by all the process streams and the other showing the total cooling required by the system. These two curves are slid together along the y-axis and where they 'pinch' together with a minimum temperature difference of, for example, 50°C, the temperature is noted. This so-called pinch temperature defines the target for optimum process design, since in a real system heat cannot be transferred from above or below this temperature. In some fuel cell systems a pinch temperature is not found, in which case a threshold is defined. Either way, pinch technology provides an excellent method for system optimisation.

Of course, other considerations need to be taken into account in system design, for example, the choice of materials for the BOP components and their mechanical layout. In the remainder of the book, various system configurations are given. These are usually drawn as process flow diagrams (PFD) showing the logical arrangement of the fuel cell stack and the BOP components. Many computer models are now available that can be used for calculating the heat and material flows around the system. Temperatures and enthalpies of the process streams can then be calculated, and these are usually the first step in system design. What normally follows is a first stage system optimisation, using pinch technology, followed by detailed design and fabrication. Many other techniques are available to the system designer, which are outside the scope of this book, and the following descriptions of fuel cell systems will indicate the opportunities for creativity in system design.

Having looked at the most important features that are common to all types of 'hot' fuel cells, we will now look at the three major types of cells in this category, starting with the phosphoric acid fuel cell (PAFC).

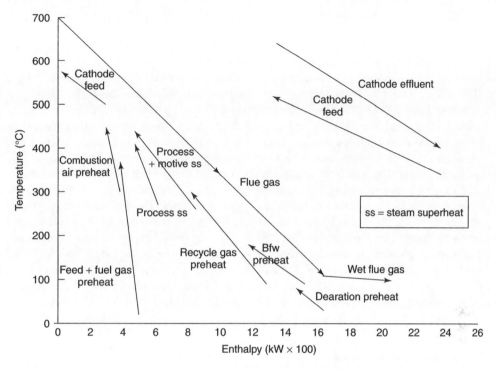

Figure 7.5 Hot and cold heating curves for a conceptual 3.25 MW MCFC system with high-pressure steam generation.

7.3 The Phosphoric Acid Fuel Cell (PAFC)

7.3.1 How it works

The PAFC works in a similar fashion to the PEM fuel cell described in Chapter 4. The PAFC uses a proton-conducting electrolyte, and the reactions occurring on the anode and cathode are those given in Figure 1.3. In the PAFC, the electrochemical reactions take place on highly dispersed electrocatalyst particles supported on carbon black. As with the PEM fuel cells, platinum (Pt) or Pt alloys are used as the catalyst at both electrodes. The electrolyte is an inorganic acid, concentrated phosphoric acid (100%) which, like the membranes in the PEM cells, will conduct protons.

The electrolyte

Phosphoric acid (H_3PO_4) is the only common inorganic acid that has enough thermal stability, chemical, and electrochemical stability, and low enough volatility (above about 150°C) to be considered as an electrolyte for fuel cells. Most importantly, phosphoric acid is tolerant to CO_2 in the fuel and oxidant, unlike the alkaline fuel cell. Phosphoric acid was therefore chosen by the American company, United Technologies (later the spin-off

ONSI Corporation), back in the 1970s as the preferred electrolyte for terrestrial fuel cell power plants.

Phosphoric acid is a colourless viscous hygroscopic liquid. In the PAFC, it is contained by capillary action (it has a contact angle $>90°$) within the pores of a matrix made of silicon carbide particles held together with a small amount of polytetrafluoroethylene (PTFE). The pure 100% phosphoric acid, used in fuel cells since the early 1980s, has a freezing point of 42°C, so to avoid stresses developing because of freezing and re-thawing, PAFC stacks are usually maintained above this temperature once they have been commissioned. Although the vapour pressure is low, some acid is lost during normal fuel cell operation over long periods at high temperature. The loss depends on the operating conditions (especially gas flow velocities and current density). It is therefore necessary to replenish electrolyte during operation, or ensure that sufficient reserve of acid is in the matrix at the start of operation to last the projected lifetime. The SiC matrix comprising particles of about 1 μm is 0.1 to 0.2 mm thick, which is thin enough to allow reasonably low ohmic losses (i.e. high cell voltages) whilst having sufficient mechanical strength and the ability to prevent crossover of reactant gases from one side of the cell to the other. This latter property is a challenge for all liquid-based electrolyte fuel cells (see also the MCFC later). Under some conditions, the pressure difference between anode and cathode can rise considerably, depending on the design of the system. The SiC matrix presently used is not robust enough to stand pressure differences greater than 100 to 200 mbar.

The electrodes and catalysts

Like the PEM fuel cell, the PAFC uses gas diffusion electrodes. In the mid-1960s, the porous electrodes used in the PAFC were PTFE-bonded Pt black, and the loadings were about 9 mg Pt cm^{-2} on each electrode. Since then, Pt supported on carbon has replaced Pt black as the electrocatalyst, as for the PEMFC, as shown in Figure 4.6. The carbon is bonded with PTFE (about 30–50 wt%) to form an electrode-support structure. The carbon has important functions:

- To disperse the Pt catalyst to ensure good utilisation of the catalytic metal,
- To provide micropores in the electrode for maximum gas diffusion to the catalyst and for electrode/electrolyte interface
- To increase the electrical conductivity of the catalyst.

By using carbon to disperse the platinum, a dramatic reduction in Pt loading has also been achieved over the last two decades – the loadings are currently about 0.10 mg Pt cm^{-2} in the anode and about 0.50 mg Pt cm^{-2} in the cathode. The activity of the Pt catalyst depends on the type of catalyst, its crystallite size, and specific surface area. Small crystallites and high surface areas generally lead to high catalyst activity. The low Pt loadings that can now be achieved result in part from the small crystallite sizes, down to around 2 nm, and high surface areas, up to 100 m^{2}g^{-1}. (See Figure 1.6.)

The PTFE binds the carbon black particles together to form an integral (but porous) structure, which is supported on a porous carbon paper substrate. The carbon paper serves as a structural support for the electrocatalyst layer, in addition to acting as the current collector. A typical carbon paper used in PAFCs has an initial porosity of about 90%, which is reduced to about 60% by impregnation with 40 wt% PTFE. This wet-proof carbon paper contains macropores of 3- to 50-μm diameter (median pore diameter of about 12.5 μm) and micropores with a median pore diameter of about 3.4 nm for gas permeability. The composite structure consisting of a carbon black/PTFE layer on carbon paper substrate forms a stable, three-phase interface in the fuel cell, with electrolyte on one side (electrocatalyst side) and the reactant gas environment on the other side of the carbon paper.

The choice of carbon is important, as is the method of dispersing the platinum, and much know-how is proprietary to the fuel cell developers. Nevertheless, it is known that heat treatment in nitrogen to very high temperatures (1000–2000°C) is found to improve the corrosion resistance of carbons in the PAFC. The development of the catalysts for PAFC fuel cells has been reviewed extensively (e.g. Kordesch, 1979; Appleby, 1984; and Kinoshita, 1988) and lifetimes of 40 000 h are now expected. However, electrode performance does decay with time. This is due primarily to the sintering (or agglomeration) of Pt catalyst particles and the obstruction of gases through the porous structure caused by electrolyte flooding. During operation, the platinum particles have the tendency to migrate to the surface of the carbon and to agglomerate into larger particles, thereby decreasing the active surface area. The rate of this sintering phenomenon depends mainly on the operating temperature. An unusual difficulty is that corrosion of carbon becomes a problem at high cell voltages (above ~0.8 V). For practical applications, low current densities, with cell voltages above 0.8 V, and hot idling at open circuit potential are therefore best avoided with the PAFC.

The stack

The PAFC stack consists of a repeating arrangement of a ribbed bipolar plate, the anode, electrolyte matrix, and cathode. In a similar manner to that described for the PEM cell, the ribbed bipolar plate serves to separate the individual cells and electrically connect them in series, whilst providing the gas supply to the anode and the cathode, respectively, as shown in Figures 1.9 and 1.10. Several designs for the bipolar plate and ancillary stack components are being used by fuel cell developers, and these aspects are described in detail elsewhere (Appleby and Foulkes, 1993). A typical PAFC stack may contain 50 or more cells connected in series to obtain the practical voltage level required.

The bipolar plates used in early PAFCs consisted of a single piece of graphite with gas channels machined on either side, as shown in Figure 7.6. Machining channels in graphite is an expensive though feasible method of manufacture. Newer manufacturing methods and designs of bipolar plates are now being used. One approach is to build up the plate in several layers. In 'multi-component bipolar plates', a thin impervious carbon plate serves to separate the reactant gases in adjacent cells in the stack, and separate porous plates with ribbed channels are used for directing gas flow. This arrangement is known as the ribbed substrate construction and is shown in Figure 7.7.

The ribbed substrate has some key advantages:

- Flat surfaces between catalyst layer and substrate promote better and uniform gas diffusion to the electrode.

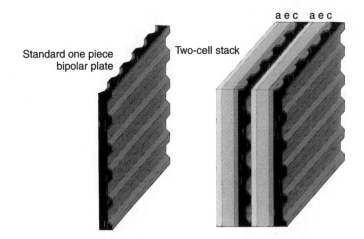

Figure 7.6   Standard one-piece ribbed bipolar plate.

- It is amenable to a continuous manufacturing process since the ribs on each substrate run in only one direction.
- Phosphoric acid can be stored in the substrate, thereby increasing the lifetime of the stack.

Stack cooling and manifolding

In PAFC stacks, provision must be included to remove the heat generated during cell operation. This can be done by either liquid (water/steam or a dielectric fluid) or gas (air) coolants that are routed through cooling channels or pipes located in the cell stack, usually about every fifth cell. Liquid cooling requires complex manifolds and connections, but better heat removal is achieved than that with air cooling. The advantage of gas cooling is its simplicity, reliability, and relatively low cost. However, the size of the cell is limited, and the air-cooling passages are much larger than those needed for liquid cooling. Water cooling is the most popular method, therefore, and is applied to the ONSI PAFC systems.

Water cooling can be done with either boiling water or pressurised water. Boiling water cooling uses the latent heat of vaporisation of water to remove the heat from the cells. Since the average temperature of the cells is around 180 to 200°C, this means that the temperature of the cooling water is about 150 to 180°C. Quite uniform temperatures in the stack can be achieved using boiling water cooling, leading to increased cell efficiency. If pressurised water is used as the cooling medium, the heat is only removed from the stack by the heat capacity of the cooling water, so the cooling is not so efficient as with boiling water. Nevertheless, pressurised water gives a better overall performance than using oil (dielectric) cooling or air cooling, though these may be preferred for smaller systems.

The main disadvantage of water cooling is that water treatment is needed to prevent corrosion of cooling pipes and blockages developing in the cooling loops. The water quality required is similar to that used for boiler feed water in conventional thermal power stations. Although not difficult to achieve with ion-exchange resins, such water treatment

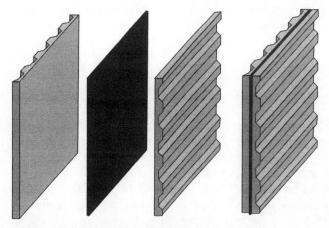

Bipolar plate made from two pieces of porous, ribbed substrate and one sheet of conductive, impermeable material such as carbon. The porous ribbed substrates hold phosphoric acid.

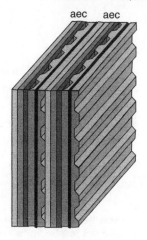

Two-cell stack. The electrolyte can now be thinner, as the ribbed substrates also hold electrolyte.

Figure 7.7 Cell interconnections made from ribbed substrates.

will add to the capital cost of PAFC systems, and for this reason water cooling is preferred only for units in the 100-kW class and above, such as the ONSI PC25 200-kW systems.

All PAFC stacks are fitted with manifolds that are usually attached to the outside of the stacks (*external manifolds*).[4] We shall see later that an alternative *internal manifold* arrangement is preferred by some MCFC developers. Inlet and outlet manifolds simply enable fuel gas and oxidant to be fed to each cell of a particular stack. Careful design of inlet fuel manifold enables the fuel gas to be supplied uniformly to each cell. This is beneficial in minimising temperature variations within the stack thereby ensuring long lifetimes. Often a stack is made of several sub-stacks arranged with the plates horizontal,

---

[4] Look back at Figures 1.12 and 1.13.

mounted on top of each other with separate fuel supplies to each sub-stack. If the fuel cell stack is to be operated at high pressure, the whole fuel cell assembly has to be located within a pressure vessel filled with nitrogen gas at a pressure slightly above that of the reactants.

7.3.2 Performance of the PAFC

A typical performance (voltage–current) curve for the PAFC is like that shown in Figure 3.1, and follows the usual form for medium- to low-temperature cells. The operating current density of PAFC stacks is usually in the range of 150 to 400 mA cm^{-2}. When operating at atmospheric pressure, this gives a cell voltage of between 600 and 800 mV. As with the PEMFC, the major polarisation occurs at the cathode, and like the PEM cell, the polarisation is greater with air (typically 560 mV at 300 mA cm^{-2}) than with pure oxygen (typically 480 mV at 300 mA cm^{-2}) because of dilution of the reactant. The anode exhibits very low polarisation (-4 mV per 100 mA cm^{-2}) on pure hydrogen, which increases when carbon monoxide is present in the fuel gas. The ohmic loss in PAFCs is also relatively small, amounting to about 12 mV per 100 mA cm^{-2}.

Operating the PAFC at pressure

Cell performance for any fuel cell is a function of pressure, temperature, reactant gas composition and utilisation. It is well known that an increase in the cell operating pressure enhances the performance of all fuel cells, including PAFCs. It was shown in Chapter 2, Section 2.5.4, that for a reversible fuel cell the increase in voltage resulting from a change in system pressure from P_1 to P_2 is given by the formula

$$\Delta V = \frac{RT}{4F} \ln\left(\frac{P_2}{P_1}\right)$$

However, this 'Nernstian' voltage change is not the only benefit of higher pressure. At the temperature of the PAFC, higher pressure also decreases the activation polarisation at the cathode, because of the increased oxygen and product water partial pressures. If the partial pressure of water is allowed to increase, a lower phosphoric acid concentration will result. This will increase ionic conductivity slightly and bring about a higher exchange current density. The important beneficial effect of increasing the exchange current density has been discussed in detail in Section 3.4.2. It results in further reduction to the activation polarisation. The greater conductivity also leads to a reduction in ohmic losses. The result is that the increase in voltage is much higher than this equation would suggest. Quoting experimental data collected over some period, Hirschenhofer et al. (1998) suggest that the formula

$$\Delta V = 63.5 \ln\left(\frac{P_2}{P_1}\right) \qquad [7.12]$$

is a reasonable approximation for a temperature range of 177°C $< T <$ 218°C and a pressure range of 1 bar $< P <$ 10 bar.

The effect of temperature

In Chapter 2, Section 2.3, we showed that for a hydrogen fuel cell the reversible voltage decreases as the temperature increases. Over the possible temperature range of the PAFC, the effect is a decrease of $0.27\,\text{mV}\,°\text{C}^{-1}$ under standard conditions (product being water vapour). However, as discussed in Chapter 3, an increase in temperature has a beneficial effect on cell performance because activation polarisation, mass transfer polarisation, and ohmic losses are reduced. The kinetics for the reduction of oxygen on platinum improves as the cell temperature increases. Again, quoting experimental data collected over some period, Hirschenhofer et al. (1998) suggest that at a mid-range operating load ($\sim\!250\,\text{mA}\,\text{cm}^{-2}$), the voltage gain ($\Delta V_T$) with increasing temperature of pure hydrogen and air is given by

$$\Delta V_T = 1.15(T_2 - T_1)\,\text{mV} \qquad\qquad [7.13]$$

The data suggests that equation 7.13 is reasonably valid for a temperature range of $180°\text{C} < T < 250°\text{C}$. It is apparent from this equation that each degree increase in cell temperature should produce a performance increase of $1.15\,\text{mV}$. Other data indicate that the coefficient for equation 7.13 may actually be in the range of 0.55 to 0.75, rather than 1.15.

Although temperature has only a minimal effect on the H_2 oxidation reaction at the anode, it is important in terms of anode poisoning. Figure 7.8 shows that increasing the cell temperature results in increased anode tolerance to CO poisoning. This increased tolerance is a result of reduced CO adsorption. A strong temperature effect is also seen for simulated coal gas (SCG in Figure 7.8).

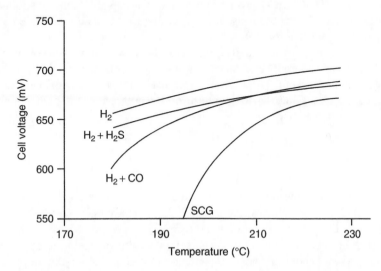

Figure 7.8 Effect of temperature on PAFC cell voltage for different fuels: H_2, $H_2 + 200\,\text{ppm}\,H_2S$ and simulated coal gas (Jalan et al., 1990).

Effects of fuel and oxidant utilisation

As mentioned in Section 7.2, fuel and oxidant utilisations are important operating parameters for fuel cells such as the PAFC. In a fuel gas that is obtained, for example, by steam reforming of natural gas (see next chapter) the carbon dioxide and unreacted hydrocarbons (e.g. methane) are electrochemically inert and act as diluents. Because the anode reaction is nearly reversible, the fuel composition and hydrogen utilisation generally do not strongly influence cell performance. The RT term in equation 7.4 is clearly lower than for the MCFC and SOFC. Further discussion is given in Hirschenhofer et al. (1998).

On the cathode side, the use of air with ~21% oxygen instead of pure oxygen results in a decrease in the current density of about a factor of three at constant electrode potential. The polarisation at the cathode increases with an increasing oxygen utilisation. Low utilisations therefore, particularly oxygen utilisation, yield high performance. As mentioned in the previous section, the drawback of low utilisations is the poor fuel usage and choice of operating utilisations requires a careful balance of all system and stack aspects. State-of-the-art PAFC systems employ utilisations of typically 85% and 50% for the fuel and oxidant, respectively.

Effects of carbon monoxide and sulphur

As with the platinum anode catalyst in the PEM fuel cell, the anode of the PAFC may be poisoned by carbon monoxide in the fuel gas. The CO occupies catalyst sites. Such CO is produced by steam reforming and for the PAFC the level that the anode can tolerate is dependent on the temperature of the cell. The higher the temperature, the greater is the tolerance for CO. The absorption of CO on the anode electrocatalyst is reversible and CO will be desorbed if the temperature is raised. Any CO has some effect on the PAFC performance, but the effect is not nearly so important as in the PEMFC. At a working temperature above 190°C, a CO level of up to 1% is acceptable, but some quote a level of 0.5% as the target. The methods used to reduce the CO levels are discussed in the next chapter, especially in Section 8.4.9.

Sulphur in the fuel stream, usually present as H_2S, will similarly poison the anode of a PAFC. State-of-the-art PAFC stacks are able to tolerate around 50 ppm of sulphur in the fuel. Sulphur poisoning does not affect the cathode, and poisoned anodes can be reactivated by increasing the temperature or by polarisation at high potentials (i.e. operating cathode potentials).

7.3.3 Recent developments in PAFC

Until recently the PAFC was the only fuel cell technology that could be said to be available commercially. Systems are now available that meet market specifications, and they are supplied with guarantees. An example has already been shown in Figure 1.16, which illustrates the ONSI Corporation 200-kW CHP system. Many of these systems have now run for several years, and so there is a wealth of operating experience from which developers and end users can draw (Brenscheidt et al. 1998). One important aspect that has come from field trials of the early PAFC plants is the reliability of the stack and the quality of power produced by the systems. The attribute of high power quality and reliability is leading to systems being preferred for so-called premium power applications, such as in banks, in hospitals, and in computing facilities. There are now over 65 MW of demonstrators, worldwide, that have been tested, are being tested, or are being fabricated. Most of the plants are in the 50- to 200-kW capacity range, but large plants of 1 MW and 5 MW have also been built. The largest plant operated to date has been that built by International Fuel Cells and Toshiba for Tokyo Electric Power. This has achieved 11 MW of grid quality AC power. Major efforts in the US and Japan are now concentrated on improving PAFCs for stationary dispersed power and on-site cogeneration (CHP) plants. The major industrial developers are the United Technologies spin-off ONSI Corporation, recently renamed UTC Fuel Cells in the US and Fuji Electric, Toshiba, and Mitsubishi Electric Companies in Japan.

Phosphoric acid electrode/electrolyte technology has now reached a level of maturity where developers and users are focusing their resources on producing commercial capacity, multi-unit demonstrations and pre-prototype installations. Cell components are being manufactured at scale and in large quantities. However, the technology is still too costly to be economical compared with alternative power generation systems, except perhaps in niche premium power applications. There is a need to increase the power density of the cells and to reduce costs, both of which are inextricably linked. System optimisation is also a key issue. For further information on general technical aspects, the interested reader should consult other references, for example, Appleby and Foulkes (1993), Blomen et al. (1993), and Hirschenhofer et al. (1998). Much of the recent technology developments are proprietary but the following gives an indication of progress made during the last few years.

During the early 1990s, the goal of the American company IFC was to design and demonstrate a large stack with a power density of $0.188\,W\,cm^{-2}$, 40 000 h useful life, and a stack cost of less than \$400/kW. A conceptual design of an improved technology stack operating at 8.2 atm and 207°C was produced. The stack would be composed of $355 \times 1\,m^2$ cells and would produce over 1 MW DC power in the same physical envelope as the 670-kW stack used in the 11-MW PAFC plant built for Tokyo Electric Power. The improvements made to the design were tested in single cells, and in subscale and full size short stacks. The results of these tests were outstanding. The power density goal was exceeded with $0.323\,W\,cm^{-2}$ being achieved in single cells operating at $645\,mA\,cm^{-2}$ and up to 0.66 V per cell. In stacks, cell performances of $0.307\,W\,cm^{-2}$ have been achieved, with an average of 0.71 V per cell at $431\,mA\,cm^{-2}$. In comparison, the Tokyo Electric Power Company's 11-MW power plant, in 1991, had an average cell performance of approximately 0.75 V per cell at $190\,mA\,cm^{-2}$. The performance degradation rate of the advanced stacks was less than 4 mV per 1000 h during a 4500-h test. The results from this program represent the highest performance of full size phosphoric acid cells and short stacks published to date.

Mitsubishi Electric Corporation has also demonstrated improved performances in single cells of 0.65 mV at $300\,mA\,cm^{-2}$. Component improvements by Mitsubishi have resulted in the lowest PAFC degradation rate publicly acknowledged, 2 mV per 1000 h for 10 000 h at 200 to $250\,mA\,cm^{-2}$ in a short stack with $3600\,cm^2$ area cells.

Catalyst development is still an important aspect of the PAFC. Transition metal (e.g. iron or cobalt) organic macrocycles have been evaluated as cathode electrocatalysts in

PAFCs (Bett, 1985). Another approach has been to alloy Pt with transition metals such as Ti, Cr, V, Zr, and Ta. Johnson Matthey Technology Centre have also obtained a 50% improvement in cathode electrocatalyst performance using platinum alloyed with nickel (Buchanan et al., 1992). Work has been done by Giner Inc. and others to make the anode electrocatalysts more tolerant to carbon monoxide and sulphur.

Other recent significant developments in PAFC technology are improvements in gas diffusion electrode construction and tests on materials that offer better carbon corrosion protection. Of course, many improvements can be made in the system design, with better BOP components such as the reformer, shift reactors, heat exchangers, and burners. Much of this is covered in the chapters that follow. For example, Figure 8.4 shows a schematic arrangement of the essential components in a PAFC system. The actual fuel cell stack is a small part of the total system.

PAFC systems have been under development for many years, and some of the earliest demonstrations were in the 1970s under the 'Target' programme funded by the American Gas Association. Many organisations have been involved in the development of PAFCs but the thrust of commercialisation is now borne mainly by two companies, UTC Fuel Cells (formerly trading under the name ONSI or International Fuel Cells) who are based in Connecticut, USA, and Fuji Electric in Japan. Fuji Electric have been developing PAFCs since the 1980s and since the 1990s have been supplying 50-kW and 100-kW systems, worldwide. Fuji have supplied over 100 such systems, which have proved their reliability and durability. Current research at Fuji is focused on improving performance by developing better reforming catalysts and on reducing costs especially of the BOP equipment.

UTC fuel cells is the leading manufacturer of stationary fuel cell systems and have supplied several hundred of their model PC25™ systems worldwide. A photograph of this was shown in Chapter 1, Figure 1.16. The units are finding niche markets in high value and prestigious locations such as a post office facility in Alaska, a science centre in Japan, the New York City police department and the National Bank of Omaha. The latter application is especially interesting as the several fuel cells are linked together with other generation equipment to create an ultra-reliable power system for what is a critical load. This has been supplied by the company Sure Power Corporation who have guaranteed 99.9999% reliability of the whole power system, thereby capitalising on the reliability and robustness of the PAFC technology. There are no single points of failure and the Sure Power system is extremely fault tolerant. The approach has been to run the PAFC alongside at least one other power generation technology thereby building in some redundancy of equipment, which has to be balanced against the resulting complexity and needs of the load.

Two recent entrants into the PAFC business are worth recording. The first is Bharat Heavy Electrical Ltd. of India. They only began R&D of PAFCs in 1987 and built their first stack in 1991. More recently they built a 50-kW system consisting of two 25-kW stacks, and this has run for over 500 h. The second company is Caltex Oil Corp. of South Korea. They also have constructed a 50-kW system (Yang et al., 2002). Alongside this development, we should note the work done by Japanese organisations such as Sanyo, Toshiba, and Mitsubishi Electric, who have all built PAFC stacks in the 1980s and early 1990s. As technical progress has mushroomed in PEMFC technology, many organisations have shelved their activities in PAFCs, but it remains to be seen which technology will achieve commercial success in the long run.

7.4 The Molten Carbonate Fuel Cell (MCFC)

7.4.1 How it works

The electrolyte of the molten carbonate fuel cell is a molten mixture of alkali metal carbonates – usually a binary mixture of lithium and potassium, or lithium and sodium carbonates, which is retained in a ceramic matrix of $LiAlO_2$. At the high operating temperatures (typically 600–700°C) the alkali carbonates form a highly conductive molten salt, with carbonate CO_3^{2-} ions providing ionic conduction. This is shown schematically in Figure 7.9, which also shows the anode and cathode reactions. Note that unlike all the other fuel cells, carbon dioxide needs to be supplied to the cathode as well as oxygen, and this becomes converted to carbonate ions, which provide the means of ion transfer between the cathode and the anode. At the anode, the carbonate ions are converted back into CO_2. There is therefore a net transfer of CO_2 from cathode to anode; one mole of CO_2 is transferred along with two Faradays of charge or two moles of electrons. (Note that the requirement for CO_2 to be supplied to the MCFC contrasts with the alkaline (electrolyte) fuel cell (AFC) we considered in Chapter 5, where CO_2 must be excluded). The overall reaction of the MCFC is therefore

$$H_2 + \tfrac{1}{2}O_2 + CO_2 \text{ (cathode)} \rightarrow H_2O + CO_2 \text{ (anode)} \qquad [7.14]$$

The Nernst reversible potential for an MCFC, taking into account the transfer of CO_2, is given by the equation

$$E = E^0 + \frac{RT}{2F} \ln \left(\frac{P_{H_2} . P_{O_2}^{\frac{1}{2}}}{P_{H_2O}} \right) + \frac{RT}{2F} \ln \left(\frac{P_{CO_{2c}}}{P_{CO_{2a}}} \right) \qquad [7.15]$$

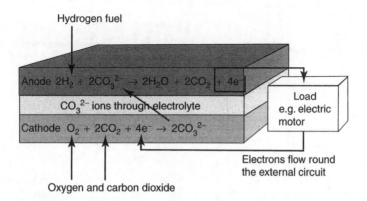

Hydrogen fuel

Anode $2H_2 + 2CO_3^{2-} \rightarrow 2H_2O + 2CO_2 + 4e^-$

CO_3^{2-} ions through electrolyte

Cathode $O_2 + 2CO_2 + 4e^- \rightarrow 2CO_3^{2-}$

Load e.g. electric motor

Electrons flow round the external circuit

Oxygen and carbon dioxide

Figure 7.9 The anode and the cathode reaction for a molten carbonate fuel cell using hydrogen fuel. Note that the product water is at the anode, and that both carbon dioxide and oxygen need to be supplied to the cathode.

where the sub-subscripts a and c refer to the anode and cathode gas compartments, respectively. When the partial pressures of CO_2 are identical at the anode and cathode, and the electrolyte is invariant, the cell potential depends only on the partial pressures of H_2, O_2, and H_2O. Usually, the CO_2 partial pressures are different in the two electrode compartments and the cell potential is therefore affected accordingly, as indicated by equation 7.15.

It is usual practice in an MCFC system that the CO_2 generated at the cell anodes is recycled externally to the cathodes where it is consumed. This might at first seem an added complication and a disadvantage for this type of cell, but this can be done by feeding the anode exhaust gas to a combustor (burner), which converts any unused hydrogen or fuel gas into water and CO_2. The exhaust gas from the combustor is then mixed with fresh air and fed to the cathode inlet, as is shown in Figure 7.10. This process is no more complex than for other 'hot' fuel cells, as the process also serves to preheat the reactant air, burn the unused fuel, and bring the waste heat into one stream for use in a bottoming cycle or for other purposes.

Another less commonly applied method is to use some type of device, such as a membrane separator, that will separate the CO_2 from the anode exit gas and will allow it to be transferred to the cathode inlet gas (a 'CO_2 transfer device'). The advantage of this method is that any unused fuel gas can be recycled to the anode inlet or used for other purposes. Another alternative to both these methods is that the CO_2 could be supplied from an external source, and this may be appropriate where a ready supply of CO_2 is available.

At the operating temperatures of MCFCs, nickel (anode) and nickel oxide (cathode), respectively, are adequate catalysts to promote the two electrochemical reactions. Unlike the PAFC or the PEMFC, noble metals are not required. Other important differences between the MCFC and the PAFC and PEMFC are the abilities to electrochemically convert carbon monoxide directly and to internally reform hydrocarbon fuels. If carbon monoxide was to be fed to the MCFC as fuel, the reactions at each electrode shown in Figure 7.11 would occur.

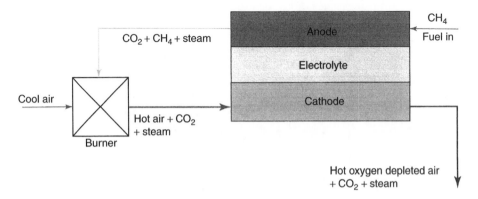

Figure 7.10 Adding carbon dioxide to the cathode gas stream need not add to the overall system complexity.

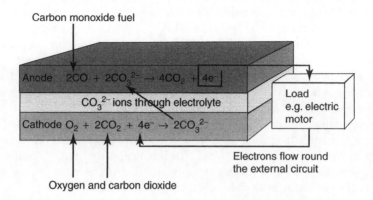

Figure 7.11 The anode and cathode reaction for a molten carbonate fuel cell using carbon monoxide fuel.

The electromotiveforce (EMF) of the carbon monoxide fuel cell is calculated in exactly the same way as for the hydrogen fuel cell, as described in the first section of Chapter 2. Figure 7.10 shows that two electrons are released for each molecule of CO, just as two electrons are released for each molecule of H_2. Thus the formula for the 'no loss', reversible OCV is identical, that is,

$$E = \frac{-\Delta \bar{g}_f}{2F}$$

The method of calculating $\Delta \bar{g}_f$ is given in Appendix 1, and it so happens that at 650°C the values for hydrogen and carbon monoxide are remarkably similar, as is shown in Table 7.1 below.

In practical applications, it is most unlikely that pure CO would be used as fuel. It is more likely that the fuel gas would contain both H_2O and CO, and in such cases the electrochemical oxidation of the CO would probably proceed via the water-gas shift reaction (equation 7.3), a fast reaction that occurs on the nickel anode electrocatalyst. The shift reaction converts CO and steam to hydrogen, which then oxidises rapidly on the anode. The two reactions (direct oxidation of CO or shift reaction and then the oxidation of H_2) are entirely equivalent.

Table 7.1 Values of $\Delta \bar{g}_f$ and E for hydrogen and carbon monoxide fuel cells at 650°C

Fuel	$\Delta \bar{g}_f$ kJ mol^{-1}	E (V)
H_2	−197	1.02
CO	−201	1.04

Unlike the PEMFC, AFC, and PAFC, the MCFC operates at a temperature high enough to enable internal reforming to be achieved. This is a particularly strong feature of both the MCFC and, as we shall see later, the SOFC. In internal reforming, steam is added to the fuel gas before it enters the stack. Inside the stack, the fuel and steam react in the presence of a suitable catalyst according to reactions such as 7.1 and 7.2. Heat for the endothermic reforming reactions is supplied by the electrochemical reactions of the cell. Internal reforming is discussed in some detail in the following chapter.

The high operating temperature of MCFCs provides the opportunity for achieving higher overall system efficiencies and greater flexibility in the use of available fuels compared with the low temperature types. Unfortunately, the higher temperatures also place severe demands on the corrosion stability and life of cell components, particularly in the aggressive environment of the molten carbonate electrolyte.

7.4.2 Implications of using a molten carbonate electrolyte

The PAFC and MCFC are similar types of fuel cell in that they both use a liquid electrolyte that is immobilised in a porous matrix. We have seen that in the PAFC, PTFE serves as a binder and wet-proofing agent to maintain the integrity of the electrode structure and to establish a stable electrolyte/gas interface in the porous electrode. The phosphoric acid is retained in a matrix of PTFE and SiC sandwiched between the anode and cathode. There are no materials available that are stable enough for use at MCFC temperatures that are comparable to PTFE. Thus, a different approach is needed to establish a stable electrolyte/gas interface in MCFC porous electrodes. The MCFC relies on a balance in capillary pressures to establish the electrolyte interfacial boundaries in the porous electrodes (Maru and Marianowski, 1976 and Mitteldorf and Wilemski, 1984). This is illustrated in Figure 7.12.

By properly co-ordinating the pore diameters in the electrodes with those of the electrolyte matrix, which contains the smallest pores, the electrolyte distribution shown in Figure 7.12 is established. This arrangement allows the electrolyte matrix to remain completely filled with molten carbonate, while the porous electrodes are partially filled, depending on their pore size distributions. The electrolyte matrix has the smallest pores and will be completely filled. The larger electrode pores will be partially filled, in inverse proportion to the pore size – the larger the pore, the less they are filled. Electrolyte management, that is the control over the optimum distribution of molten carbonate electrolyte in the different cell components, is critical for achieving high performance and endurance with MCFCs. This feature is very specific to this type of fuel cell. Various undesirable processes (i.e. consumption by corrosion reactions, potential driven migration, creepage of salt and salt vaporisation) occur, all of which contribute to the redistribution of molten carbonate in MCFCs. These aspects are discussed by Maru et al. (1986) and Kunz (1987).

7.4.3 Cell components in the MCFC

In the early days of the MCFC, the electrode materials used were, in many cases, precious metals, but the technology soon evolved during the 1960s and the 1970s saw the use of nickel-based alloys at the anode and oxides at the cathode. Since the mid-1970s, the materials for the state-of-the-art electrodes and electrolyte structure (molten

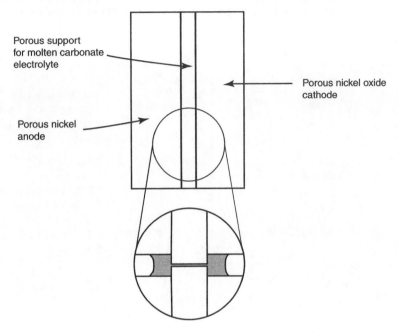

Porous support
for molten carbonate
electrolyte

Porous nickel oxide
cathode

Porous nickel
anode

Figure 7.12 Dynamic equilibrium in porous MCFC cell elements (porous electrodes are depicted with pores covered by a thin film of electrolyte).

carbonate/$LiAlO_2$) have remained essentially unchanged. A major development in the 1980s has been the evolution in the technology for fabrication of electrolyte structures. Several reviews of developments in cell components for MCFCs have been published, for example, by Maru et al. (1984), Petri and Benjamin (1986), and Hirschenhofer et al. (1998). Table 7.2 summarises the evolution of some of the principal materials of the MCFC since the 1960s.

Electrolyte

State-of-the-art MCFC electrolytes contain typically 60 wt% carbonate constrained in a matrix of 40 wt% $LiAlO_2$. The γ form of $LiAlO_2$ is the most stable in the MCFC electrolyte and is used in the form of fibres of <1-μm diameter. Other materials (e.g. larger size particles of $LiAlO_2$) may be added and many details are proprietary. Until the 1990s, the matrix was often fabricated into a tile by a hot pressing process, and it is still often referred to as the electrolyte 'tile'. Nowadays, the matrix is invariably made using tape-casting methods commonly employed in the ceramics and electronics industry. The process involves dispersing the ceramic materials in a 'solvent'. This contains dissolved binders (historically organic compounds), plasticisers, and additives to achieve the desired viscosity and rheology of the resulting mixture or 'slip'. This material is then cast in the form of a thin film over a moving smooth surface, and the required thickness is obtained by shearing with an adjustable blade device. After drying the slip, this material

Table 7.2 Evolution of cell component technology for molten carbonate fuel cells (Hirschenhofer et al., 1998)

Component	Ca. 1965	Ca. 1975	Current status
Anode	Pt, Pd, or Ni	Ni – 10 wt% Cr	Ni–Cr/Ni–Al 3–6 μm pore size 45–70% initial porosity 0.20–1.5-mm thickness 0.1–1 m^2/g

Cathode	Ag$_2$O or
	  lithiated NiO
		Lithiated NiO	Lithiated NiO
			7–15 μm pore size
			70–80% initial porosity
			60–65% after lithiation and
			  oxidation
			0.5–1-mm thickness
			0.5 m^2/g

Electrolyte Support	MgO
	10–20 m^2/g
		Mixture of
		  α-β-γ-
		  LiAlO$_2$
		  0.1–12 m^2/g
			α-LiAlO$_2$$\beta$-LiAlO$_2$
			0.5–1-mm thickness

Electrolyte[a]	52 Li-48 Na 43.5 Li-31.5 Na-25 K 'Paste'	62 Li–38 K ~60–65 wt% Hot press 'tile' 1.8-mm thickness	62 Li–38 K 50 Li–50 Na ~50 wt% Tape cast 0.5–1-mm thickness

[a]Figures in this row are mole percent unless stated otherwise.

is then heated further in air and any organic binder is burnt out at 250 to 300°C. The semi-stiff 'green' structure is then assembled into the stack structure. Tape casting of the electrolyte and other components provides a means of producing large-area components. The methods can also be applied to the cathode and anode materials and fabrication of stacks of electrode area up to 1 m^2 is now easily achieved.

The ohmic resistance of the MCFC electrolyte, and especially the ceramic matrix, has an important and large effect on the operating voltage, compared with most other fuel cells. Under typical MCFC operating conditions Yuh et al. (1992) found that the electrolyte accounted for 70% of the ohmic losses. Furthermore, the losses were dependent on the thickness of the electrolyte according to the formula

$$\Delta V = 0.533t$$

where t is the thickness in centimetre. From this relationship, it can be seen that a fuel cell with an electrolyte structure of 0.025-cm thickness would operate at a cell voltage 82 mV higher than an identical cell with a structure of 0.18-cm thickness. Using tape-casting methods, electrolyte matrices can now be made quite thin (0.25–0.5 mm), which has an advantage in reducing the ohmic resistance. Note that this is similar in thickness to the electrolyte of the PAFC and the resulting large ohmic resistances limit the current densities of these technologies compared with, for example, the PEMFC and the SOFC. There is a trade-off between low resistance and long-term stability, which is obtained with thicker materials. This fundamental issue in utilising liquid electrolytes limits the power densities obtainable with the PAFC and MCFC. With the MCFC, the typical power density at 650°C is $0.16 \, \text{W cm}^{-2}$.

Finally, in considering the MCFC electrolyte, we must point out an important difference between the MCFC and all other types of fuel cell. The final preparation of the cell is carried out once the stack components are assembled. So, layers of electrodes, electrolyte and matrix, and the various non-porous components (current collectors and bipolar plates) are assembled together, and the whole package is heated slowly up to the fuel cell temperature. As the carbonate reaches its melt temperature (over 450°C) it becomes absorbed into the ceramic matrix. This absorption process results in a significant shrinkage of the stack components, and provision must be made for this in the mechanical design of the stack assembly. In addition, a reducing gas must be supplied to the anode side of the cell as the package is heated to the operating temperature to ensure that the nickel anode remains in the reduced state. The net result is that the MCFC stack takes a long time to condition. Whereas a PEM fuel cell will turn on from cold in seconds, an MCFC stack takes typically 14 h or more. The other point to note is that every time the MCFC stack is heated and cooled through the electrolyte melt temperature, stresses are set up within the electrolyte support or tile. Heating and cooling of MCFC stacks must therefore be carried out slowly to avoid cracking the electrolyte tile and the resulting permanent damage caused by fuel *crossover*. In addition, it is necessary to protect the MCFC anode during shut downs from reoxidation by purging the anode compartment with inert gas. Thermal cycling of MCFC stacks is therefore best avoided and MCFC systems are ideally suited to applications that need continuous power.

Anodes

As indicated in Table 7.2, state-of-the-art MCFC anodes are made of a porous sintered Ni-Cr/Ni-Al alloy. These are usually made with a thickness of 0.4 to 0.8 mm with a porosity of between 55 and 75%. Fabrication is by hot pressing finely divided powder or by tape-casting a slurry of the powdered material, which is subsequently sintered. Chromium (usually 10–20%) is added to the basic nickel component to reduce the sintering of the nickel during cell operation. This can be a major problem in the MCFC anode, leading to growth in pore sizes, loss of surface area, and mechanical deformation under compressive load in the stack. This can result in performance decay in the MCFC, through redistribution of carbonate from the electrolyte. Unfortunately, the chromium added to anodes also reacts with lithium from the electrolyte with time, thereby exacerbating the loss of electrolyte. This can be overcome to some extent by the addition of aluminium, which improves creep resistance in the anode and reduces electrolyte loss. This is believed to be due to the formation of $LiAlO_2$ within the nickel particles. Although Ni-Cr/Ni-Al alloy anodes have achieved commercially acceptable stability, the cost is relatively high and developers are investigating alternative materials. Partial substitution of the nickel with copper, for example, can go some way in reducing the materials' costs. Complete substitution of the Ni by Cu is not feasible, however, as Cu exhibits more creep than Ni. In an attempt to improve the tolerance to sulphur in the fuel stream, various ceramic anodes are also being investigated. $LiFeO_2$, with and without dopings of Mn and Nb, for example, have been tested.

The anode of the MCFC needs to provide more than just electro-catalytic activity. Because the anode reaction is relatively fast at MCFC temperatures, a high surface area is not required, compared with the cathode. Partial flooding of anode with molten carbonate is therefore acceptable, and this is used to good effect to act as a reservoir for carbonate, much in the same way that the porous carbon substrate does in the PAFC. The partial flooding of the anode also provides a means for replenishing carbonate in a stack during prolonged use.

In some earlier MCFC stacks, a layer or 'bubble barrier' was located between the anode and the electrolyte. This consisted of a thin layer of Ni or $LiAlO_2$ containing only small uniform pores. It served to prevent a flow of electrolyte to the anode and at the same time reduced the risk of gas crossover. As we have seen before, this latter problem is common to all liquid fuel cell systems in which an excess of pressure on one side of the cell may cause the gas to cross the electrolyte. Nowadays, using a tape cast structure it is possible to control the pore distribution of anode materials during manufacture so that small pores are found closest to the electrolyte and larger pores nearer the fuel gas channels. Long-term electrolyte loss is, however, still a significant problem with the MCFC and a totally satisfactory solution to electrolyte management is yet to be achieved.

Cathodes

One of the major problems with the MCFC is that the nickel oxide state-of-the-art cathode material has a small, but significant, solubility in molten carbonates. Through dissolution, some nickel ions are formed in the electrolyte. These then tend to diffuse into the electrolyte towards the anode. As the nickel ions move towards the chemically reducing conditions at the anode (hydrogen is present from the fuel gas), metallic nickel can precipitate out in the electrolyte. This precipitation of nickel can cause internal short-circuits of the fuel cell with subsequent loss of power. Furthermore, the precipitated nickel can act as a sink for nickel ions, which promotes the further dissolution of nickel from the cathode. The phenomenon of nickel dissolution becomes worse at high CO_2 partial pressures because of the reaction

$$NiO + CO_2 \rightarrow Ni^{2+} + CO_3^{2-} \qquad [7.16]$$

It has been found that this problem is reduced if the more basic, rather than acidic, carbonates are used in the electrolyte. The basicity of the common alkali metal carbonates is (basic) $Li_2CO_3 > Na_2CO_3 > K_2CO_3$ (acidic). The lowest dissolution rates have been found for the eutectic mixtures 62% Li_2CO_3 + 38% K_2CO_3 and 52% Li_2CO_3 + 48% Na_2CO_3. The addition of some alkaline earth oxides (CaO SrO and BaO) has also been found to be beneficial.

With state-of-the-art nickel oxide cathodes, nickel dissolution can be minimised by (1) using a basic carbonate; (2) operating at atmospheric pressure and keeping the CO_2

partial pressure in the cathode compartment low; and (3) using a relatively thick electrolyte matrix to increase the Ni^{2+} diffusion path. By these means, cell lifetimes of 40 000 h have been demonstrated under atmospheric pressure operation. For operation at higher pressure alternative materials have been investigated, the most studied being $LiCoO_2$ and $LiFeO_2$. Of these two materials, $LiCoO_2$ has the lower dissolution rate, being an order of magnitude lower than NiO at atmospheric pressure. Dissolution of $LiCO_2$ also shows a lower dependency on CO_2 partial pressure than NiO.

Non-porous components

The bipolar plates for the MCFC are usually fabricated from thin sheets of stainless steel. The anode side of the plate is coated with nickel. This is stable in the reducing environment of the anode, it provides a conducting path for current collection, and it is not wetted by electrolyte that may migrate out of the anode. Gas-tight sealing of the cell is achieved by allowing the electrolyte from the matrix to contact the bipolar plate at the edge of each cell outside the electrochemically active area. To avoid corrosion of the stainless steel in this 'wet-seal' area, the bipolar plate is coated with a thin layer of aluminium. This provides a protective layer of $LiAlO_2$ after reaction of Al with Li_2CO_3. There are many designs of bipolar plate, depending on whether the gases are manifolded externally or internally. Some designs of bipolar plate have been developed especially for internal reforming, such that the reforming catalyst can be incorporated within the anode gas flow field. (See Figure 8.3 in the next chapter.)

7.4.4 Stack configuration and sealing

The stack configuration for the MCFC is very different from those described in previous chapters for the PEM, AFC, and PAFC, although there are, of course, similarities. The most important difference is in the aspect of sealing. As described in the previous section, the MCFC stack is composed of various porous components (matrix and electrodes) and non-porous components (current collector/bipolar plate). In assembling and sealing these components, it is important to ensure good flow distribution of gases between individual cells, uniform distribution within each cell, and good thermal management to reduce temperature gradients throughout the stack. Several proprietary designs have been developed for all the stack components but there are some generic aspects that are described below with examples taken from real systems.

Manifolding

Reactant gases need to be supplied in parallel to all cells in the same stack via common manifolds. Some stack designs rely on external manifolds, whereas others use internal manifolds.

The basic arrangement for *external manifolding* is as shown in Chapter 1, Figures 1.12 and 1.13. The electrodes are about the same area as the bipolar plates, and the reactant gases are fed in and removed from the appropriate faces of the fuel cell stack. One advantage of external manifolding is its simplicity, enabling a low-pressure drop in the manifold and good flow distribution between cells. A disadvantage is that the two gas flows are at right angles to each other – cross-flow – and this can cause uneven temperature distribution over the face of the electrodes. Other disadvantages have been gas leakage and migration ('ion-pumping') of electrolyte. Each external manifold must have an insulating gasket to form a seal with the edges of the stack. This is usually zirconia felt, which provides a small amount of elasticity to ensure a good seal. Note that most stack developers arrange the cells to lie horizontal and the fuel and the oxidants are supplied to the sides of the stack. An alternative arrangement pioneered by MTU in their 'hot module' (see Figure 7.17) is to have vertically mounted cells with the anode inlet manifold located underneath the stack. In this way, sealing with the gasket is enhanced by the weight of the whole stack.

Internal manifolding refers to a means of gas distribution among the cells through channels or ducts within the stack itself, penetrating the separator plates. This is also illustrated in Chapter 1, Figure 1.14. An important advantage of internal manifolding is that there is much more flexibility in the direction of flow of the gases. For even temperature distribution, co-flow or counter-flow can be used, as was discussed in Section 7.2.2. The ducts are formed by holes in each separator plate that line up with each other once the stack components are assembled. The separator plates are designed with various internal geometries (flow inserts, corrugations, etc.), which, as well as providing the walls of the internal manifold, also control the flow distribution across each plate. This is shown in Figure 7.13c. Internal manifolding offers a great deal of flexibility in stack design, particularly with respect to flow configuration. The main disadvantages are associated with the more complex design of the bipolar plate needed.

In internal manifolding, the preferred method of sealing is to use the electrolyte matrix itself as a sealant. The electrolyte matrix forms a wet seal in the manifold areas around the gas ducts. In internally manifolded stacks, the entire periphery of the cell may be sealed in this way. It is possible to seal the manifolds with separate gaskets and to extend the matrix only over the active cell area, but this method is rarely used.

7.4.5 Internal reforming

The concept of internal reforming for the MCFC and SOFC has been mentioned already and will be discussed further in the next chapter. For the MCFC, it presents particular challenges. The principal variations of internal reforming configuration are shown in Figure 8.3 in the next chapter. These are described in detail elsewhere (Dicks, 1998). Direct internal reforming (DIR) offers a high cell performance advantage compared with indirect internal reforming (IIR). The reasons for this are quite subtle but it should be noted that with DIR the major product from the reforming reaction, hydrogen, is consumed directly by the electrochemical reaction. Therefore, by allowing the reforming reaction and the anode electrochemical reactions to be coupled, the reforming reaction is shifted in the forward direction. *This leads to a higher conversion of hydrocarbon than would normally be expected from reforming equilibrium conditions at the reaction temperature.* At high fuel utilisations in the DIR-MCFC, almost 100% of the methane is converted into hydrogen at 650°C. This compares with a typical conversion of 85%, which would be predicted from the equilibrium of the simple reforming of a steam/methane (2:1 by

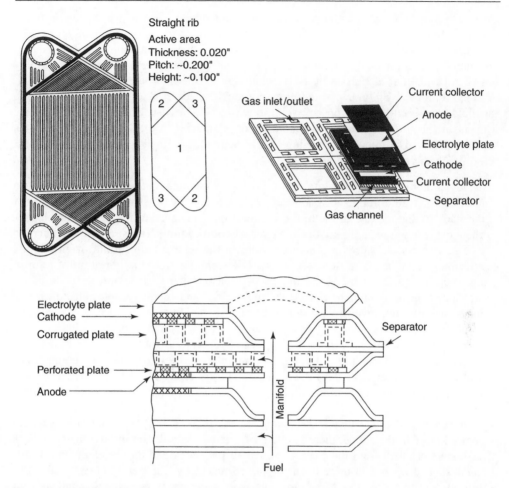

Figure 7.13 Examples of practical separator plate designs with internal manifolding: (a) IMHEX design of ECN, (b) multiple cell stack of Hitachi, and (c) the cross section of the wet-seal area in an internally manifolded MCFC stack.

vol.) mixture at the same temperature. Internal reforming eliminates the cost of an external reformer and system efficiency is improved, but at the expense of a potentially more complex cell configuration. This provides developers of MCFC with a choice of an external reforming or internal reforming approach. Some developers have adopted a combination of the two methods.

Internal reforming can only be carried out in an MCFC stack if a supported metal catalyst is incorporated. This is because although nickel is a good reforming catalyst, the conventional low surface area porous nickel anode has insufficient catalytic activity in itself to support the steam reforming reaction at the 650°C operating temperature. We shall see that this is not the case in the SOFC, in which complete internal reforming may be carried out on the SOFC anode. In the DIR-MCFC, the reforming catalyst needs to be close to the anode for the reaction to occur at a sufficient rate. Several groups demonstrated internal reforming in the MCFC in the 1960s and identified the major problem areas to be associated with catalyst degradation, caused by carbon deposition, sintering, and catalyst poisoning by alkali from the electrolyte. Most recently, development of internal reforming has been carried out in Europe by BG Technology in a BCN (Dutch Fuel Cell Corporation) led, European Union supported programme (Kortbeek et al., 1998). Others active in this field have been Haldor Topsoe, working with MTU Friedrichshafen in Germany and Mitsubishi Electric Corporation in Japan. Other companies in the US, Japan, and Korea are now involved in internal reforming MCFC technology. Key requirements for MCFC reforming catalysts are as follows:

- *Sustained activity to achieve the desired cell performance and lifetime.* For use in an MCFC, steam reforming catalysts need to provide sufficient activity for the lifetime of the stack so that the rate of the reforming reaction is matched to the rate of the electro-chemical reaction, which may decline over a period of time. The strongly endothermic reforming reaction causes a pronounced dip in the temperature profile of an internal reforming cell. This is most severe in the DIR configuration. Optimisation of reform-ing catalyst activity is therefore important to ensure that such temperature variations are kept to a minimum, to reduce thermal stress, and thereby contribute towards a long stack life. Improvements in temperature distribution across the stack may also be achieved through the recycling of either anode gas, or cathode gas, or both.
- *Resistance to poisons in the fuel.* Most raw hydrocarbon fuels that may be used in MCFC systems (including natural gas) contain impurities (e.g. sulphur compounds) that are harmful for both the MCFC anode and the reforming catalyst. The tolerance of most reforming catalysts to sulphur is very low, typically in the parts per billion (ppb) range.
- *Resistance to alkali/carbonate poisoning.* In the case of DIR, in which catalysts are located close to the anode, there is a risk of catalyst degradation through reaction with carbonate or alkali from the electrolyte. Alkali poisoning and the development of DIR catalysts is described in more detail in the references by Clark et al. (1997) and Dicks (1998), and tolerance to degradation by alkali materials is the biggest challenge for MCFC catalysts. In contrast, many commercial catalysts are available that can be used in IIR-MCFC systems, and it is therefore not surprising that most MCFC developers have adopted the less demanding IIR approach. Commonly, supported nickel catalysts have been used, although more recently researchers have tested supported ruthenium. Poisoning of DIR MCFC reforming catalysts is now known to occur through two prin-cipal routes: creep of liquid molten carbonate along the metallic cell components and transport in the gas phase in the form of alkali hydroxyl species. These are illustrated schematically in Figure 7.14, which also indicates one possible solution to the prob-lem, that is, the insertion of a protective porous shield between the anode and the catalyst.

7.4.6 Performance of MCFCs

The operating conditions for an MCFC are selected essentially on the same basis as those for the PAFC. The stack size, efficiency, voltage level, load requirement, and cost are

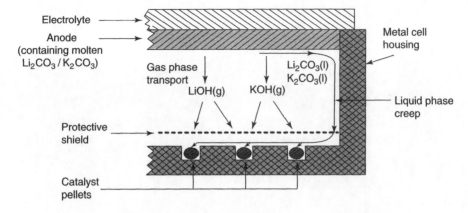

Figure 7.14 Alkali transport mechanisms in the DIR-MCFC.

all important and a trade-off between these factors is usually sought. The performance (V/I) curve is defined by cell pressure, temperature, gas composition, and utilisation. State-of-the-art MCFCs generally operate in the range of 100 to 200 mA cm^{-2} at 750 to 900 mV per cell. Figure 7.15 shows the progress made over the last 30 years in improving the performance of generic MCFCs.

As with the PAFC, there is significant polarisation of the cathode in the MCFC. This is illustrated in Figure 7.16, which shows typical cathode performance curves obtained at 650°C with an oxidant composition (12.6% O_2/18.4% CO_2/69% N_2) typically used

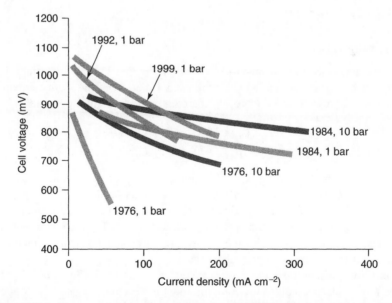

Figure 7.15 Progress in generic performance of MCFCs on reformate gas and air.

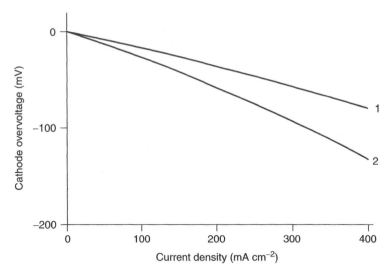

Figure 7.16   Influence of oxidant gas composition on MCFC cathode polarisation at 650°C (a) 33% O_2/67% CO_2 and (b) 12.6% O_2/18.4% CO_2/69% N_2 (Bregoli and Kunz, 1982).

in MCFCs and a baseline composition (33% O_2/67% CO_2). The baseline composition contains the reactants, O_2 and CO_2, in the stoichiometric ratio that is needed in the electrochemical reaction at the cathode (see Figure 7.9). With this gas composition, little or no diffusion limitations occur in the cathode because the reactants are provided primarily by bulk flow. The other (more realistic) gas composition, which contains a substantial fraction of nitrogen, yields a cathode performance that is limited by gas diffusion and from dilution by an inert gas.

Influence of pressure

There is a performance improvement to be made by increasing the operating pressure of the MCFC. We have already shown in Section 2.5.4 that, from the Nernst equation, for a change in system pressure from P_1 to P_2, the change in reversible potential is given by

$$\Delta V = \frac{RT}{4F} \ln\left(\frac{P_2}{P_1}\right)$$

From this it can be shown that a fivefold increase in pressure should yield, at 650°C, a gain in OCV of 32 mV, and a tenfold pressure increase should give an increase of 46 mV. In practice the increase is somewhat greater because of the effect of reduced cathode polarisation. Increasing the operating pressure of MCFCs results in enhanced cell voltages because of the increase in the partial pressure of the reactants, increase in gas solubilities, and increase in mass transport rates.

As was shown when considering the PEMFC at pressure in Section 4.7.2, there are important power costs involved in compressing the reactant gases. Also opposing the benefits of increased pressure are the effects on undesirable side reactions such as carbon deposition (Boudouard reaction). Higher pressure also *inhibits* the steam reformation reaction of equation 7.1, which is a disadvantage if internal reforming is being used. These effects, as will be described in the next chapter, can be minimised by increasing the steam content of the fuel stream. In practice, the benefits of pressurised operation are significant only up to about 5 bar. Above this there are disadvantages brought about by the system design constraints.

The problem of *differential pressure* is another factor to consider. To reduce the risk of gas crossover between the anode and the cathode in the MCFC, the difference in pressure between the two sides of each cell should be kept as low as possible. For safety reasons the cathode is usually maintained at a slightly higher (a few millibars) pressure than the anode. The ceramic matrix that constrains the electrolyte in the MCFC is a fragile material that is also susceptible to cracking if subjected to stresses induced either through thermal cycling, temperature variations, or excessive pressure differences between the anode and the cathode. The minimisation of pressure difference between anode and cathode compartments in stacks has always been a concern of system designers, since the recycling of anode burn-off gas to the cathode is normally required and there are inevitably pressure losses associated with this transfer of gases. (See Figure 7.10.) The problems associated with control of small pressure differences have also mitigated against running the stacks at elevated pressures even though there may be advantages from an efficiency point of view. Since provision of gas-tight sealing in the MCFC is achieved through the use of the molten carbonate itself as a sealing medium, this also means that the difference in pressure between the cell compartments and the outside of the stack has to be minimised even when running at elevated pressures. Therefore, any pressurised stack must be enclosed within a pressure vessel in which the stack is surrounded by a non-reactive pressurising gas – usually nitrogen.

Another issue relating to the choice of pressure is that, as we have shown in Section 7.3, an improvement in the overall efficiency of fuel systems can be achieved by combining high-temperature fuel cells with gas turbines. Gas turbines require pressurised (typically 5 bar) hot exhaust gas. Solid oxide fuel cells are very suitable for this application, as they can run in a pressurised mode and have a high exhaust gas temperature. MCFCs could also be combined with GTs but the exhaust temperature is lower. In addition, for the reason described above, the molten carbonate system is not so amenable to pressurisation. Such systems are thus unlikely to be developed.

Generally, it is thought that it is uneconomical to pressurise MCFC systems less than 1 MW. More discussion of the influence of pressure is to be found in Hirschenhofer et al. (1998) and in Selman in Blomen et al. (1993).

Influence of temperature

Simple thermodynamic calculations indicate that the reversible potential of MCFCs should decrease with increasing temperature. This is brought about by the Gibbs free energy changes explained in Chapter 2 *and* a change in equilibrium gas composition at the anode with temperature. The main reason for the latter is that the shift reaction (equation 7.3) achieves rapid equilibrium at the anode of the MCFC and the gas composition depends on the equilibrium for this reaction. The equilibrium constant (K) for the shift reaction increases with temperature and the equilibrium composition changes with temperature and utilisation to affect the cell voltage, as illustrated in Table 7.3.

Table 7.3 Equilibrium composition for fuel gas and reversible cell potential calculated using the Nernst equation and assuming initial anode gas composition of 77.5% H_2/19.4% CO_2/3.1% H_2O at 1 atmosphere and cathode composition of 30% O_2, 60% CO_2, /10% N_2

Parameter	Temperature (K)		
	800	900	1000
P_{H_2}	0.669	0.649	0.6543
P_{CO_2}	0.088	0.068	0.053
P_{CO}	0.106	0.126	0.141
P_{H_2O}	0.138	0.157	0.172
E (volts)	1.155	1.143	1.133
K	0.247	0.48	0.711

Under real operating conditions, the influence of temperature is actually dominated by the cathode polarisation. As temperature is increased, the cathode polarisation is considerably reduced. The net effect is that the operating voltage of the MCFC actually increases with temperature. However, above 650°C this effect is very slight, only about 0.25 mV per °C. Since higher temperatures also increase the rate of all the undesired processes, particularly electrolyte evaporation and material corrosion, 650°C is generally regarded as an optimum operating temperature (Hirschenhofer et al., 1998).

7.4.7 Practical MCFC systems

Molten carbonate fuel cell technology is being actively developed in the USA, Asia, and Europe. An example system is the 250-kW 'Hot Module' of MTU Friedrichshafen shown in Figure 7.17.

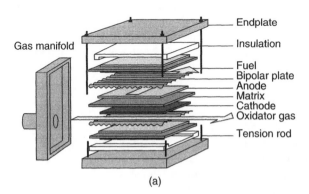

(a)

Figure 7.17 Example MCFC system, the MTU 'Hot Module' cogeneration system showing (a) stack construction, (b) simplified flow diagram, and (c) early demonstration units under construction.

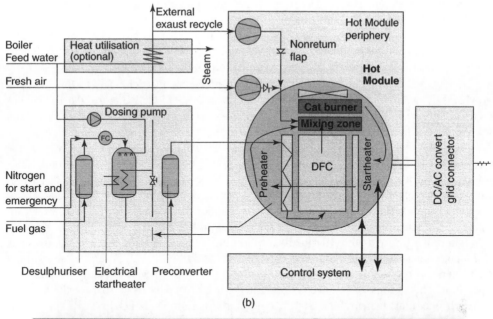

(b)

(c)

Figure 7.17 (*continued*).

Currently, one industrial corporation is actively pursuing the commercialisation of MCFCs in the US: Fuel Cell Energy, formerly known as Energy Research Corporation. Europe and Japan have several developers of the technology: MTU Friedrichshafen (Germany) who are themselves a partner of Fuel Cell Energy, Ansaldo (Italy), Hitachi, Ishikawajima-Harima Heavy Industries, Mitsubishi Electric Corporation, and Toshiba Corporation.[5] Organisations in Korea have also entered the MCFC field. The technical status of MCFC is exemplified by a 2-MW demonstration by Fuel Cell Energy, the MTU 'hot module' and the 1-MW pilot plant that has been operated by the Central Research Institute for Electrical Power Industries (CRIEPI) in Japan.

A review carried out by Dicks and Siddle (1999) indicated that the MCFC was a few years away from being truly commercial. Since then good progress towards commercialisation has been made by Fuel Cell Energy who have set up a manufacturing facility in Connecticut capable of producing up to two hundred 250-kW systems per year. They are already marketing early units and have a backlog of orders. To this end, Fuel Cell Energy have recently signed partnerships with Caterpillar Inc. for distribution of systems throughout the US, and with Marubeni Corporation, a Japanese trading company, for sales in Japan, China, Southeast Asia, and Australia. In addition to the 250-kW systems, Fuel Cell Energy is promoting its products for operation at up to 10-MW capacity for baseload operation. This is important since, as remarked earlier, it takes 14 to 16 h for the systems to reach operating temperature from a cold start. In Europe, MTU Friedrichshafen has an agreement with Fuel Cell Energy to develop and market 'Hot Modules' of 250 kW in Europe. Several field trials of the 250-kW systems are currently underway. The first was located at the University of Bielefeld in Germany, followed by an installation in the hospital at Bad Neustadt/Saale (Figure 7.18). Fuel Cell Energy has tested systems in Alabama at the Mercedes Benz production facility in Tuscaloosa and at the Los Angeles department of Water and Power.

The current status of the MTU system is summarised in Table 7.4. The system is designed for industrial and commercial cogeneration applications, such as hospitals, and the philosophy behind the design of the 'Hot Module' was as follows:

- all hot components are integrated into one common vessel;
- common thermal insulation with internal air recycle for best temperature levelling;
- minimum flow resistance and pressure differences;
- horizontal fuel cell block with internal reforming;
- gravity-sealed fuel manifold;
- simple and elegant mechanics;
- standard truck transportable up to 400 kW and beyond.

The system design for the hot module has all the processes that need to be run at elevated temperature located within the hot module, whereas systems such as power conditioning, natural gas compression, and so on are located outside the hot module.

Ansaldo Fuel Cells SpA (AFCo), an Italian company, is currently the other main developer of MCFC in Europe. The company was set up by Ansaldo Ricerche Srl, a

---

[5] It is worth remarking that significant MCFC development programmes in the US, the Netherlands, and Japan have ended. These are a reflection partly of the fragile nature of the emerging global industry, the perceived technical problems associated with MCFC compared with some other fuel cell technologies, and the difficulty of sustaining funding in new technologies.

Figure 7.18 Early MTU "Hot Module" cogeneration system as installed in the hospital at Bielefeld. (Pictures reproduced by kind permission of MTU Friedrichshafen.)

Table 7.4 Current status of MTU Friedrichshafen MCFC

Power	279 kW (250 kW net AC)
No. of cells	292
Efficiency	49% LHV
Temperature of available heat	450°C
Stack degradation	1%/1000 h in a 16 000 15 × 15 cm lab scale test stack
Fuel quality	Normal pipeline gas used

Finmeccanica company with over 10 years experience in the development of fuel cell technology. The company is aiming to commercialise their indigenous European MCFC technology for on-site dispersed energy production at low-medium sizes (from 50 kW to some MW). The AFCo Series 500 unit is based on a 100-kW 'proof of concept' power plant that was run at a site near Milan (Italy) during 1998–1999. At the heart of the Series 500 are two electrochemical modules, each formed by two stacks integrated with their relevant auxiliary systems. The plant is skid-mounted and the stacks, integrated reformers, and catalytic burners are located inside a pressurised vessel, while all other auxiliary systems (including power conditioning and control systems) are installed on a separate skid. The concept is therefore similar in philosophy to the MTU 'Hot Module' in that all the elements that run at elevated temperatures are integrated into one unit, but there are significant differences in detail between the two systems. Presently, the first-of-a-kind unit is under development and contracts are underway for the demonstration of a stack module fed by a biomass gasifier, diesel fuel, and landfill-gas.

Current trials of MCFC power plant show that electrical efficiencies of up to 50% (LHV) can be achieved. However, projected electrical and overall efficiencies of 55% and at least 80% are anticipated for the MCFC in the long term. Emissions from these plants are predicted to be negligible with regard to SO_x and particulates, and below 10 ppm NO_x. Noise levels are below 65 dbA. In many cogeneration applications, these attributes should give the MCFC an edge over competing technologies such as engine and turbine-based power generation systems.

The Dutch Energy Research Foundation, ECN, brought in other industrial partners to develop an advanced DIR-MCFC system for the European cogeneration market during the late 1990s. Although this development programme has now ended, the consortium developed several advanced system concepts. These included a novel method of stack connection illustrated in Figure 7.19. In this arrangement, shown for three stacks but applicable to MCFC systems of two or more stacks, stacks are connected together in series on the cathode side, and in parallel on the anode side, with recycle of the anode gas applied. Calculations have shown that the system eliminates the need for all major heat exchangers and provides a high system efficiency (over 50% HHV).

Another novel system concept from the ECN project is the so-called SMARTER™ stack, shown schematically in Figure 7.20. In this stack configuration, two types of cells of slightly different geometries are embodied in the one stack. The two cells differ in that anode recycle is applied to one and single-pass operation is applied to the other. Again,

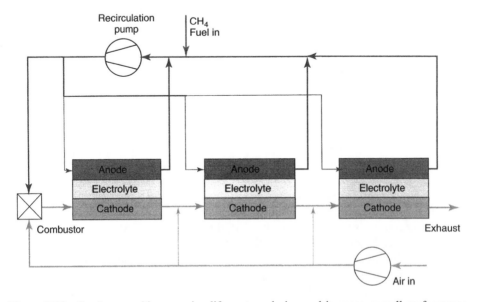

Figure 7.19   Stack networking can simplify system design and increase overall performance.

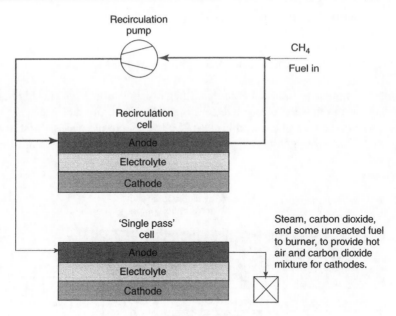

Figure 7.20   The SMARTER™ stack concept.

this arrangement shows an efficiency advantage of 5 to 10 percentage points. Interestingly, both the series connection and SMARTER™ stack concepts can be applied to SOFC as well as MCFC systems. The recirculation of the anode gas in these systems has three important advantages.

1. No preheating of the fuel supply is needed.
2. The fuel entering the cell has already been partially reformed, and so there is less stress on any direct internal reformation catalyst.
3. The gas flow rates are larger, so there is higher heat capacity in the system, and so the temperature changes through each stack are reduced, giving more even performance.

For more information on the other developers of MCFC, the review by Dicks and Siddle (1999) should be consulted. There are few large developers and in Japan the main support comes from Ishikawajima-Heavy Industries (IHI) who recently reported the construction of two 300-kW systems. Development is continuing and IHI are carrying out long-term tests on a 10-kW natural–gas fuelled system. Thus, although the emphasis in the MCFC community appears to be on getting systems out for field trials, there are a number of technical issues that remain to be dealt with.

7.5 The Solid Oxide Fuel Cell

7.5.1 How it works

The SOFC is a complete solid-state device that uses an oxide ion-conducting ceramic material as the electrolyte. It is therefore simpler in concept than all the other fuel cell systems described, as only two phases (gas and solid) are required. The electrolyte management issues that arise with the PAFC and MCFC do not occur and the high operating temperatures mean that precious metal electrocatalysts are not needed. As with the MCFC, both hydrogen and carbon monoxide can act as fuels in the SOFC, as shown in Figure 7.21.[6]

The SOFC is similar to the MCFC in that a negatively charged ion ($O^=$) is transferred from the cathode through the electrolyte to the anode. Thus, product water is formed at the anode. Development can be traced back to 1899 when Nernst was the first to describe zirconia[7] (ZrO_2) as an oxygen ion conductor. Until recently, SOFCs have all been based

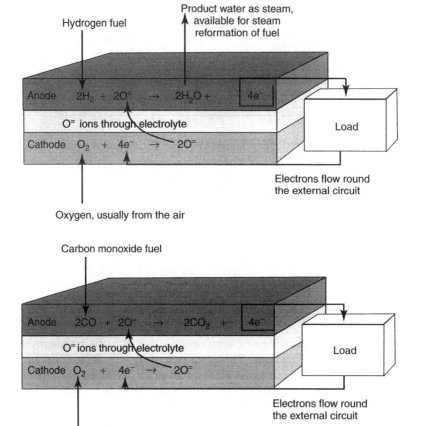

Figure 7.21 The separate anode and cathode reactions for the SOFC, when using hydrogen and carbon monoxide fuel.

---

[6] The high temperatures and presence of steam also means that CO oxidation producing hydrogen, via the shift reaction (equation 7.3), invariably occurs in practical systems, as with the MCFC. The use of the CO may thus be more indirect, but just as useful, as shown in Figure 7.21.

[7] Zirconia is zirconium oxide, yttria is yttrium oxide, etc.

on an electrolyte of zirconia stabilised with the addition of a small percentage of yttria (Y_2O_3). Above a temperature of about 800°C, zirconia becomes a conductor of oxygen ions ($O^=$), and typically the state-of-the-art zirconia-based SOFC operates between 800 and 1100°C. This is the highest operating temperature of all fuel cells, which presents both challenges for the construction and durability, and also opportunities, for example, in combined cycle (bottoming cycle) applications.

The anode of the SOFC is usually a zirconia cermet (an intimate mixture of ceramic and metal). The metallic component is nickel, chosen amongst other things because of its high electronic conductivity and stability under chemically reducing and part-reducing conditions. The presence of nickel can be used to advantage as an internal reforming catalyst, and it is possible to carry out internal reforming in the SOFC directly on the anode (Pointon, 1997). The material for the cathode has been something of a challenge. In the early days of development noble metals were used, but have fallen out of favour on cost grounds. Most SOFC cathodes are now made from electronically conducting oxides or mixed electronically conducting and ion-conducting ceramics. The most common cathode material of the latter type is strontium-doped lanthanum manganite.

As can be seen from Figure 7.21, unlike the MCFC, the SOFC requires no carbon dioxide recycling, which leads to system simplification.[8] The absence of carbon dioxide at the cathode means that the open circuit cell voltage is given by the simple form of the Nernst equation (given in Chapter 2 as 2.8). However, one disadvantage of the SOFC compared with the MCFC is that at the higher operating temperature the Gibbs free energy of formation of water is less negative. This means that the OCV of the SOFC at 1000°C is about 100 mV lower than the MCFC at 650°C. (See Chapter 2, particularly Figure 2.4 and Table 2.2.) This could lead to lower efficiencies for the SOFC. However, in practice the effect is offset at least in part by the lower internal resistance of the SOFC and the use of thinner electrolytes. The result of this is that the SOFC can be operated at relatively high current densities (up to $1000\,mA\,cm^{-2}$).

7.5.2 SOFC components

Electrolyte

Zirconia doped with 8 to 10 mole % yttria (yttria-stabilised zirconia (YSZ)) is still the most effective electrolyte for the high-temperature SOFC, although several others have been investigated (Steele, 1994) including Bi_2O_3, CeO_2, and Ta_2O_5. Zirconia is highly stable in both the reducing and oxidising environments that are to be found at the anode and cathode of the fuel cell, respectively. The ability to conduct $O^=$ ions is brought about by the fluorite crystal structure of zirconia in which some of the Zr^{4+} ions are replaced with Y^{3+} ions. When this ion exchange occurs, a number of oxide-ion sites become vacant because of three $O^=$ ions replacing four $O^=$ ions. Oxide-ion transport occurs between vacancies located at tetrahedral sites in the perovskite lattice, and the process is now well understood at the atomic and molecular level (Kahn et al., 1998). The ionic conductivity of YSZ ($0.02\,S\,cm^{-1}$ at 800°C and $0.1\,S\,cm^{-1}$ at 1000°C) is comparable with that of

---

[8] Though as we noted in Figure 7.10, the simplification might be very slight.

liquid electrolytes, and it can be made very thin (25–50 μm) ensuring that the ohmic loss in the SOFC is comparable with other fuel cell types. A small amount of alumina may be added to the YSZ to improve its mechanical stability, and tetragonal phase zirconia has also been added to YSZ to strengthen the electrolyte structure so that thinner materials can be made.

Thin electrolyte structures of about 40-μm thickness can be fabricated by 'electrochemical vapour deposition' (EVD), as well as by tape casting and other ceramic processing techniques. The EVD process was pioneered by Siemens Westinghouse to produce thin layers of refractory oxides suitable for the electrolyte, anode, and interconnection in their tubular SOFC design (see below). However, it is now only used for fabrication of the electrolyte. In this technique, the starting material is a tube of cathode material. The appropriate metal chloride vapour to form the electrolyte is introduced on one side of the tube surface and an oxygen/steam mixture on the other side. The gas environments on both sides of the tube act to form two galvanic couples. The net result is the formation of a dense and uniform metal oxide layer on the tube in which the deposition rate is controlled by the diffusion rate of ionic species and the concentration of electronic charge carriers.

Zirconia-based electrolytes are suitable for SOFCs because they exhibit pure anionic conductivity. Some materials, such as CeO_2 and Bi_2O_3, show higher oxygen ion conductivities than YSZ but they are less stable at low oxygen partial pressures as found at the anode of the SOFC. This gives rise to defect oxide formation and increased electronic conductivity and thus internal electric currents, which lower the cell potential (see Section 3.5). Some good progress has been made in recent years in stabilising ceria by the addition of gadolinium (Sahibzada et al., 1997), and adding zinc to lanthanum-doped Bi_2O_3. Most recently, other materials have been produced with enhanced oxide-ion conductivity at temperatures lower than that required by zirconia. The most notable of these is the system LaSrGaMgO (LSGM) (Feng et al., 1996 and Ishihara et al., 1994). This material is a superior oxide-ion electrolyte that provides performance at 800°C comparable to YSZ at 1000°C, as shown in Figures 7.22 and 7.23.

Figure 7.22 compares the conductivities of some of the more common electrolyte materials. To ensure that the total internal resistance (that is, electrolyte and electrodes) of an SOFC is sufficiently small, in Figure 7.22 the target value for the area specific resistivity (ASR) of the electrolyte is set at $0.15 \, V \, cm^2$. Films of oxide electrolytes can be reliably produced using cheap, conventional ceramic fabrication routes at thicknesses down to ~15 mm. It follows that the specific conductivity of the electrolyte must exceed $10^{-2} \, S \, cm^{-1}$. Figure 7.22 shows that this is achieved at 500°C for the electrolyte $Ce_{0.9}Gd_{0.1}O_{1.95}$ and at 700°C for the electrolyte $(ZrO_2)_{0.9} \, (Y_2O_3)_{0.1}$. Although the electrolyte $Bi_2V_{0.9}Cu_{0.1}O_{5.35}$ exhibits higher conductivities, it is not stable in the reducing environment imposed by the fuel in the anode compartment of an SOFC.

Anode

The anode of state-of-the-art SOFCs is a cermet made of metallic nickel and a YSZ skeleton. The zirconia serves to inhibit sintering of the metal particles and provides a thermal expansion coefficient comparable to that of the electrolyte. The anode has a high porosity (20–40%) so that mass transport of reactant and product gases is not

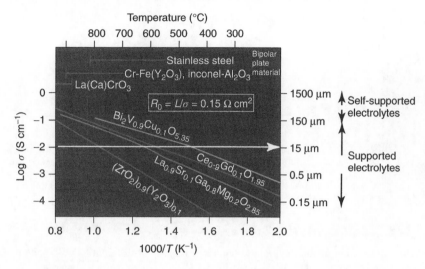

Figure 7.22 Specific conductivity versus reciprocal temperature for selected solid-oxide electrolytes.

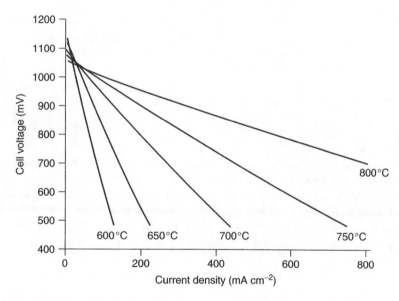

Figure 7.23 Typical single cell performance of LSGM electrolyte (500-m thick).

inhibited. There is some ohmic polarisation loss at the interface between the anode and the electrolyte and several developers are investigating bi-layer anodes in an attempt to reduce this. Often, a small amount of ceria is added to the anode cermet. This improves the tolerance of the anodes to temperature cycling and redox (changing of the anode gas

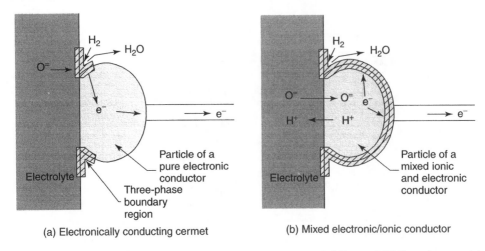

(a) Electronically conducting cermet   (b) Mixed electronic/ionic conductor

Figure 7.24 Illustration of the three-phase boundary regions of different SOFC anode materials. Similar extension of the boundary is obtained in mixed conducting cathode materials.

from a reducing feed gas to an oxidising gas, and vice versa). Control of the particle size of the YSZ can also improve the stability of the anode under redox conditions. Most recently the attention of the developers has been directed towards novel ceramic anodes that especially promote the direct oxidation of methane (see Section 8.4.6). Examples are Gd-doped ceria mixed with Zr and Y, and various TiO_2-based systems. In such anodes, there is mixed conductivity for both electrons and oxygen ions. A further advantage of using mixed conductors as anodes is that they can provide a means of extending the three-phase boundary between reactant-anode electrolyte, as shown in Figure 7.24.

Cathode

Similar to the anode, the cathode is a porous structure that must allow rapid mass transport of reactant and product gases. Strontium-doped lanthanum manganite $(La_{0.84}Sr_{0.16})MnO_3$, a p-type semiconductor, is most commonly used for the cathode material. Although adequate for most SOFCs, other materials may be used, particularly attractive being p-type conducting perovskite structures that exhibit mixed ionic and electronic conductivity. This is especially important for lower-temperature operation since the polarisation of the cathode increases significantly as the SOFC temperature is lowered. It is in cells operating at around 650°C that the advantages of using mixed conducting oxides become apparent. As well as the perovskites, lanthanum strontium ferrite, lanthanum strontium cobalite, and n-type semiconductors are better electrocatalysts than the state-of-the-art lanthanum strontium manganite, because they are mixed conductors (Han, 1993).

Interconnect material

The 'interconnect' is the means by which connection is achieved between neighbouring fuel cells. In planar fuel cell terminology this is the bipolar plate, but the arrangement is different for tubular geometries as will be described in the next section. Metals can be used as the interconnect, but these tend to be expensive 'inconel' type stainless steels, particularly for stacks that need to operate at 800 to 1000°C. Conventional steels also have a mismatch in thermal expansion coefficient with the YSZ electrolyte. To overcome this, Siemens and others have tried to develop new alloys, such as the Cr-$5Fe$-$1Y_2O_3$ Siemens/Plansee alloy. Unfortunately, such alloys can poison the cathode with chromium. Under the cathode atmosphere, chromium can evaporate from the interconnect and deposit at the $(La,Sr)MnO_3(LSM)/YSZ/gas$ three-phase boundary resulting in rapid deactivation of the cathode. An advantage for the low-temperature SOFC is that cheaper materials may be used, such as austenitic steels, which do not contain chromium. Metal interconnects also tend to form oxide coatings, which can limit their electrical conductivity and act as a barrier to mass transport.

An alternative, and one that is favoured for the tubular design, is the use of a ceramic material for the interconnect, lanthanum chromite being the preferred choice. The electronic conductivity of this material is enhanced when some of the lanthanum is substituted by magnesium or other alkaline earth elements. A fuller discussion of the effect of dopants is given in the recent review by Yokokawa et al. (2001). Unfortunately, the material needs to be sintered to quite high temperatures (1625°C) to produce a dense phase. This exposes one of the major problems with the SOFC – that of fabrication. All the cell components need to be compatible with respect to chemical stability and mechanical compliance (i.e. similar thermal expansion coefficients). The various layers need also to be deposited in such a way that good adherence is achieved without degrading the material due to the use of too high a sintering temperature. Many of the methods of fabrication are proprietary and considerable research is being carried out in this field.

Sealing materials

A key issue with SOFCs is the method of sealing the ceramic components to obtain gas-tightness, particularly with planar SOFCs. The wet-seal arrangement used in the MCFC cannot be used, and the wide temperature ranges involved posed particular problems. The usual approach has been to use glasses that have transition temperatures close to the operating temperature of the cell. These materials soften as the cells are heated up and form a seal all around the cell. This is important in planar stack designs in which, for example, a number of cells may be assembled in one layer (see below). A particular problem is the migration of silica from such glasses, especially onto the anodes, causing degradation in cell performance. For all-ceramic stacks, glass ceramics have been used, but migration of the silica component can still be a problem on both the anode and cathode sides.

7.5.3 Practical design and stacking arrangements for the SOFC

Tubular design

The tubular SOFC was pioneered by the US Westinghouse Electric Corporation (now Siemens Westinghouse Power Corporation or SWPC) in the late 1970s. The original design used a porous calcia-stabilised zirconia support tube, 1 to 2 mm thick onto which the

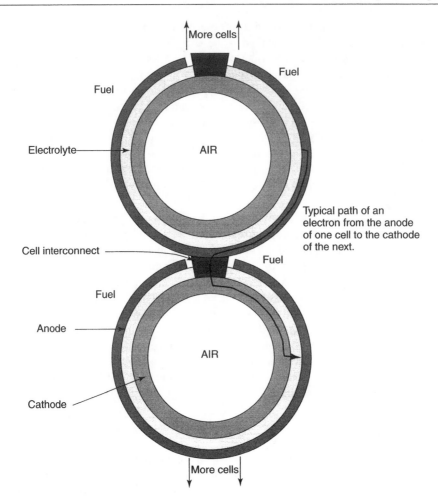

Figure 7.25 End view of tubular type solid oxide fuel cell produced by Siemens Westinghouse. The electrolyte and the anode are built onto the air cathode.

cylindrical anodes were deposited. By a process of masking, the electrolyte, interconnect, and finally the fuel electrode were deposited on top of the anode. The process was reversed in the early 1980s so that the air electrode became the first layer to be deposited on the zirconia tube, and the fuel electrode was on the outside. This tube became the norm for the next 15 years. The problem with the tubular design from the start has been the low power density and the high fabrication costs. The low power density is a result of the long path for electrical power through each cell and the large voids within the stack structure, as shown in Figure 7.25. The high costs are due to the method of electrolyte and electrode deposition – electrochemical vapour deposition (EVD), which is a batch process carried out in a vacuum chamber. Most recently, the zirconia support tube has been eliminated from the design and the tubes are now made from air-electrode materials, resulting in

Figure 7.26 Small stack of 24 tubular SOFCs. Each tube has a diameter of 2.2 cm and is about 150 cm long. (Photograph reproduced by kind permission of Siemens Westinghouse Power Corporation).

an air-electrode-support (AES) onto which the electrolyte is deposited by EVD, followed by plasma spraying of the anode. Further details of the fabrication methods are given elsewhere, (Hirschenhofer et al., 1998 and Yokokawa et al., 2001). The arrangement of the AES cell is shown in Figure 7.25. The present generation of Siemens Westinghouse SOFC tubes are 150 cm long and 2.2 cm in diameter. They are arranged in series/parallel stacks of 24 tubes as shown in Figure 7.26.

One great advantage of the tubular design of SOFC is that high-temperature gas-tight seals are eliminated. The way this is done is shown in Figure 7.27. Each tube is fabricated like a large test tube, sealed at one end. Fuel flows along the outside of the tube, towards the open end. Air is fed through a thin alumina air supply tube located centrally inside each tubular fuel cell. Heat generated within the cell brings the air up to the operating temperature. The air then flows through the fuel cell back-up to the open end. At this point air and unused fuel from the anode exhaust mix are instantly combusted and so the cell exit is above 1000°C. This combustion provides additional heat to preheat the air supply tube. Thus the tubular SOFC has a built-in air preheat and anode exhaust gas combustor and does not require high-temperature seals. Finally, by allowing imperfect sealing around the tubes, some recirculation of the anode product gas occurs, allowing internal reforming of fuel gas on the SOFC anode, as the anode product contains steam.

Several other organisations, notably Mitsubishi Heavy Industries and TOTO in Japan and also Adelan Ltd in the United Kingdom have been developing tubular SOFC designs.

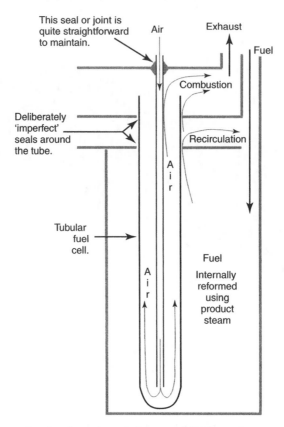

Figure 7.27 Diagram showing how the tubular-type SOFC can be constructed with (almost) no seals.

TOTO have adopted another method of cell fabrication – that of wet-sintering. The aim of this is to overcome the cost of the EVD process. Although much progress has been made with this recently, the Siemens Westinghouse SOFC design remains the most advanced of the tubular SOFC types. Figures 7.26 and 7.28 show stacks built up of tubular fuel cells and Figure 7.29 shows a 100-kW demonstration that has been built and operated in the Netherlands.

Planar design

Alternatives to the tubular SOFC have been developed for many years, notably several types of planar configuration and a monolithic design. The planar configurations more closely resemble the stacking arrangements described for the PAFC and PEMFC. This bipolar or flat-plate structure enables a simple series electrical connection between cells without the long current path through the tubular cell shown in Figure 7.24. The bipolar flat-plate design thus results in lower ohmic losses than in the tubular arrangement. This

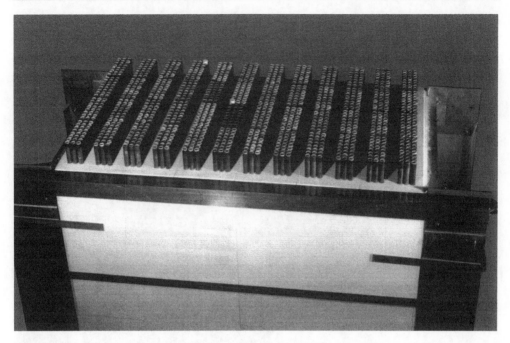

Figure 7.28 Larger stack made from bundles of 24 SOFC tubes. There are 1152 cells, and this stack has a power of about 200 kW. (Photograph reproduced by permission of Siemens Westinghouse.)

leads to a superior stack performance and a much higher power density. Another advantage of the planar design is that low-cost fabrication methods such as screen printing and tape casting can be used.

One of the major disadvantages of the planar design is the need for gas-tight sealing around the edge of the cell components. Using compressive seals this is difficult to achieve, and glass ceramics have been developed in an attempt to improve high-temperature sealing. Similarly, thermal stresses at the interfaces between the different cell and stack materials tend to cause mechanical degradation, so thermal robustness is important. Particularly challenging is the brittleness of planar SOFCs in tension. The tensile strength of zirconia SOFCs is only about 20% of their compressive strength. Thermal cycling is also a problem for the planar SOFC. Finally, the issue of thermal stresses and fabrication of very thin components places a major constraint on the scale-up of planar SOFCs. Until recently planar SOFCs could be manufactured only in sizes up to 5 × 5 cm. Now 10 × 10 cm planar cells are routinely made (Huijsmans et al., 1999). Such cells may be assembled into a stack by building them into a window-frame arrangement such that several cells are located in one stack layer.

Despite their fundamental attraction in terms of power density and efficiency compared with tubular designs, several organisations have abandoned development of the planar SOFC design because of all the inherent technical problems. These have included Siemens in Europe who had built 20-kW stacks by 1999. However, planar technology is being

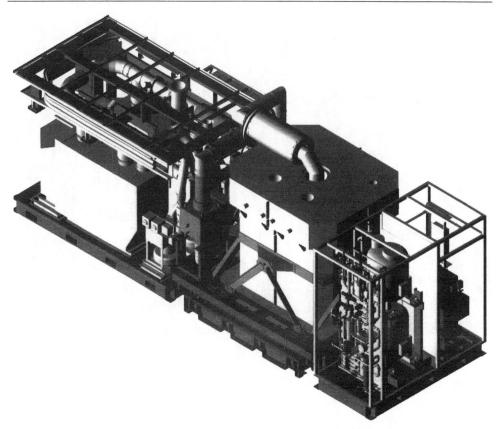

Figure 7.29 Part of a 100-kW SOFC combined heat and power unit. The fuel cell stack is the central unit. The fuel processing system is on the small skid at the front, and the thermal management/electrical/control system is at the rear. The unit is about $8.5 \times 3 \times 3$ m. (Diagram reproduced by kind permission of Siemens Westinghouse Power Corporation.)

carried forward by companies such as Ceramic Fuel Cells Ltd in Australia, Sulzer Hexis, and ECN in Europe, Allied Signal, and SOFCo (Khankar et al., 1998) in the US, Global Thermoelectric in Canada, Rolls Royce in the United Kingdom and by companies such as Murata Manufacturing in Japan (Badwal et al., 1998, Foger et al., 1999).

Sulzer Hexis are targeting the small-scale cogeneration market with their unique planar stack arrangement (Figure 7.30). In this design, pipeline natural gas is desulphurised and fed to a small pre-reformer or partial oxidation unit located beneath the stack. Future designs will incorporate pre-reforming within the stack. The fuel gas is then fed into the centre of a cylindrical stack made of layers of circular SOFCs interspersed with air manifolds. As shown in Figure 7.30, the design of the bipolar plate ensures that the reactant air is preheated. Unreacted fuel is burnt to heat water for the domestic heating system.

Allied Signal, SOFCo, Ztek, Global Thermoelectric Rolls Royce and the various Japanese developers of SOFC are all at the laboratory stage with stacks of up to about 1 kW

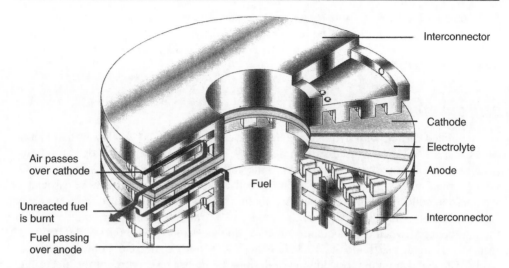

Figure 7.30   Ring-type solid oxide fuel cell with metal cell interconnects. The fuel gas comes up the central tube and out over the anode. Air is blown in from the outside over the cathode. After passing over both electrodes the gas streams mix, and the remaining fuel is burnt to provide more heat. (Diagram reproduced by kind permission of Sulzer Hexis Ltd).

being demonstrated. The Allied Signal SOFC is a flat-plate concept that involves stacking high-performance thin-electrolyte cells with lightweight metallic interconnect assemblies. Each cell comprises a relatively thick anode that supports a thin film electrolyte. This SOFC design can be operated at temperatures between 600 and 800°C. Single cells used in this design contain supported thin electrolytes. Numerous multi-cell stacks have been assembled and performance tested, demonstrating power densities of over 650 mW cm^{-2} at 800°C. Allied Signal are targeting small-scale applications for their technology (Minh et al., 1998). Similarly, SOFCo has demonstrated several sub-kilowatt scale stacks at an operating temperature of 850 to 900°C with performance degradations of 0.5% per 1000 h. Zetek have built 1-kW stacks and have a concept for a 100-kW class system of 25-kW module, and Global Thermoelectric are working with Delphi Automotive and BMW on a kilowatt-class system to provide auxiliary power for vehicles. Further discussion of SOFCs operating at lower temperatures is provided in Section 7.5.6.

Integrated planar SOFC and flat-tube high-power density designs

The integrated planar SOFC configuration of Rolls Royce seeks to retain the specific advantages of both the tubular and planar arrangements (Day, 2000). It has many similar features to the original Westinghouse tubular design and consists of an assembly of small planar SOFCs fabricated on a ceramic housing. The housing serves as a manifold for the fuel gas, with a novel sealing arrangement. Rather than using a bipolar plate, the cells are connected by an inter-connector fabricated onto the cell housing. Most recently, Siemens-Westinghouse have proposed a flat-tube development that aims to increase the power density of stacks by reducing the number or length of cells required to achieve a given output. The effect of this should be to reduce the required footprint of plant, with a resultant lowering of costs. The arrangement is similar to that in Figure 7.25, except that the circular tubes are more square in section, with less space and more contact area between each cell.

7.5.4 SOFC performance

As mentioned earlier, with hydrogen as the fuel, the OCV of SOFCs is lower than that of MCFCs and PAFCs (see discussion in Section 7.2). However, the higher operating temperature of SOFCs reduces polarisation at the cathode. So, the voltage losses in SOFCs are governed mainly by ohmic losses in the cell components, including those associated with current collection. The contribution to ohmic polarisation in a tubular cell is typically some 45% from cathode, 18% from the anode, 12% from the electrolyte, and 25% from the interconnect, when these components have thickness of 2.2, 0.1, 0.04 and 0.085 mm, respectively, and resistivities at $1000°C$ of 0.013, 3×10^{-6}, 10, and 1 ohm cm, respectively. The cathode ohmic loss dominates despite the higher resistivities of the electrolyte and cell interconnection because of the short conduction path through these components and the long current path in the plane of the cathode – see Figure 7.25. As we have seen in the previous section, SOFCs are being developed with various materials and designs and are being made by several different methods. SOFCs, especially the planar type, may well develop considerably over the next few years and the performance characteristics are likely to change. It could be that the development will lead to a standardisation of cell types, as has been achieved already with PAFC and MCFC. On the other hand, several quite different types could come into use, as with the PEMFC. The following section gives general indications of the effect of pressure and temperature on SOFC performance. For further discussion, the reader is referred to Hirschenhofer et al. (1998) and the proceedings of the European SOFC Forums.

Influence of pressure

SOFCs, like all fuel cell types, show an enhanced performance with increasing cell pressure. Unlike low- and medium-temperature cells, the improvement is mainly due to the increase in the Nernst potential. We showed in Section 2.5.4 that the voltage change for an increase in pressure from P_1 to P_2 follows very closely the theoretical equation

$$\Delta V = 0.027 \ln \left(\frac{P_2}{P_1} \right)$$

The relationship is borne out in practice, and Siemens Westinghouse in conjunction with Ontario Hydro Technologies has tested AES cells at pressures up to 15 atm, on both hydrogen and natural gas (Singhal, 1997). Operation at higher pressure is particularly advantageous when using the SOFC in a combined cycle system with a gas turbine. In other cases, as with the PEMFC, the power costs involved in compressing the reactants make the benefits marginal.

Influence of temperature

The temperature of an SOFC has a very marked effect on its performance, though the details will vary greatly between cell types and materials used. The predominant effect is that higher temperatures increase the conductivity of the materials, and this reduces the ohmic losses within the cell. As we saw in Chapter 3, ohmic losses are the most important type of loss in the SOFC.

For SOFC-combined cycle and hybrid systems, it is beneficial to keep the operating temperature of the SOFC high. For other applications, such as cogeneration, and possible transport applications (the SOFC is being developed by BMW for use as an auxiliary power supply for vehicles), it is an advantage to operate at lower temperatures, as the higher temperatures bring material and construction difficulties. Unfortunately, as Figure 7.23 clearly shows, the performance decreases substantially for SOFCs as the temperature is lowered. Indeed, if zirconia (the standard electrolyte material) is used in place of the LGSM electrolyte of Figure 7.23, the performance falls off at even higher temperatures, and operation at about 900 to 1000°C is required.

As was mentioned in the section on cell interconnects, one of the main advantages of operating at lower temperatures is the possibility of using cheaper construction materials and methods. Making electrolytes and electrodes that work well at lower temperatures is a major focus of current SOFC research.

7.5.5 SOFC combined cycles, novel system designs and hybrid systems

In Section 7.2.3, we discussed how a high-temperature fuel cell could be combined with a steam turbine in a bottoming cycle. The ability to use both gas turbines and steam turbines in a combined cycle with an SOFC has been known in concept for many years. However, it is only recently that pressurised operation of SOFC stacks has been demonstrated for prolonged periods, making the SOFC/GT combined cycle system feasible practically. Pioneered by Siemens Westinghouse in their SureCell™ concept, the ideas of combined SOFC/GT are now being explored by other developers. Figure 7.31 shows the design of the Siemens Westinghouse concept at the 300-kW scale, and the essential process features are shown in Figure 7.32. These systems, and variations on them, are further described in the literature. (Veyo and Forbes, 1998; Bevc, 1997; Fry et al., 1997; and Hassman, 2001.)

The first complete SOFC/GT system was delivered by Siemens Westinghouse to Southern Californian Edison in May 2000. A second hybrid system was built for the Canadian company Ontario Hydro, and further units are being built for customers in Europe. In addition, a utility consortium led by Energie Baden-Württemberg (EnBW) and shared by EDF, Gaz de France and the Austrian utility TIWAG jointly with SWPC will build the first 1-MW scale all-electric hybrid cycle power plant in Europe. Only one other project of this size is planned for the demonstration phase and this will be located in the United States. The expected performance data for both the 300-kW class and 1-MW-class hybrid systems are shown in Table 7.5. On the basis of the 220-kW systems, the performance of the 300-kW systems should be achieved, and with larger systems it is envisaged that higher pressures would be possible. With the SOFC running at 7 bar, for example, a hybrid system of 2-MW to 20-MW scale would deliver 60 to 70% efficiency. However, a more

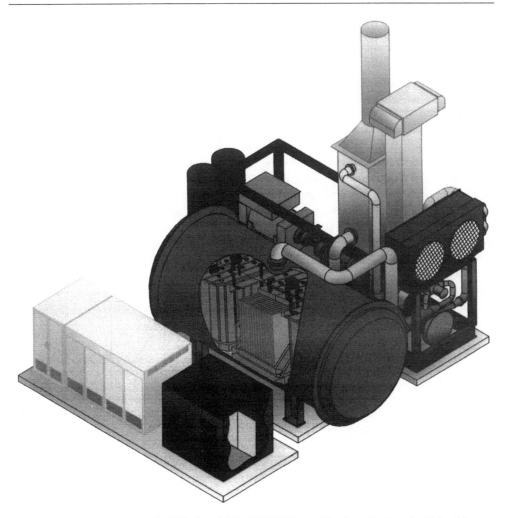

Figure 7.31 Design of a 300-kW class MW SOFC/GT combined cycle plant built by Siemens Westinghouse. The SOFC is shown in the middle of the diagram, and operates at about 10 bar inside the cylindrical pressure vessel. The gas turbine, compressor, and alternator are behind the fuel cell.

sophisticated GT would be needed and the added investment may not be justified for the few percentage points gained in system efficiency.

Siemens Westinghouse has now expanded its production facilities to cope with the demand for increased numbers of cells for its increasing number of demonstration systems. However, they are not the only company interested in hybrid systems; it is worthwhile remarking that there are many opportunities for novel system design, and there is scope for considerable creativity by the systems engineer. There are many examples in the literature. For instance, by using the concept of series stack connection mentioned first in

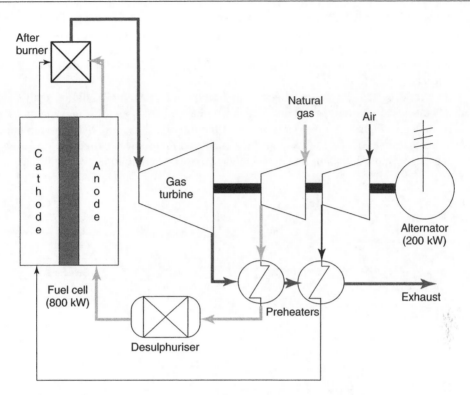

Figure 7.32 System diagram for the 300-kW class SOFC/GT combined cycle system shown in Figure 7.31.

Table 7.5 Expected performance of Siemens Westinghouse 300-kW and 1-MW class hybrid systems

	300 kW	1 MW
Electrical net AC efficiency	>55%	>55%, approaching 60%
SOFC AC power	244 kW	805 kW
Gas turbine AC power	65 kW	220 kW
Total net AC power	300 kW	1014 kW
Pressure ratio of turbine compressor	3–4	3–4
Emissions: CO_2	<350 kg MWh^{-1}	<350 kg MWh^{-1}
Nox	<0.5 ppm	<0.5 ppm
CO	0 ppm	0 ppm
Sox	0 ppm	0 ppm
Particulates	0 ppm	0 ppm
Ground noise level (5 m from housing)	<75 dBa	<75 dBa

connection with the MCFC (Section 7.4.6), a 'multi-stage' SOFC concept 'UltraFuelCell' has been developed by the US Department of Energy and described by Hirschenhofer (1998). However, perhaps one of the most novel ideas is the hybrid or combined system concept described by Siemens Westinghouse (Vollmar et al., 1999) and BG Technology (Mescal, 1999).

This hybrid system is one in which both SOFC and PEM fuel cells are combined. The advantages of each type of fuel cell are enhanced by operating in synergy. The system shown in Figure 7.33 works as follows: the internal reforming SOFC is run under conditions that give low fuel utilisation. This enables a high power output for a relatively low stack size. Unused fuel appears in the anode exhaust where it undergoes shift reaction,

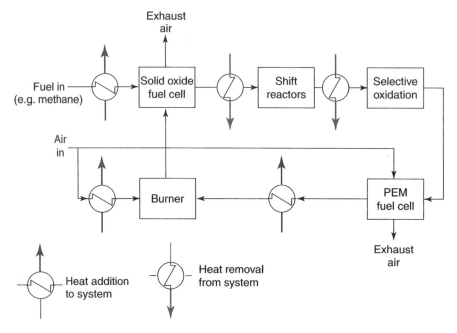

Figure 7.33 SOFC-PEM hybrid system (Mescal, 1999). See also Table 7.6.

Table 7.6 Summary of output powers for the hybrid system shown in Figure 7.33 (Mescal, 1999)

SOFC stack power	369.3 kW
PEM stack power	146.7 kW
Turbine power	100.3 kW
Compressor power	−100.8 kW
Net power output	515.5 kW
Electrical output	489.7 kW
Overall efficiency (%) LHV	61

followed by a process stage when the final traces of carbon monoxide are removed. At this stage, the gas comprises mainly hydrogen and carbon dioxide, with some steam. This gas, once it is cooled, is suitable for use as a fuel in the PEM stack. The use of two stacks of different types for power generation results in a high overall electrical efficiency. The system becomes particularly attractive if an economic analysis is carried out. Preliminary calculations show that the system is more cost-effective than an SOFC-only system because of the anticipated relatively low cost of the PEM stack. On the other hand, the system has a much higher efficiency than a natural gas fuelled PEM-only system could achieve. We also see in the following chapter that the fuel processing technology for running a PEMFC from natural gas is complex, bulky, and expensive. How much better to use an SOFC as the fuel processor!

7.5.6 Intermediate temperature SOFCs

During the 1970s and 1980s, support for the development of SOFCs came from large generating equipment manufacturers such as Westinghouse, ABB, and GEC. The focus was on developing large systems, thereby benefiting from volume production of the cells. High temperatures were required (and still are for SOFC hybrid systems). By 1990 it was becoming recognised that if smaller systems were to be built there would be cost advantages in reducing the operating temperature, provided the internal resistance of the cell and the electrode kinetics were adequate and that internal reforming could be carried out if possible. More recently, the development of smaller units has been supported by a paradigm shift in the power generation business in which increased emphasis is being placed on dispersed power systems, embedded generation, and cogeneration. This shift in the business environment has been brought about by a liberalisation of the established energy markets in developed countries. To that we should add the need in many underdeveloped areas of the world for small remote-area power supplies. Finally, there has been a growing interest recently shown by some automotive companies for additional on-board power in vehicles to support auxiliary equipment such as air-conditioning, electrically operated steering, and other high-power electrical loads. Thus, we are finding that the standard 14-V systems that have fared well for many years are being superseded by 42-V systems able to deliver more power to the growing number of vehicle accessories.

We have already discussed some materials that could be used as low-temperature SOFC electrolytes. Referring again to Figure 7.22, if we assume that the electrolyte should not contribute more than $0.15\,V cm^2$ to the total cell area specific resistance, then for a thickness (L) of $15\,\mu m$, the associated specific ionic conductivity (σ) of the electrolyte should exceed $10^{-2}\,S cm^{-1}$ (since $\sigma = L/ASR = 0.0015/0.15$).

Clearly, yttria-stabilised zirconia has sufficient ionic conductivity to meet the target at around $700°C$, and for ceria-gadolinium oxide (CGO) the temperature is around $500°C$. This assumes that electrolytes as thin as $15\,\mu m$ can of course be made. In fact, such thin films can be made by a variety of techniques, but because the material is so fragile it is necessary to provide support for the electrolyte in such cells. In other words, the thin electrolyte does not have the mechanical strength to support the electrodes as is the case with conventional high-temperature planar SOFCs. Most manufacturers have opted for supporting the electrolyte on relatively thick anodes, and these are often referred to as thick-film PEN (Positive electrode–Electrolyte–Negative electrode) structures. Anode-supported cells can be sintered at temperatures of around 1400°C to produce a dense thin electrolyte whilst maintaining a reasonable permeability for the Ni-YSZ anode.

One of the problems with anode-supported cells is that any difference in thermal expansion between anode and electrolyte becomes more significant than in conventional high-temperature SOFCs. For this reason many developers use porous nickel cermet anodes with interfacial regions made of Ni/YSZ doped with ceria. Operating at temperatures below about 700°C means that metallic bipolar plates can be used, and the lower the temperature, the less exotic the steel needs to be. Ferritic stainless steels can be used below about 600°C, and these have the advantage that they have a low thermal expansion coefficient. Conventional doped LSM-YSZ cathodes can be used but there is much development in progress to improve cathode materials as the cathode overpotentials become more significant as the cell temperatures are lowered. A recent review of cathode materials has been published by Ralph (2001).

Sealing of IT-SOFC stacks is also an issue, as with conventional planar stacks. As has been noted before, organisations such as Sulzer Hexis and Global Thermoelectric have addressed this issue by opting for a circular design of cell in which the fuel and air are fed via a manifold to the centre of the PEN structure. The air and fuel gases are distributed over the cathode and anode, and the flow rates are adjusted to ensure almost complete utilisation of the fuel by the time it leaves the stack where any excess fuel is combusted. It is also possible to use compliant gaskets in cells operating near 500°C, which gives a much greater flexibility in stack design than when using glass seals. At Imperial College, researchers have capitalised on the close thermal expansion of ferritic steels and YSZ electrolytes to make thick-film PEN structures supported on a porous stainless steel foil (Oishi et al., 2002). The structures are claimed to be very robust and able to withstand the temperature cycles that are expected in IT-SOFC applications.

The search for new materials for IT-SOFCs is currently a key activity. Ceria materials, as noted earlier, are more conductive than the traditional YSZ except that ceria is unstable at elevated temperatures in reducing atmospheres. Fortunately, if the temperature of operation is as low as 500°C, then the increased electronic conductivity caused by reduction of Ce^{4+} to Ce^{3+} ions is relatively low and then this issue appears not to be such a big concern (Steele, 2001). Lanthanum gallate electrolytes are also being investigated for IT-SOFC applications, and time will tell which material offers the best prospects, both technically and economically, for commercial systems.

References

Appleby J. (1984) in *Proceedings of the Workshop on the Electrochemistry of Carbon*, S. Sarangapani, J.R. Akridge, and B. Schumm (eds) The Electrochemical Society, Pennington, NJ, p. 251.
Appleby A.J. and Foulkes F.R. (1993) *A Fuel Cell Handbook*, 2nd ed., Krieger Publishing Co., New York.
Badwal S.P.S. and Foger K. (1998) "SOFC development at Ceramic Fuel Cell Limited." *Proceedings of the Third European Solid Oxide Fuel Cell Forum*, pp. 95–104.
Bett J.A.S., Kunz H.R., Smith S.W., and Van Dine L.L. (1985) "Investigation of Alloy Catalysts and Redox Catalysts for Phosphoric Acid Electrochemical Systems," FCR-7157F, prepared by International Fuel Cells for Los Alamos National Laboratory under Contract No. 9-X13-D6271-1.

Bevc F. (1997) "Advances in solid oxide fuel cells and integrated power plants", *Proceedings of the Institution of Mechanical Engineers (London)*, **211**, Part A, 359–366.

Blomen L., Leo J.M.J., and Mugerwa M. (1993) *Fuel Cell Systems*, Plenum Publishing, New York.

Bregoli L.J. and Kunz H.R. (1982) "The effect of thickness on the performance of molten carbonate fuel cell cathodes", *Journal of the Electrochemical Society*, **129**(12), 2711–2715.

Brenscheidt T., Janowitz K., Salge H.J., Wendt H., and Brammer F. (1998) "Performance of ONSI PC25 PAFC cogeneration plant", *International Journal of Hydrogen Energy*, **23**(1), 53–56.

Buchanan J.S., Hards G.A., Keck L., and Potter R.J. (1992) "Investigation into the superior oxygen reduction activity of platinum alloy phosphoric acid fuel cell catalysts," *Fuel Cell Seminar Abstracts*, Tucson, Arizona, U.S.

Clarke S.H., Dicks A.L., Pointon K, Smith T.A., and Swann A. (1997) "Catalytic aspects of the steam reforming of hydrocarbons in internal reforming fuel cells", *Catalysis Today*, **38**, 411–423.

Day, M.J. (2000) in *4th European SOFC Forum*, A.J. McEvoy (ed.) (European Fuel Cell Forum, Oberrohrdorf, Switzerland, pp. 133–140.

Dicks A.L. (1998) "Advances in catalysts for internal reforming in high temperature fuel cells", *Journal of Power Sources*, **71**, 111–122.

Dicks A.L. and Siddle A. (1999) "Assessment of Commercial Prospects of Molten carbonate Fuel Cells", ETSU Report No.F/03/00168/REP, AEA Technology, Harwell, UK.

Feng M., Goodenough J., Huang K., and Milliken C. (1996) *Journal of Power Sources*, **64**, 47.

Foger K., Godfrey B., and Pham T. (1999) "Development of 25 kW SOFC system", *Fuel Cells Bulletin No*, ISSN 1350–4789, Elsevier Science, Amsterdam, pp. 9–11.

Fry M.R., Watson H., and Hatchman J.C. (1997) "Design of a prototype fuel cell/composite cycle power station", *Proceedings of the Institution of Mechanical Engineers (London)*, **211**, Part A, 171–180.

Gardner F.J. (1997) "Thermodynamic processes in solid oxide and other fuel cells", *Proceedings of the Institution of Mechanical Engineers (London)*, **211**, Part A, 367–380.

Han P. et al. (1993) "Novel oxide fuel cells operating at 600–800°C", *An EPRI/GRI Fuel Cell Workshop on Fuel Cell Technology Research and Development*, New Orleans, LA.

Hassman K. (2001) "SOFC power plants, Westinghouse approach", *Fuel Cells*, **1**(1), 78–84.

Hirschenhofer J.H., Stauffer D.B., and Engelman R.R. (1998) *Fuel Cell Handbook Fourth Edition*, Report no. DOE/FETC-99/1076, Parsons Corporation, for U.S. Department of Energy.

Huijsmans J.P.P., Huiberts R.C., and Christie G.M. (1999) "Production line for planar SOFC ceramics: from laboratory to pre-pilot scale manufacturing", *Fuel Cells Bulletin No.*, ISSN 1464–2859, Elsevier Science, Amsterdam, pp. 5–7.

Ishihara T., Matsuda H., and Takita Y. (1994) "Doped $LaGaO_3$ perovskite type oxide as a new oxide ionic conductor", *Journal of the American Chemical Society*, **116**(9), 3801–3803.

Jalan V., Poirier J., Desai M., and Morrisean B. (1990) "Development of CO and H_2S tolerant PAFC anode catalysts." *Proceedings of the Second Annual Fuel Cell Contractors Review Meeting*, (1990).

Kahn M.S., Islam S., and Bates D.R. (1998) "Cation doping and oxygen diffusion in zirconia: a combined simulation and molecular dynamics study", *Journal of Material Chemistry*, **8**(10), 2299–2307.

Khankar A., Elangovan S, Hartvigsen J., Rowley D., and Tharp M. (1998) "Recent progress in SOFCo's planar SOFC development." *Proceedings of the Fuel Cell Seminar*, November, 16–19, Palm Springs, Calif., US.

Kinoshita K. (1988) *Carbon: Electrochemical and Physicochemical Properties*, Wiley Interscience, New York.

Kunz H.R. (1987) "Transport of electrolyte in molten carbonate fuel cells", *Journal of the Electrochemical Society*, **134**(1), 105–113.

Kordesch K.V. (1979) *"Survey of Carbon and Its Role in Phosphoric Acid Fuel Cells,"* BNL 51418, prepared for Brookhaven National Laboratory.

Kortbeek P.J. and Ottervanger R. (1998) "Developing advanced DIR-MCFC co-generation for clean and competitive power", *Abstracts of the Fuel Cell Seminar*, Palm Springs, California.

Maru H.C. and Marianowski L.G. (1976) Extended abstracts, abstract #31. Fall Meeting of the Electrochemical Society, October, 17–22, Las Vegas, NV, p. 82.

Maru H.C., Pigeaud A., Chamberlin R., and Wilemski G. (1986) in *Proceedings of the Symposium on Electrochemical Modeling of Battery, Fuel Cell, and Photoenergy Conversion Systems*, J.R. Selman and H.C. Maru (eds) The Electrochemical Society, Pennington, NJ, p. 398.

Maru H.C., Paetsch L., and Pigeaud A. (1984) in *Proceedings of the Symposium on Molten Carbonate Fuel Cell Technology*, R.J. Selman and T.D. Claar (eds) The Electrochemical Society, Pennington, NJ, p. 20.

Mescal C.M. (1999) "*SOFC-PEM Hybrid Fuel Cell Systems*," ETSU Report No. F/03/00177/REP, AEA Technology, Harwell, UK.

Minh H., Anumakonda A., Chung B., Doshi R., Ferall J., Lear G., Montgomery K., Ong E., Schipper L., and Yamanis J. (1998) "High performance, reduced-temperature solid oxide fuel cell technology." *Proceedings of the Fuel Cell Seminar*, November Palm Springs, CA, US, pp. 16–19.

Mitteldorf J. and Wilemski G. (1984) "Film thickness and distribution of electrolyte in porous fuel cell components", *Journal of the Electrochemical Society*, **131**(8), 1784–1788.

Oishi N., Rudkin R.A., Steele B.C.H., and Brandon N.P. (2002) "Thick film stainless steel supported IT-SOFCS for operation at 500–600°C", *Scientific Advances in Fuel Cells*, Elsevier Science Ltd., Amsterdam.

Petri R.J. and Benjamin T.G. (1986) in *Proceedings of the 21st Intersociety Energy Conversion Engineering Conference*, Volume 2, American Chemical Society, Washington, DC, p. 1156.

Pointon K.D. (1997) "*Review of Work on Internal Reforming in the Solid Oxide Fuel Cell*," ETSUreport F/01/00121/REP, AEA Technology, Harwell UK.

Ralph J.M., Schoeler A.C., and Krumpelt M. (2001) "Materials for lower temperature SOFC", *Journal of Material Science*, **36**, 1161–1172.

Sahibzada M., Rudkin R.A., Steele B.C.H., Metcalfe I., and Kilner J.A. (1997) "Evaluation of PEN structures incorporating supported thick film Ce 0.9 Gd 0.1 O1.95 electrolytes." *Proceedings of the fifth International Symposium on Solid Oxide Fuel Cells (SOFC-V)*, pp. 244–253.

Singhal S.C. (1997) "Recent progress in tubular solid oxide fuel cell technology", *Proceedings of the Fifth International Symposium on Solid Oxide Fuel Cells (SOFC-V)*, The Electrochemical Society, Pennington, NJ, U.S.

Steele B.C.H. (1994) "State-of-the-art SOFC ceramic materials." *Proceedings of the First European Solid Oxide Fuel Cell Forum*, Vol. 1, October, 3–7, Lucerne, Switzerland, pp. 375–397.

Steele B.C.H. and Heinzel A. (2001) "Materials for fuel cell technologies", *Nature*, **414**, 345–352.

Veyo S.E. and Forbes C.A. (1998) "Demonstrations based on Westinghouse's prototype commercial AES design." *Proceedings of the Third European Solid Oxide Fuel Cell Forum*, pp. 79–86.

Vollmar H.-E., Maiaer C.-U., Nolscher C., Merklein T., and Pippinger M. (1999) "Innovative concepts for the co-production of electricity and syngas with solid oxide fuel cells." *Abstracts from the 6th Grove Fuel Cell Symposium*, September 13–16, Elsevier Science, London to be published in Journal of Power Sources.

Yang J. C., Park Y. S., Seo S. H., Lee H. J. and Noh J. S. (2002) Development of a 50 kW PAFC power generation system, *Journal of Power Sources*, **106**(1–2), 68–75.

Yokokawa H., Sakai N., Terushia H., and Wamaji K. (2001) "Recent developments in solid oxide fuel cell materials", *Fuel Cells*, **1**(2), 117–131.

Yuh C., Farooque, Johnsen R. (1992) "Understanding of carbonate fuel cell resistances in MCFCs." *Proceedings of the Fourth Annual Fuel Cells Contractors Review Meeting*, U.S.DOE/METC, pp. 53–57.

8

Fuelling Fuel Cells

8.1 Introduction

The fuel cell types that have been described in detail so far employ hydrogen as the preferred fuel, on account of its high reactivity for the electrochemical anode reaction, and because the oxidation of hydrogen produces water, which is environmentally benign. Vehicles employing proton exchange membrane (PEM) fuel cells running on hydrogen may therefore be termed *zero-emission*, because the only emission from the vehicle is water. Unfortunately, hydrogen does not occur naturally as a gaseous fuel, and so for practical fuel cell systems it usually has to be generated from whatever fuel source is locally available. Table 8.1 gives basic chemical and physical data on hydrogen and some of these other fuels that might be considered for fuel cells.

The next section deals with the characteristics of the various primary fossil fuels that can be used for the generation of hydrogen for fuel cell systems, ranging from petroleum and natural gas to coal. Biofuels are also a possible source of hydrogen gas, and their possibilities are discussed in Section 8.3. Sections 8.4 to 8.6 explain how such primary fuels may be converted to hydrogen, using different chemical conversion technologies. For stationary fuel cell power plants, there are good arguments for carrying out the chemical conversion of fuel as close to the fuel cell stack as possible. The principal reason for this is that heat, which is generated in the fuel cell stack, may be used for part of the fuel processing. Heat integration in such systems is therefore an important part of system design and this is dealt with in Section 8.5. For transportation applications, the generation of hydrogen on board vehicles presents particular system design considerations, and these are dealt with in Section 8.6.

Another commonly used way of producing hydrogen is to use electricity to electrolyse water – the reverse of a fuel cell. Since the purpose of a fuel cell is to produce electricity, this may at first seem perverse. However, in many cases it is in fact a very convenient and efficient way of providing hydrogen for mobile fuel cells. Electrolysers are briefly discussed in Section 8.7.

Fuel Cell Systems Explained, Second Edition James Larminie and Andrew Dicks
© 2003 John Wiley & Sons, Ltd   ISBN: 0-470-84857-X

Table 8.1 Some properties of hydrogen and other fuels considered for fuel cell systems

	Hydrogen H_2	Methane CH_4	Ammonia NH_3	Methanol CH_3OH	Ethanol C_2H_5OH	Gasoline[a] C_8H_{18}
Molecular weight	2.016	16.04	17.03	32.04	46.07	114.2
Freezing point (°C)	−259.2	−182.5	−77.7	−98.8	−114.1	−56.8
Boiling point (°C)	−252.77	−161.5	−33.4	64.7	78.3	125.7
Net enthalpy of combustion at 25°C (kJ mol^{-1})	241.8	802.5	316.3	638.5	1275.9	5512.0
Heat of vaporisation (kJ kg^{-1})	445.6	510	1371	1129	839.3	368.1
Liquid density (kg m^{-3})	77	425	674	786	789	702
Specific heat at STP (J mol^{-1} K^{-1})	28.8	34.1	36.4	76.6	112.4	188.9
Flammability limits in air (%)	4–77	4–16	15–28	6–36	4–19	1–6
Autoignition temperature in air (°C)	571	632	651	464	423	220

[a]Gasoline is a blend of hydrocarbons and varies with producer, application, and season. N-octane is reasonably representative of properties except vapour pressure, which is intentionally raised by introducing light fractions.

A promising way of producing hydrogen that is not yet commercially used, but may well be soon, is using biological methods to break down the fuel – fossil or bio. Some of these methods are based on enzymes, others on bacteria, and still others use light. These methods, and their prospects, are outlined in Section 8.8.

Although hydrogen will normally be generated as required by the fuel processors described in Sections 8.4 to 8.6, there are times when hydrogen is produced in large central plants, or by electrolysers, and is stored and transported for fuel cell use. There is already something of an infrastructure for producing, storing, and supplying hydrogen, as it is widely used in the chemical industry, in petroleum refining and in ammonia manufacture, for example. There are those who see hydrogen being widely used in this way as an energy vector (a method of storing and transporting energy) in the future, when we rely less on fossil fuels and more on renewable sources of energy. However, for certain applications of fuel cells this may be a suitable way of providing fuel even now. This is especially so in the case of small, portable, low-power fuel cell systems. In these circumstances the special and difficult problems involved with transporting and storing hydrogen come to the fore.

The problem of storing hydrogen is a complex one, not least because some of the ways of doing it are so radically different. However, two distinct groups of methods can be identified. In one the hydrogen is stored simply as hydrogen – either compressed, or liquefied, or held in some kind of 'absorber'. The possible methods of doing this are explained in Section 8.9.2. This section also addresses the important issue of hydrogen safety.

In the second group of hydrogen storage methods, the hydrogen is produced in large chemical plants and is then used to produce hydrogen-rich chemicals or man-made fuels. Among these are ammonia and methanol, whose characteristics are also given in Table 8.1.

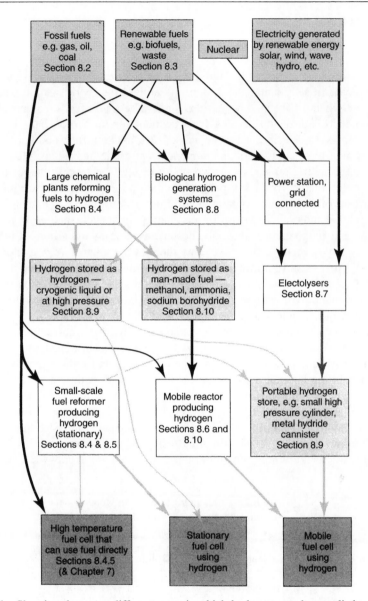

Figure 8.1 Showing the many different ways is which hydrogen can be supplied to fuel cells.

These 'hydrogen-carrier' compounds can be made to give up their hydrogen much more easily than fossil fuels and can be used in mobile systems. The most important of these compounds, and the ways they could be used, are explained in Section 8.10.

We see then that there are many possible ways of fuelling a fuel cell, and the options are varied. Figure 8.1 is an attempt to show how these different possibilities fit together and where the information you may require can be found in this chapter.

8.2 Fossil Fuels

8.2.1 Petroleum

Petroleum is a mixture of gaseous, liquid, and solid hydrocarbon-based chemical compounds that occur in sedimentary rock deposits around the world. Crude petroleum has little value, but when it is refined it provides high-value liquid feeds, solvents, lubricants, and other products. Petroleum-derived fuels account for up to one half of the world's total energy supply, and include gasoline, diesel fuel, aviation fuel, and kerosene. Simple distillation is able to separate various components of crude petroleum into generic fractions of different boiling ranges, as shown in Table 8.2. The proportions of the different fractions that are obtained from any particular crude oil depend on the origin of supply.

Each fraction of petroleum contains a different proportion of chemical compounds, these being normal and branched paraffins or alkanes, monocyclic and polycyclic paraffins (naphthenes), and mononuclear and polynuclear aromatic hydrocarbons. Light naphthas contain principally normal alkanes and some monocyclic alkanes (e.g. cyclopentane and cyclohexane). As we move through from low boiling to higher boiling fractions, the proportion of these low molecular weight alkanes falls and the proportion of polycyclic alkanes and aromatic hydrocarbons increases. The various fractions can be used without further refining, but chemical modification is made to some fractions by the oil companies. Hydro-refining, or hydrogenation, for example, is routinely carried out by refiners to reduce the aromatic content of some fractions. Gasoline is also doped with various compounds to improve engine lubrication, reduce corrosion, and reduce the risk of 'knock' or pre-ignition in the internal combustion engine. It is also common to blend petroleum fractions and to carry out hydro-refining to ensure sufficiently high 'octane' rating for different types of engine. In recent years, oxygenates have been added to petroleum for the automobile to improve combustion characteristics in engines. More details of such petrochemical engineering are to be found in other textbooks on the subject of fuel technology. For fuel cells, we need to know something of the physical and combustion characteristics

Table 8.2 Crude petroleum is a mixture of compounds that can be separated into different generic boiling fractions

Petroleum fraction	Boiling range (°C)
Light naphtha	−1−150 ·
Gasoline	−1−180
Heavy naphtha	150−205
Kerosene	205−260
Stove oil	205−290
Light gas oil	260−315
Heavy gas oil	315−425
Lubricating oil	>400
Vacuum gas oil	425−625
Residue	>600

of the fuel, but it is also very important to understand the chemical composition of the fuel. It is the chemical composition that largely determines the type of fuel processing, which may be used for generating hydrogen, as will become apparent in the following sections. It will be seen later that different technologies have to be used to convert the various fraction types into hydrogen for fuel cell systems. Especially of importance, when the fuel is converted catalytically, is the various trace compounds that may be present, since they may act as poisons for the conversion catalysts, and indeed also for the fuel cell stack. The most significant trace compounds in fossil fuels are mainly the organic compounds containing sulphur, nitrogen or oxygen, and organo-metallic compounds, which include various porphyrins. The removal of contaminant sulphur compounds from raw fuels is dealt with in Section 8.4.

The gasoline fraction of petroleum, together with the heavier diesel, is widely distributed as a fuel for vehicles – ranging from passenger cars to heavy-duty trucks and buses. Because the infrastructure for delivering such fuels is mature, there are good arguments for fuelling fuel cell vehicles (FCVs) using a similar fuel. However, 'well-to-wheel' techno-economic studies indicate that the benefits of using gasoline in FCVs, compared with internal combustion engines, are not as high as those from using hydrogen or even methanol.[1] If such fuel is used for future FCVs, it will be of simpler composition than the present vehicle fuels, since there are no reasons for adding components for anti-knock or lubrication purposes. It is also expected that in the future petroleum fuels will have a much lower sulphur content than those currently distributed.

8.2.2 Petroleum in mixtures: tar sands, oil shales, gas hydrates, and LPG

There are various deposits of naturally occurring petroleum substances that can be found as solid or near-solid material in sandstone at depths that are usually less than 2000 m. Some may be found as an outcrop on the surface. Huge amounts of such tar sands can be found in various parts of the USA and Canada, but the high bitumen content (very high molecular weight) makes recovery not as attractive as more conventional petroleum deposits. Similarly, oil shales comprise a significant and largely untapped source of petroleum materials. Oil shales are compact laminated rocks of sedimentary origin in which organic petroleum is locked. The oil can be obtained from the rock by distillation. Of the worldwide shale oil deposits estimated to be over 20×10^{12} barrels, only a small fraction (3×10^{12} barrels) is easily recoverable by conventional technologies. Much of the oil locked inside oil shales is of high molecular weight and bituminous in nature. A discussion of the processing of such materials is outside the scope of this book.

In natural petroleum reservoirs, where the prevailing pressures are high and temperatures low (under permafrost, for example), methane and other normally gaseous hydrocarbons form ice-like hydrogen-bonded complexes (clathrates) with water. Such complexes are known as gas hydrates. Methane hydrates have also been considered as a method for transporting natural gas from remote fields. Other low molecular weight hydrocarbons, such as propane and butane, are often found associated in crude petroleum. When the crude petroleum material is distilled, these emerge as a gaseous product that

---

[1] See Chapter 11.

is marketed as Liquefied Petroleum Gas or LPG. It is obtained as a refinery by-product and is used extensively for applications as diverse as camping gas stoves and as a vehicle fuel. LPG is also attractive for fuel cell systems that may be used for remote stationary power sources where there is no pipeline source of gas, and possibly also for some vehicle applications.

8.2.3 Coal and coal gases

Coal is the most abundant of all fossil fuels, and chemically it is the most complex, being formed from the compaction and induration of various plant remains similar to those in peat. Coal is classified according to the kinds of inherent plant material (coal type), the degree of metamorphism (coal rank), and the degree of impurities (coal grade). It is worth pointing out that apart from combustion, further processing of coal to produce liquids, gases, and coke is highly dependent on the properties of the raw coal material. For example, primary coking coals are those that have 20 to 30% volatile organic matter. Bituminous coal is best suited for carbonisation (heating in air to temperatures of 750 to 1500°C to form 'coal gas' or 'town gas'), whereas there are several processes for gasification of a variety of coal types.

Coal carbonisation was the original method for producing 'town gas' in the nineteenth century. Simple carbonisation or distillation of coal yielded gas (a mixture of mainly hydrogen and carbon oxides), organic liquids (tars and phenolics), and a residual coke. Carried out in retorts of various types, carbonisation has been superseded, for large-scale production, by various coal gasification processes. In gasification, coal is usually reacted with steam and oxygen (or air) at high temperatures. The products of primary coal gasification are mainly gases, together with smaller amounts of liquids and solids. The relative proportion of products depends on the type of coal, the temperature and pressure of the reaction, and the relative amounts of steam or oxygen injected into the gasifier. Further processing of the raw gasifier product gas can be carried out, for example, to increase the methane content or alter the hydrogen/carbon monoxide ratio depending on what syngas is required.

The numerous coal gasification systems available today can be classified as one of three basic types: (1) moving bed; (2) fluidised bed; and (3) entrained bed. All three types use steam and either air or oxygen, to partially oxidise coal into a gaseous product. The moving-bed gasifiers produce a low-temperature (450–650°C) gas containing methane and ethane arising from devolatilisation of the coal, together with a hydrocarbon liquid stream containing naphtha, tars, oils, and phenolic liquids. Entrained-bed gasifiers produce gas at high temperature (>1200°C), virtually no devolatilisation products in the product gas stream, and much lower amounts of liquid hydrocarbons. In fact, the entrained-bed gas product is composed almost entirely of hydrogen, carbon monoxide, and carbon dioxide. The product gas from the fluidised-bed gasifier falls somewhere between these two other reactor types in composition and temperature (925–1050°C). In all of these gasifiers, the heat required for gasification (the reaction of coal and steam) is effectively provided by the partial oxidation of the coal. The temperature, and therefore composition, of the product gas is dependent upon the amount of oxidant and steam, as well as the design of the reactor that each process utilises. Gasifier product gases invariably contain contaminants

Table 8.3 Typical coal gas compositions (mole percent basis)

	BG-Lurgi gasifier (non-slagging)	BG-Lurgi slagging gasifier	Moving bed O_2-blown (Lurgi)	Fluidised bed (Winkler)	Entrained bed O_2-blown (Texaco)	Entrained bed air-blown	Entrained bed O_2-blown (Shell)
Coal	Pittsburg 8	Pittsburg 8	Illinois no.6	Texas Lignite	Illinois no.6	Illinois no.6	Illinois no.6
Ar	Trace	Trace	Trace	0.7	0.9	Trace	1.1
CH_4	8.5	7.2	3.3	4.6	0.1	1.0	
C_2H_6	0.7	0.1	0.1				
C_2H_4	0.3	0.2	0.2				
H_2	29.1	39.0	21.0	28.3	30.3	9.0	26.7
CO	18.0	55.5	5.8	33.1	39.6	16	63.1
CO_2	31.1	3.9	11.8	15.5	10.8	6	1.5
N_2	2.4	4.0	0.2	0.6	0.7	62	4.1
NH_4			0.4	0.1	0.1		
H_2O			61.8	16.8	16.5	5.0	2.0
H_2S			0.5	0.2	1.0		1.3
Total	100.0	100.0	100.0	100.0	100.0	100.0	100.0

that need to be removed before they can be used for fuel cells. Contaminant removal is therefore essential, and methods of gas cleanup are described in Section 8.4. Table 8.3 shows the compositions of some raw coal gases produced by some of the leading types of coal gasifiers. Fuel cell developers have for many years recognised the benefits of fuelling fuel cells with the gases produced from such coal gasifiers, an example study being that of Brown et al. (1996).

8.2.4 Natural gas

Natural gas is the combustible gas that occurs in porous rocks in the earth's crust. It is found with or close to crude oil reserves, but may also occur alone in separate reservoirs. Most commonly, it forms a gas cap trapped between liquid petroleum and an impervious rock layer (cap rock) in a petroleum reservoir. If the pressure is high enough, the gas will be intimately mixed with or dissolved in the crude petroleum.

Chemically, natural gas comprises a mixture of hydrocarbons of low boiling point. Methane is the component usually present in the greatest concentration, with smaller amounts of ethane, propane, and so on. In addition to hydrocarbons, natural gas contains various quantities of nitrogen, carbon dioxide, and traces of other gases such as helium (often present in commercially recoverable quantities). Sulphur is also present to a greater or lesser extent, mostly in the form of hydrogen sulphide. The overall composition varies according to the source of the natural gas, and there are also seasonal variations. Natural gas is often described as being dry or lean (containing mostly methane), wet (containing considerable concentrations of higher molecular weight hydrocarbons), sour gas (with significant levels of H_2S), sweet gas (low in H_2S), and casing-head gas (derived from

Table 8.4 Typical compositions of natural gases from different geographic regions

Component	North Sea	Qatar	Netherlands	Pakistan	Ekofisk	Indonesia
CH_4	94.86	76.6	81.4	93.48	85.5	84.88
C_2H_6	3.90	12.59	2.9	0.24	8.36	7.54
C_3H_8		2.38	0.4	0.24	2.85	1.60
i-C_4H_{10}	0.15	0.11		0.04	0.86	0.03
n-C_4H_{10}		0.21	0.1	0.06		0.12
C_{5+}		0.02		0.41	0.22	1.82
N_2	0.79	0.24	14.2	4.02	0.43	4.0
S	4 ppm	1.02	1 ppm	N/A	30 ppm	2 ppm

Values are % by volume unless otherwise stated

an oil well by extraction at the surface). Table 8.4 shows some typical compositions of natural gases from different regions around the world.

Some processing of the natural gas may be carried out close to the point of extraction before it enters a transmission system. Examples are the bulk removal of sulphur (sweetening), removal of high molecular weight hydrocarbons, nitrogen, acid gases, liquid water, and liquid hydrocarbons. Nevertheless, there are wide variations in the composition of natural gas fed to transmission systems around the world. UK natural gas, for example, varies in composition according to the field from which it is supplied. Since natural gas composition does vary so much in composition, even with the season, it is common to enrich natural gas in some geographic regions by blending in mixtures of ethane, propane, and butane, for example. The requirements for this are usually dictated by the agreements between the gas supplier and the gas distributor. In addition, to meet the need for peak demand, enrichment with mixtures of propane/air or butane/air is also practised. The latter has important consequences for gas processing in fuel cell systems. For example, the proprietary CRG catalyst, widely used around the world for steam reforming of natural gas, will only tolerate a small percentage of oxygen in the feed gas.

Natural gas has no distinctive odour (except for very sour gases), and for safety reasons, pipeline companies and utilities commonly odorise the gas either as the gas enters the transmission system, or within local distribution zones. Various odorants may be used, the most common being mixtures of thiophenes and mercaptans. Tetrahydrothiophene (THT) is widely used throughout Europe and in the US (as Pennwall's odorant), whereas in the UK a cocktail of compounds is used comprising ethyl mercaptan, tertiary butyl mercaptan, and diethyl sulphide.

8.3 Bio-Fuels

Biomatter or biomass is a catch-all term for natural organic material associated with living organisms, including terrestrial and marine vegetable matter – everything from algae to trees, together with animal tissue and manure. On a global basis it is estimated that over 150 gigatonnes of vegetable biomatter are generated annually. The term 'biomass' is more often associated with production, expressed in tonnes per hectare, often also expressed in tonnes per hectare annual yield (t ha^{-1}). This yield ranges from about 13 t ha^{-1} for water hyacinth to 120 t ha^{-1} for napier grass. In view of its high energy content, biomass represents an important source of renewable fuel. This can be obtained by one of several routes:

• Direct combustion
• Conversion to biogas via pyrolysis, hydrogasification, or anaerobic digestion
• Conversion to ethanol via fermentation
• Conversion to syngas thermochemically, and then to methanol or ammonia
• Conversion to liquid hydrocarbons by hydrogenation, or via the syngas/Fischer–Tropsch route.

Another source of bio-fuel is from municipal waste. Again, apart from direct pyrolysis or incineration, gaseous fuels arising from landfill sites and other forms of refuse digestion can form a useful source of renewable energy, well suited to fuel cell systems. In the United Kingdom, approximately 80% of household waste is tipped, 10% is incinerated, and 5% is composted. The combustible component of such waste may be extracted chemically by conversion to gases, liquid distillates, and char by anaerobic pyrolysis or biochemically by anaerobic digestion using methane-forming bacteria. Anaerobic digestion, as currently practised, requires a wet waste of relatively high nitrogen content, and the nitrogen/carbon ratio of about 0.03 is increased to 0.07 by the addition of animal manure, sewage sludge or other nitrogen-rich waste. Anaerobic digesters can be made at a relatively small scale (a few kilowatts), compared with pyrolysis gasifiers that normally only become attractive at the megawatt scale.

Biogases produced from biomass, landfill, or anaerobic digestion contain mixtures of methane, carbon dioxide, and nitrogen, together with various other organic materials. The compositions vary widely and, for example in the case of landfill, depend on the age of the site. A new site tends to produce gas with a high heating value, and this tends to decrease over a period of time. Some compositions of biogases are given in Table 8.5.

There is a particular attraction for using biogases in fuel cell systems. Most biogases have low heating values, caused by the relatively high levels of carbon oxides and nitrogen, making them unattractive for use in gas engines. However, this drawback is not an issue for fuel cells, particularly the molten carbonate (electrolyte) fuel cell (MCFC) and solid oxide fuel cell (SOFC), which are able to handle very high concentrations of carbon oxides. This is also the case, but to a lesser extent for the phosphoric acid (electrolyte) fuel cell (PAFC). Developers have been assessing the use of biogases in such fuel cells for some time (Warren et al., 1986 and Langnickel, 1999).

Bioliquids such as methanol and ethanol are also attractive bio-fuels for some fuel cell systems. Methanol is proposed as a fuel for FCVs. Methanol can be synthesised from syngas that may be derived from biomass or natural gas. Ethanol can be produced directly by fermentation of biomass. Alcohols are attractive also because of the ease with which they can be reformed into hydrogen-rich gas. This is necessary if the reforming is to be carried out onboard a fuel cell vehicle.

Table 8.5 Example compositions of biogases

	Biogas[a]	Biogas[b]	Biogas[c]	Biogas[d]	Landfill gas[e]
Source	Agricultural sludge		Agricultural sludge	Brewery effluent	
Methane (vol %)	55–65	55–70	50–70	65–75	57
Ethane (vol %)		0			
Propane (vol %)		0			
Carbon dioxide (vol %)	33–43	30–45	30–40	25–35	37
Nitrogen (vol %)	2–1	0–2	small		6
Hydrogen sulphide (ppm)	<2000	~500	small	<5000	
Ammonia (ppm)	<1000	~100		<1	
Hydrogen (vol %)			small		
Water, relative moisture	80				
Higher heating value (MJ nm^{-3})		23.3	>20		
Density (kg nm^{-3})		1.16			

[a]Paper BP-12 20th World Gas conference 1997.
[b]Combined Utilization of Biogas and Natural Gas, J. Jemsen, S. Tafdrup, and Johannes Chrisensen, Paper BO-06, 20th World Gas Conference, 1997.
[c]Renewable Energy World, March 1999, page 75.
[d]Caddet renewable energy newsletter, July 1999 pages 14–16 (Biogas used in Toshiba 200-kW phosphoric acid fuel cell).
[e]Caddet renewable energy Technical Brochure No. 32 (1996).

8.4 The Basics of Fuel Processing

8.4.1 Fuel cell requirements

Fuel processing may be defined as the conversion of the raw primary fuel, supplied to a fuel cell system, into the fuel gas required by the stack. Each type of fuel cell stack has some particular fuel requirements, summarised in Table 8.6. Essentially, the lower the operating temperature of the stack, the more stringent are the requirements, and the greater the demand placed on fuel processing. For example, fuel fed to a PAFC needs to be hydrogen-rich and contain less than about 0.5% carbon monoxide. The fuel fed to a PEM fuel cell needs to be essentially carbon monoxide free, while both the MCFC and the SOFC are capable of utilising carbon monoxide through the water-gas shift reaction that occurs within the cell. Additionally, SOFCs and internal reforming MCFCs can utilise methane within the fuel cell, whereas PAFCs and PEMs cannot. It is not widely known that PEMFCs can utilise some hydrocarbons, such as propane, directly, although the performance is poor.

Considerable research has been carried out in the field of fuel processing and reviews of some of the key technologies are readily available (Dicks, 1996). The following sections are intended to provide a basic explanation of the various technologies. Some detailed design of individual reactors and systems are proprietary, of course, but there is a wealth of information also available from various organisations involved in the development

Table 8.6 The fuel requirements for the principal types of fuel cells

Gas species	PEM fuel cell	AFC	PAFC	MCFC	SOFC
H_2	Fuel	Fuel	Fuel	Fuel	Fuel
CO	Poison (>10 ppm)	Poison	Poison (>0.5%)	Fuel[a]	Fuel[a]
CH_4	Diluent	Diluent	Diluent	Diluent[b]	Diluent[b]
CO_2 and H_2O	Diluent	Poison[c]	Diluent	Diluent	Diluent
S (as H_2S and COS)	Few studies, to date	Unknown	Poison (>50 ppm)	Poison (>0.5 ppm)	Poison (>1.0 ppm)

[a]In reality CO reacts with H_2O producing H_2 and CO_2 via the shift reaction 8.3 and CH_4 with H_2O reforms to H_2 and CO faster than reacting as a fuel at the electrode.
[b]A fuel in the internal reforming MCFC and SOFC.
[c]The fact that CO_2 is a poison for the alkaline fuel cell more or less rules out its use with reformed fuels.

of fuel cell systems. Much of what is discussed here may be regarded as the practical application of chemical engineering to fuel cell systems.

8.4.2 Desulphurisation

Natural gas and petroleum liquids contain organic sulphur compounds that normally have to be removed before any further fuel processing can be carried out. Even if suphur levels in fuels are below 0.2 ppm, some deactivation of steam reforming catalysts can occur. Shift catalysts are even more intolerant to sulphur (Farrauto, 2001), and to ensure adequate lifetimes of fuel processors the desulphurisation step is very important. Even if the fuel processor catalysts were tolerant to some sulphur, it has been shown that levels of only 1 ppb are enough to permanently poison a PEM anode catalyst.

The sulphur compounds found in fuels vary. In the case of natural gas, the only sulphur compounds may be the odorants that have been added by the utility company for safety reasons. With petroleum fractions, the compounds may be highly aromatic in nature. Gasoline currently contains some 300 to 400 ppm of sulphur as organic compounds. In the drive to reduce emissions from vehicles, regulations are being introduced to limit the sulphur in gasoline and diesel fuels. For example in the United States, the proposed limit is 30 to 80 ppm for the year 2004. As levels are reduced, the methods used to remove or desulphurise such fuels may need to be adapted.

For these reasons the design of a desulphurisation system must be undertaken with care. If the fuel cell plant has a source of hydrogen-rich gas (usually from the reformer exit), it is common practice to recycle a small amount of this back to a hydrodesulphurisation (HDS) reactor. In this reactor, any organic sulphur-containing compounds are converted, over a supported nickel-molybdenum oxide or cobalt-molybdenum oxide catalyst, into hydrogen sulphide via hydrogenolysis reactions of the type

$$(C_2H_5)_2S + 2H_2 \rightarrow 2C_2H_6 + H_2S \qquad [8.1]$$

The rate of hydrogenolysis increases with increasing temperature and under operating temperatures of 300 to 400°C, and in the presence of excess hydrogen, the reaction is normally complete. It is worth pointing out that the lighter sulphur compounds easily undergo hydrogenolysis, whilst thiophene (C_4H_4S) and tetrahydrothiophene (THT) ($C_4H_8O_2S$) react with more difficulty, with a much slower reaction rate. The H_2S that is formed by such reactions is subsequently absorbed onto a bed of zinc oxide, forming zinc sulphide:

$$H_2S + ZnO \rightarrow ZnS + H_2O \qquad\qquad [8.2]$$

HDS is practised widely in industry and commercial catalysts and absorbents are available. The operating conditions and feed gas composition determine the choice between nickel or cobalt catalysts. For example, nickel is a more powerful hydrogenation catalyst than cobalt and can cause more hydrocracking. Such undesirable hydrocracking reactions can occur over HDS catalysts if they become reduced to the metals by the hydrogen-rich gas, and the concentrations of organic sulphur compounds are very low. Alkene hydrogenation can also cause problems if such compounds are present in significant quantities. Such problems may occur with some liquid petroleum feedstocks but are unlikely to happen with natural gases. The optimum temperature for most HDS catalysts is between 350°C and 400°C, and the catalyst and zinc oxide may be placed in the same vessel. A variation on the HDS process, known as the PURASPEC™ process, is marketed by Synetix, part of the ICI group.

HDS as a means of removing sulphur to very low levels is ideally suited to PEM or PAFC systems. Unfortunately, HDS cannot easily be applied to internal reforming MCFC or SOFC systems, since there is no hydrogen-rich stream to feed to the HDS reactor. At least one developer of high temperature fuel cells is overcoming this problem by building a reformer reactor into their system for the sole purpose of generating hydrogen for HDS. Other developers have opted for removing sulphur from the feed gases using an absorbent. Activated carbon is especially suitable for small systems and can be impregnated with metallic promoters to enhance the absorption of specific materials such as hydrogen sulphide. Molecular sieves may also be used, and Okada et al. (1992) have developed a mixed oxide absorbent system for fuel cells. The absorption capacity of materials such as active carbon is, however, quite low, and the beds of absorbent need replacing at regular intervals. This may be a serious economic disadvantage for large systems.

Some organisations claim to have developed sulphur-tolerant reforming or partial oxidation catalysts. Argonne National Laboratory, for example, has developed a sulphur-tolerant catalyst for its autothermal diesel reformer, but this is only intended to cope with the relatively low levels of sulphur found in commercial low-sulphur diesel fuel. If partial oxidation is used to process the fuel, rather than steam reforming, it is likely that the product gas will still contain a few ppm of sulphur and removing this before applying the gas to a PEM fuel cell anode is an essential but non-trivial additional process step. Zinc oxide, although regenerable in theory, degrades over time. There is a need, therefore, to develop desulphurisation systems that can operate at reformer temperatures and with high concentrations of steam in the fuel stream. To address this issue, Mcdermott in the United States has been developing a regenerable zinc oxide bed. In addition, if desulphurisation is to be carried out on a vehicle (in the case in which gasoline is to be reformed on board, for example) it is likely that conventional zinc oxide pellets will not be robust enough. Argonne National Labs, for example, are therefore developing a monolith that is coated with zinc oxide absorbant. This would also have the advantage of a lower weight compared to the conventional pelleted zinc oxide absorbants.

8.4.3 Steam reforming

Steam reforming is a mature technology, practised industrially on a large scale for hydrogen production and several detailed reviews of the technology have been published (Van Hook (1980); Rostrup–Nielsen (1984) and (1993), and useful data for system design is provided by Twigg (1989). The basic reforming reactions for methane and a generic hydrocarbon C_nH_m are

$$CH_4 + H_2O \rightarrow CO + 3H_2 \quad [\Delta H = 206\,\text{kJ}\,\text{mol}^{-1}] \tag{8.3}$$

$$C_nH_m + nH_2O \rightarrow nCO + (\tfrac{m}{2} + n)H_2 \tag{8.4}$$

$$CO + H_2O \rightarrow CO_2 + H_2 \quad [\Delta H = -41\,\text{kJ}\,\text{mol}^{-1}] \tag{8.5}$$

The reforming reactions (8.3 and 8.4), more correctly termed *oxygenolysis reactions*, and the associated water-gas shift reaction 8.5 are carried out normally over a supported nickel catalyst at elevated temperatures, typically above 500°C. Reactions 8.3 and 8.5 are reversible and normally reach equilibrium over an active catalyst, as at such high temperatures the rates of reaction are very fast. Over a catalyst that is active for reaction 8.3, reaction 8.5 nearly always occurs as well. The combination of the two reactions taking place means that the overall product gas is a mixture of carbon monoxide, carbon dioxide, and hydrogen, together with unconverted methane and steam. The actual composition of the product from the reformer is then governed by the temperature of the reactor (actually the outlet temperature), the operating pressure, the composition of the feed gas, and the proportion of steam fed to the reactor. Graphs and computer models using thermodynamic data are available to determine the composition of the equilibrium product gas for different operating conditions. Figure 8.2 is an example, showing the composition of the output at 1 bar.

Reaction 8.3, there are three molecules of hydrogen and one molecule of carbon monoxide produced for every molecule of methane reacted. Le Chatelier's principle therefore tells us that the equilibrium will be moved to the right (i.e. in favour of hydrogen) if the pressure in the reactor is kept low. Increasing the pressure will favour formation of methane, since moving to the left of the equilibrium reduces the number of molecules. The effect of pressure on the equilibrium position of the shift reaction (reaction 8.5) is very small.

Another feature of reactions 8.3 and 8.4 is that they are usually very *endothermic*, which means that heat needs to be supplied to the reaction to drive it forward to produce hydrogen and carbon monoxide. Higher temperatures (up to 700°C) therefore favour hydrogen formation, as shown in Figure 8.2.

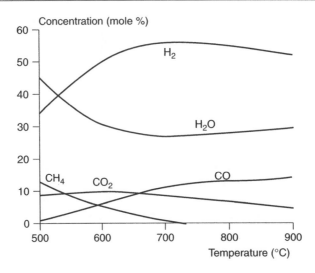

Figure 8.2 Equilibrium concentrations of steam reformation reactant gases as a function of temperature.

It is important to note at this stage that although the shift reaction 8.5 does occur at the same time as steam reforming, at the high temperatures needed for hydrogen generation, the equilibrium point for the reaction is well to the left of the equation. The result is that by no means will all the carbon monoxide be converted to carbon dioxide. For fuel cell systems that require low levels of CO, further processing will be required. This is explained in Section 8.4.9.

It is important to realise also that steam reforming is not always endothermic. For example, in the case of steam reforming a petroleum hydrocarbon such as naphtha, with the empirical formula $CH_{2.2}$, the reaction is most endothermic at the limit when the whole of the carbon is reformed to give oxides of carbon and hydrogen. This is the case when the reaction is carried out at relatively high temperatures. It is less endothermic and eventually exothermic (liberates heat) as the temperature is lowered. This is because as the temperature is lowered, the reverse of reaction 8.1 becomes favoured, that is, a competing reaction, namely, the formation of methane, starts to dominate. This effect is illustrated in Table 8.7.

The implications of this for fuel cell systems are as follows: Natural gas reforming will invariably be endothermic, and heat will need to be supplied to the reformer at sufficiently high temperature to ensure a reasonable degree of conversion. Naphtha, and similar fractions (gasoline, diesel, and logistics fuels) will also react under endothermic conditions if hydrogen is the preferred product and the operating temperature is kept high. However, if the reforming of naphtha is carried out at modest temperatures (up to 500–600°C), then the reactions will tend to yield greater concentrations of methane. More importantly, the need for external heating will diminish as the reaction becomes exothermic. This may have important consequences for the reforming of gasoline as envisaged in some automotive fuel cell systems.

Table 8.7 Examples of typical heat(s) of reaction in naphtha reforming at different temperatures, pressures, and steam/carbon ratios. (Twigg, 1989)

Pressure kPa	Temp °C	Steam ratio[a]	Reaction	ΔH (25°C) (kJ mol^{-1} CH$_{2.2}$)
2070	800	3.0	$CH_{2.2} + 3H_2O \rightarrow$ $0.2CH_4 + 0.4CO_2 + 0.4CO + 1.94H_2 + 1.81H_2O$	+102.5
2760	750	3.0	$CH_{2.2} + 3H_2O \rightarrow$ $0.35CH_4 + 0.4CO_2 + 0.25CO + 1.5H_2 + 1.95H_2O$	+75.0
31050	450	2.0	$CH_{2.2} + 2H_2O \rightarrow$ $0.75CH_4 + 0.25CO_2 + 0.14H_2 + 1.5H_2O$	−48.0

[a]Steam/carbon ratio is the ratio between the number of moles of steam and the number of moles of carbon in the steam + fuel fed to the reformer reactor.

Another type of reforming is known as dry reforming, or CO_2 reforming, which can be carried out if there is no ready source of steam:

$$CH_4 + CO_2 \rightarrow 2CO + 2H_2 \quad [\Delta H = 247\,\text{kJ mol}^{-1}] \quad\quad [8.6]$$

This reaction may occur in internal reforming fuel cells, for example the MCFC, when anode exhaust gas containing carbon dioxide and water is recycled to the fuel cell inlet. Mixed reforming is a term that is sometimes used to describe a hybrid approach in which both steam and CO_2 are used to reform the fuel. Both dry and mixed reforming have energy and environmental advantages compared with traditional steam reforming. The reactions are catalysed by nickel but deactivation due to carbon formation and nickel sintering can be particularly severe, and better catalysts are required.

Hydrocarbons such as methane are not the only fuels suitable for steam reforming. Alcohols will also react in an oxygenolysis or steam reforming reaction, for example, methanol

$$CH_3OH + H_2O \rightarrow 3H_2 + CO_2 \quad [\Delta H = 49.7\,\text{kJ mol}^{-1}] \quad\quad [8.7]$$

The mildly endothermic steam reforming of methanol is one of the reasons why methanol is finding favour with vehicle manufacturers as a possible fuel for FCVs. Little heat needs to be supplied to sustain the reaction, which will readily occur at modest temperatures (e.g. 250°C) over catalysts of mild activity such as copper supported on zinc oxide. Notice also that carbon monoxide does not feature as a principal product of methanol reforming. This makes the methanol reformate particularly suited to PEM fuel cells, where carbon monoxide, even at the ppm level, can cause substantial losses in performance because of poisoning of the platinum anode electrochemical catalyst. However, it is important to note that although carbon monoxide does not feature in reaction 8.7, this does not mean that it is not produced at all. The water-gas shift reaction of reaction 8.5 is reversible, and carbon monoxide in small quantities is produced. The result is that the carbon monoxide removal methods described in Section 8.4.9 are still needed with PEM fuel cells, though the CO levels are low enough for PAFC.

8.4.4 Carbon formation and pre-reforming

One of the most critical issues in fuel cell systems is the risk of carbon formation from the fuel gas. This can occur in several areas of the system where hot fuel gas is present. Natural gas, for example, will decompose when heated in the absence of air or steam at temperatures above about 650°C via pyrolysis reactions of the type

$$CH_4 \rightarrow C + 2H_2 \quad [\Delta H = 75\,kJ\,mol^{-1}] \qquad [8.8]$$

Similar reactions can be written out for other hydrocarbons. Higher hydrocarbons tend to decompose more easily than methane and therefore the risk of carbon formation is higher with vaporised liquid petroleum fuels than with natural gas. Another source of carbon formation is from the disproportionation of carbon monoxide via the so-called Boudouard reaction:

$$2CO \rightarrow C + CO_2 \qquad [8.9]$$

This reaction is catalysed by metals such as nickel, and therefore there is a high risk of it occurring on steam reforming catalysts that contain nickel and on nickel-containing stainless steel used for fabricating the reactors. Fortunately, there is a simple expedient to reduce the risk of carbon formation from reactions 8.8 and 8.9, and that is to add steam to the fuel stream. The principal effect of this is to promote the shift reaction 8.5, which has the effect of reducing the partial pressure of carbon monoxide in the fuel gas stream. Steam also leads to the carbon gasification reaction, which is also very fast:

$$C + H_2O \rightarrow CO + H_2 \qquad [8.10]$$

The minimum amount of steam that needs to be added to a hydrocarbon fuel gas to avoid carbon deposition may be calculated. The principle here is that it is assumed that a given fuel gas/steam mixture reacts via reactions 8.3, 8.4, and 8.5 to produce a gas that is at equilibrium with respect to reactions 8.3 and 8.5 at the particular temperature of operation. The partial pressures of carbon monoxide and carbon dioxide in this gas are then used to calculate an equilibrium constant for the Boudouard reaction 8.9. This calculated equilibrium constant is then compared with what would be expected from the thermodynamic calculation at the temperature considered. If the calculated constant is greater than the theoretical one, then carbon deposition is predicted on thermodynamic grounds. If the calculated constant is lower than theory predicts, then the gas is said to be in a safe region and carbon deposition will not occur. In practice, a steam/carbon ratio of 2.0 to 3.0 is normally employed in steam reforming systems so that carbon deposition may be avoided with a margin of safety.

A particular type of carbon formation occurs on metals, known as carburisation, leading to spalling of metal in a phenomenon known as 'metal dusting'. Again, it is important to reduce the risk of this in fuel cell systems, and some developers have used copper-coated stainless steel in their fuel gas preheaters to keep the risk to a minimum.

Carbon formation on steam reforming catalysts has been the subject of intense study over the years and is well understood. Carbon formed via the pyrolysis (8.8) and the Boudouard (8.9) reactions adopts different forms that can be identified under an electron microscope. The most damaging form of carbon is the filament that 'grows' attached to nickel crystallites within the catalyst. Such carbon formation can be very fast. If the steam flow to the reformer reactor is shut down for some reason, the consequences can be disastrous, as carbon formation occurs within seconds, leading to permanent breakdown with fouling of the catalyst and plugging of the reactor. Control of the fuel processing is therefore an important aspect of fuel cell systems, especially if the steam is obtained by recycling products from the fuel cell stack. Commercial steam reforming catalysts contain elements such as potassium and molybdenum to reduce the risk of carbon formation, which would otherwise rapidly deactivate the catalyst.

Another procedure to reduce the risk of carbon formation in a fuel cell system is to carry out some pre-reforming of the fuel gas before it is fed to the reformer reactor. Pre-reforming is a term commonly used in industry to describe the conversion of high molecular weight hydrocarbons via the steam reforming reaction at relatively low temperatures (typically 250–500°C). This process step (also known as 'sweetening' of the gas) is carried out before the main reforming reactions in a reactor, which operates adiabatically, that is, no heat is supplied to it or removed from it. The advantage of carrying out pre-reforming is that high molecular weight hydrocarbons, which are more reactive than methane, are converted into hydrogen preferentially. The gas from the exit of a pre-reformer therefore comprises mainly methane, with steam, together with small amounts of hydrogen and carbon oxides, depending on the temperature of the pre-reformer reactor.

The use of a pre-reformer in a fuel cell system is illustrated in Figure 8.3, which was designed for an SOFC demonstration. In this case, the pre-reformer was built not only to remove the higher hydrocarbons from the natural gas but also to convert some 15% of the methane into hydrogen. This amount of external pre-reforming ensured that the

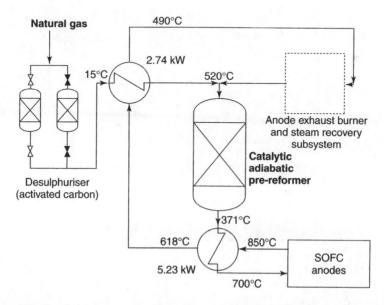

Figure 8.3 Pre-reformer system devised for a Siemens 50-kW SOFC demonstration.

anodes of the SOFC were kept in a reduced condition and that not too much thermal stress was placed on the SOFC stack by only allowing 85% reforming of the methane inside the stack.

8.4.5 Internal reforming

Fuel cell developers have for many years known that the heat required to sustain the endothermic reforming of low molecular weight hydrocarbons (e.g. natural gas) can be provided by the electrochemical reaction in the stack. This has led to various elegant internal reforming concepts that have been applied to the molten carbonate or solid oxide fuel cells, on account of their high operating temperatures.

In contrast to the steam reforming reactions (8.3 and 8.4), the fuel cell reactions are exothermic, mainly because of heat production in the cell caused by internal resistances. Under practical conditions, with a cell voltage of 0.78 V, this heat evolved amounts to 470 kJ per mole methane. The overall heat production is about twice the heat consumed by the steam reforming reaction in an internally reforming fuel cell. Hence, the cooling required by the cell, which is usually achieved by flowing excess gas through the cathode in the case of external reforming systems, will be much smaller for internal reforming systems. This has a major benefit on the electrical efficiency of the overall system.

Developers of internal reforming fuel cells have generally adopted one of two approaches, and these are usually referred to as direct (DIR) and indirect (IIR) internal reforming. They are illustrated schematically in Figure 8.4. In some cases, a combination of both approaches has been taken. A thermodynamic analysis and comparison of the two approaches to internal reforming in the MCFC has been completed by Freni and Maggio

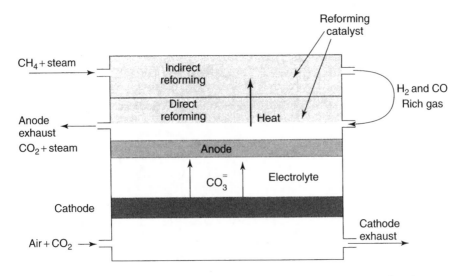

Figure 8.4 Schematic representation of direct and indirect internal reforming.

(1997). Application of internal reforming offers several further advantages compared with external reforming:

- System cost is reduced because a separate external reformer is not needed.
- With DIR, less steam is required (the anode reaction in the SOFC and MCFC produces steam).
- There is a more even distribution of hydrogen in a DIR cell that may result in a more even temperature distribution.
- The methane conversion is high, especially in DIR systems where the cell consumes hydrogen as it is produced.
- The efficiency of the system is higher. This is mainly because internal reforming provides an elegant method of cooling the stack, reducing the need for excess cathode air. This in turn lowers the requirement for air compression and recirculation.

Indirect internal reforming (IIR)

Also known as integrated reforming, this approach involves conversion of methane by reformers positioned in close thermal contact with the stack. An example of this type of arrangement alternates plate reformers with small cell packages. The reformate from each plate is fed to neighbouring cells. IIR benefits from close thermal contact between stack and reformer but suffers from the fact that heat is transferred well only from cells adjacent to the reformers and steam for the reforming must be raised separately. A variation of this type of arrangement places the reforming catalyst in the gas distribution path of each cell. With IIR, the reforming reaction and electrochemical reactions are separated.

Direct internal reforming (DIR)

In direct internal reforming, the reforming reactions are carried out within the anode compartment of the stack. This can be done by placing the reforming catalyst within the fuel cell channels in the case of the MCFC, but for the SOFC the high temperature of operation and anode nickel content of most SOFCs mean that the reactions can be performed directly on the anode. The advantage of DIR is that not only does it offer good heat transfer but there is also chemical integration – product steam from the anode electrochemical reaction can be used for the reforming without the need for recycling spent fuel. In principle, the endothermic reaction can be used to help control the temperature of the stack but this effect is not enough to completely offset the heat produced by the electrochemical reaction and management of the temperature gradients is an issue. Examples of the use of DIR are to be found in the Siemens Westinghouse tubular SOFC and in various DIR-MCFC concepts. These were described in Chapter 7.

Finally, it should be pointed out that internal reforming may be applied to several hydrocarbon fuels such as natural gas and vaporised liquids such as naphtha and kerosenes. Logistic fuels have also been demonstrated in internal reforming MCFC stacks, and coal gases are particularly attractive for internal reforming MCFC and SOFC stacks, since not only are carbon monoxide and hydrogen consumed directly as fuels but the residual methane (e.g. from the BG-Lurgi slagging gasifier – see Table 8.3) is also internally reformed.

8.4.6 Direct hydrocarbon oxidation

A particular variation on internal reforming is that of direct hydrocarbon oxidation within the fuel cell, also known as dry reforming. The Gibbs free energy change for the reaction of direct oxidation of methane to carbon dioxide and water is $-796\,kJ\,mol^{-1}$, which is very close to the enthalpy change ($\Delta H = -802.5\,kJ\,mol^{-1}$). In other words, if methane could be oxidised directly, most of the heat of reaction would be converted directly into electricity, with a maximum efficiency of

$$\frac{\Delta G}{\Delta H} = \frac{-796.5}{-802.5} \times 100 = 99.2\%$$

A great problem with direct hydrocarbon oxidation is that of carbon formation. As discussed in Section 8.4.4, this has a propensity to form on nickel-containing materials at temperatures even as low as 600°C. Nickel can be replaced by copper as Gorte (2002) has found that Cu-based cermets are stable in hydrocarbon environments as long as the temperature is limited (copper has a relatively low melting point of 1084°C, compared with 1453.0°C for nickel). Gorte has been able to show unequivocally that direct oxidation can occur on Cu-cermets, demonstrating for the first time the feasibility of this approach.

In addition, direct oxidation has also been investigated on novel ceramic anodes for SOFCs made of mixed ionic- and electronic-conducting materials. A good example is the work reported by Perry Murray et al. (1999) on ceria-doped zirconia anodes in low-temperature SOFCs. Many other materials are also being investigated and a recent review has been published by Irvine and Sauvet (2001).

Despite the attractiveness of direct hydrocarbon oxidation, development is at an early stage. If steam is present with the fuel (and it will be generated, for example, on an SOFC anode), some steam reforming may occur. It would be easy to think that this would tend to offset the great advantage of direct methane oxidation, that is, its high theoretical efficiency. In fact, there is a substantial benefit in not having to supply steam to the fuel cell. Nevertheless, perhaps it is in the area of proton-conducting fuel cells that direct oxidation really does have a long-term benefit. During developments of acid fuel cells in the 1960s, several workers reported on the direct oxidation of hydrocarbons (Baker, 1965; Smith, 1966; and Liebhafsky and Cairns, 1968). Propane, for example, will decompose on platinum catalysts at moderate temperatures (below 200°C) to form protons, electrons, and CO_2 (via reaction with water from the aqueous acid electrolyte). As in the PEM fuel cell, the protons migrate to the cathode where they are oxidised with air and electrons to form water. It may be prudent for researchers now to revisit some of this earlier work to see if it holds the key for new ways of fuelling fuel cells.

8.4.7 Partial oxidation and autothermal reforming

As an alternative to steam reforming, methane and other hydrocarbons may be converted to hydrogen for fuel cells via partial oxidation:

$$CH_4 + \tfrac{1}{2}O_2 \rightarrow CO + 2H_2 \qquad [\Delta H = -247\,kJ\,mol^{-1}] \qquad [8.11]$$

Partial oxidation can be carried out at high temperatures (typically 1200–1500°C) without a catalyst. In this case it has the advantage over catalytic processes in that materials such as sulphur compounds do not need to be removed, although the sulphur does have to be removed at a later stage (as H_2S). High-temperature partial oxidation can also handle much heavier petroleum fractions than catalytic processes and is therefore attractive for processing diesels, logistic fuels, and residual fractions. Such high-temperature partial oxidation has been carried out on a large scale by several companies but it does not scale down well, and control of the reaction is problematic. If the temperature is reduced, and a catalyst employed, then the process becomes known as catalytic partial oxidation (CPO). Catalysts for CPO tend to be supported platinum-metal or nickel based.

It should be noted that reaction 8.11 produces less hydrogen per molecule of methane than reaction 8.3. This means that partial oxidation (either non-catalytic or catalysed) is usually less efficient than steam reforming for fuel cell applications. Reaction 8.11 is effectively the summation of the steam reforming and oxidation reactions; about half the fuel that is converted into hydrogen is oxidised to provide heat for the endothermic reforming reaction. Unlike the steam reforming reaction, no heat from the fuel cell can be utilised in the reaction, and the net effect is a reduced overall system efficiency. Another disadvantage of partial oxidation is when air is used as the oxidant. This results in a lowering of the partial pressure of hydrogen at the fuel cell because of the presence of the nitrogen. This in turn results in a lowering of the Nernst potential of the cell, again resulting in a lowering of system efficiency. To offset these negative aspects, a key advantage of partial oxidation is that it does not require steam. It may therefore be considered for applications in which system simplicity is regarded as more important than high electrical conversion efficiency, for example, small-scale cogeneration, also known as micro-cogeneration.

Autothermal reforming is another commonly used term in fuel processing. This usually describes a process in which both steam and oxidant (oxygen, or more normally air) are fed with the fuel to a catalytic reactor. It can therefore be considered as a combination of POX and the steam reforming processes already described. The basic idea of autothermal reforming is that both the *endothermic* steam reforming reaction 8.3 and the *exothermic* POX reaction of 8.11 occur together, so that no heat needs to be supplied or removed from the system. However, there is some confusion in the literature between the terms *partial oxidation* and *autothermal reforming*. Joensen and Rostrup-Nielsen have recently published a review that explains the issues in some detail (Joensen and Rostrup-Nielsen, 2002). If steam is added to a fuel/oxidant mixture and passed through a bed of catalyst, the catalyst, and the operating temperature and pressure will determine to what extent the steam reforming reaction occurs. Several studies (Dvorak, 1998) have tried to determine the relative rates of the reforming and partial oxidation reactions over different catalysts and usually it is assumed that the oxidation reaction is brought to equilibrium faster than the steam reforming reaction. This has been termed the *indirect mechanism* of CPO. Dvorak et al. showed that supported nickel catalysts tend to promote this indirect CPO mechanism, whereas supported ruthenium catalysts promote a 'direct mechanism' whereby oxidation, steam reforming, and shift reactions occur in parallel. In some CPO processes, both steam and oxidant are fed with the fuel, an example being the Shell partial oxidation process (Kramer et al., 2001). This uses a proprietary reactor design containing a Pt-group catalyst over which CPO occurs at the top of the bed, the reaction being limited by mass transfer. Further down the bed, the steam reforming and shift reactions bring the gas to equilibrium. This process is now being developed by HydrogenSource in the United States. In other examples of CPO where the steam oxidation and reforming reactions operate in parallel, the rate of the reactions are not limited by mass transfer and are brought to equilibrium without gain or loss of heat. This is true autothermal reforming.

The advantages of autothermal reforming and CPO are that less steam is needed compared with conventional reforming and that all the heat for the reforming reaction is provided by partial combustion of the fuel. This means that no complex heat management engineering is required, resulting in a simple system design. As we shall see in Section 8.6, this is particularly attractive for mobile applications.

8.4.8 Hydrogen generation by pyrolysis or thermal cracking of hydrocarbons

An alternative to all the above methods of generating hydrogen is to simply heat hydrocarbon fuels in the absence of air (pyrolysis). The hydrocarbon 'cracks' or decomposes into hydrogen and solid carbon. The process is ideally suited to simple hydrocarbon fuels; otherwise various by-products may be formed. The advantage of thermal cracking is that the hydrogen that is produced is very pure. The challenge is that the carbon that is also produced has to be removed from the reactor. This can be done by switching off the supply of fuel and admitting air to the reactor to burn off the carbon as carbon dioxide. The principle of switching the flow of fuel and oxidant is simple, but there are real difficulties, not least of which are the safety implications of admitting fuel and air into a reactor at high temperatures. Control of the pyrolysis is critical, otherwise too much carbon will built up, and on a catalyst this can cause irreversible damage. If no catalyst is employed, again too much carbon may be formed. This may plug the reactor and mean that no flow of oxidising gas can be established to burn off the deposited material. Despite these substantial problems, pyrolysis is being considered seriously as an option for some fuel cell systems. Cracking of propane has been proposed recently to provide hydrogen for small PEM fuel cell systems (Ledjeff–Hey et al., 1999 and Kalk et al., 2000).

8.4.9 Further fuel processing – carbon monoxide removal

A steam reformer reactor running on natural gas and operating at atmospheric pressure with an outlet temperature of 800°C produces a gas comprising some 75% hydrogen, 15% carbon monoxide, and 10% carbon dioxide on a dry basis. For the PEM fuel cell and the PAFC, reference to Table 8.6 shows that the carbon monoxide content must be reduced to much lower levels. Similarly, even the product from a methanol reformer operating at about 200°C will have at least 0.1% carbon monoxide content, depending on the pressure and water content. This will be satisfactory for a PAFC, but is far too high for a PEMFC. The problem of reducing the carbon monoxide content of reformed gas streams is thus very important.

We have seen that the water-gas shift reaction

$$CO + H_2O \leftrightarrow CO_2 + H_2$$

takes place at the same time as the basic steam reforming reaction. However, the thermodynamics of the reaction are such that higher temperatures favour the production of carbon monoxide and shifts the equilibrium to the left ($K = -4.35$). The first approach is thus to *cool* the product gas from the steam reformer and to pass it through a reactor containing catalyst, which promotes the shift reaction. This has the effect of converting carbon monoxide into carbon dioxide. Depending on the reformate composition, more than one shift reactor may be needed to reduce the carbon monoxide level to an acceptable level for the PAFC fuel cell. The iron–chromium catalyst is found to be effective for promoting the shift reactor at relatively high temperatures (400–500°C), and this may be followed by further cooling of the gas before passing to a second, low-temperature reactor (200–250°C) containing a copper catalyst. At this temperature, the proportion of carbon monoxide present will typically be about 0.25 to 0.5%, and so these two stages of shift conversion are sufficient to decrease the carbon monoxide content to meet the needs of the PAFC. However, this level is equivalent to 2500 to 5000 ppm, which exceeds the limit for PEM fuel cells by a factor of about 100. It is similar to the CO content in the product from a methanol reformer.

Until recently, the shift reactors in fuel cell systems at the kilowatt scale have been utilising industrial iron and copper catalysts. These are easily poisoned by sulphur. They also present something of a safety hazard as the catalyst in the reduced state becomes pyrophoric when exposed to air. In addition, the size of the vessels required are substantial in relation to the stacks, the reformer reactor, and other balance of plant components. To improve the situation, some developers are working on novel catalysts that are able to operate at high space velocities and at low temperatures. Several new materials are being studied including base metal and precious metal catalysts (Hilaire, 2001). Base metals (which are lower cost) tend to have lower sulphur tolerance, less stability in air (pyrophoric), some are non-pyrophoric, and have higher activity but lower long-term stability. Precious metal catalysts, for example, Pt on Ceria developed by Nextech in the United States (Swartz, 2002) or gold on ceria (Andreeva, 2002), are non-pyrophoric, more tolerant to sulphur, and may become viable options provided costs can be kept reasonably low. Some transition metal carbides also provide sulphur tolerance and relatively high activity. There is clearly room for improvement in shift catalysts for fuel cell systems, but in the end there will be a trade-off between cost, performance, and vessel size.

For PEM fuel cells, further carbon monoxide removal is essential after the shift reactors. This is usually done in one of three ways:

- In the *Selective oxidation reactor* a small amount of air (typically around 2%) is added to the fuel stream, which then passes over a precious metal catalyst. This catalyst preferentially absorbs the carbon monoxide, rather than the hydrogen, where it reacts with the oxygen in the air. In addition to the obvious problem of cost, these units need to be very carefully controlled. There is the presence of hydrogen, carbon monoxide, and oxygen, at an elevated temperature, with a noble metal catalyst. Measures must be taken to ensure that an explosive mixture is not produced. This is a special problem in cases where the flow rate of the gas is highly variable, such as with a PEMFC on a vehicle.
- The *Methanation* of the carbon monoxide is an approach that reduces the danger of producing explosive gas mixtures. The reaction is the opposite of the steam reformation

reaction of equation 8.3, that is,

$$CO + 3H_2 \rightarrow CH_4 + H_2O \qquad (\Delta H = -206\,\text{kJ}\,\text{mol}^{-1})$$

This method has the obvious disadvantage that hydrogen is being consumed, and so the efficiency is reduced. However, the quantities involved are small – we are reducing the carbon monoxide content from about 0.25%. The methane does not poison the fuel cell, but simply acts as a diluent. Catalysts are available, which will promote this reaction so that at about 200°C the carbon monoxide levels will be less than 10 ppm. The catalysts will also ensure that any unconverted methanol is reacted to methane, hydrogen, or carbon dioxide.

- *Palladium/platinum membranes* can be used to separate and purify the hydrogen. This is a mature technology that has been used for many years to produce hydrogen of exceptional purity. However, these devices are expensive.

Note that recently workers at the National Renewable Energy Laboratory in the United States have been studying promotion of the shift reaction by bacteria. If this approach is successful, there may be less need for the final CO clean-up stage. Biological systems are referred to in Section 8.8. Another method of CO removal, which has been investigated by Honeywell and others, is the electrochemical oxidation of CO. Honeywell claims that this technique can reduce CO levels from 500 ppm down to ppb levels, but these ideas are at an early stage of development.

A further method of hydrogen purification, which can be applied, is that of Pressure Swing Absorption (PSA). In this process, the reformer product gas is passed into a reactor containing absorbent material. Hydrogen gas is preferentially absorbed on this material. After a set time, the reactor is isolated and the feed gas is diverted into a parallel reactor. At this stage the first reactor is depressurised, allowing pure hydrogen to desorb from the material. The process is repeated and the two reactors are alternately pressurised and depressurised. PSA is used in the Hyradix fuel processor.

These extra process stages add considerably to the cost and complexity of the fuel-processing systems for PAFC and PEMFC in comparison with those needed for MCFC and SOFC, a factor which we have already noted in Chapter 7.

8.5 Practical Fuel Processing – Stationary Applications

8.5.1 Conventional industrial steam reforming

Before we look at some practical fuel processor systems for fuel cells, it is worth making some remarks about industrial steam reforming plants. Steam reformers have been built for many years to provide hydrogen for oil refineries and for chemical plants. The latter have been used mainly for ammonia production for the fertiliser industry. Such large-scale reformer systems produce typically between 7 and 30 million normal cubic metres (Nm^3) of hydrogen per day. These reformers consist of a number of tubular reactors packed with pelleted catalysts operating at temperatures up to 850°C, and pressures up to 25 bar. The reactors are up to 12 m long and need to be made from expensive alloy steels to endure the high temperatures and the reducing gas conditions. Such reformers can be scaled down reasonably well to give hydrogen outputs of some 0.1 to 0.3 million $Nm^3\,day^{-1}$. So, about $4000\,Nm^3\,h^{-1}$ is the smallest efficient rate of production using such well-established technology. (Note, $1\,Nm^3$ is $1\,m^3$ at NTP.)

To give a better perspective to this, we can refer back to Chapter 2, where we noted that the enthalpy of combustion of hydrogen is $-241.8\,kJ\,mol^{-1}$ (lower heating value (LHV) basis). The reader can verify that this means that a hydrogen feed of $1\,Nm^3\,h^{-1}$ when combusted will produce about $3\,kW$ of heat energy, and therefore $200\,Nm^3\,h^{-1}$ of hydrogen when combusted will produce $600\,kW$ of energy. If the hydrogen is fed into a fuel cell system rather than being combusted, and assuming the system has an overall electrical efficiency of 40% (LHV), then clearly the energy produced by such a fuel cell fed with $200\,Nm^3\,h^{-1}$ of hydrogen would be $0.4 \times 600 = 240\,kW$. The reader can also work out how many hydrogen-powered fuel cell vehicles this amount of hydrogen would support, assuming so many hours on the road each day.[2] Scaling down, we can therefore say that $10\,Nm^3\,h^{-1}$ of hydrogen is enough to supply a 5-kW stationary fuel cell power system having an electrical efficiency of 40%. Unfortunately, scaling down conventional tubular reformers to this extent has several disadvantages – the systems are expensive because of the need to run at high temperatures and pressures, and they are large in terms of footprint area and weight. For these reasons, simply scaling down the mature and well-proven designs for producing hydrogen in a large-scale plant is not sensible, and special systems need to be produced for the much smaller demands of fuel cell systems.

8.5.2 System designs for natural gas fed PEMFC and PAFC plants with steam reformers

In most stationary fuel cell plants, natural gas is the fuel of choice. It is widely available, clean, and the processing of natural gas is generally straightforward. For the PEM fuel cell and the PAFC, steam reforming is the preferred option, ensuring high overall fuel conversion efficiencies. In both PEMFC and PAFC systems the sulphur removal can be achieved by HDS.

A fuel processor for a stationary fuel cell system supplied with natural gas may therefore consist of an HDS unit for sulphur removal, steam reformer for conversion to hydrogen and carbon monoxide, and stages of high-temperature and low-temperature shift for conversion of carbon monoxide. For a PEMFC, further carbon monoxide removal will be necessary. With such systems, a degree of process integration is required, whereby heat from the fuel cell is utilised for various pre-heating duties. This is because the various chemical processes (desulphurisation, steam reforming, high temperature shift reaction, low-temperature shift reaction) take place at different temperatures. There are thus a number of temperature changes to effect. The minimum required temperature changes are as follows:

- initial heating of the dry fuel gas to $\sim$300°C prior to desulphurisation,
- further heating prior to steam reforming at 600°C or more,
- cooling of the reformer product gas to $\sim$400°C for the high temperature shift reaction,
- further cooling to $\sim$200°C for the low-temperature shift reaction,

---

[2] It has been estimated that $10\,000\,Nm^3\,day^{-1}$ would be enough to supply a fleet of 280 PEMFC buses and 9000 PEMFC cars (Ogden, 2001).

- possible heating or cooling prior to entry to the fuel cell (depending on the fuel cell),
- heating of water to produce the steam needed in the steam reformer.

In addition to these six temperature changes, the steam reforming process requires heat at high temperature. This is usually provided by burning the anode exhaust gas from the fuel cell stack, which always contains some unconverted fuel. The reformer will incorporate a burner in which this unused fuel is burnt. To obtain the high temperatures required by the reformer, it helps if the fuel gas from the fuel cell anode exhaust is further heated. Similarly, if the air needed by the burner is preheated it will be more effective. We can see that in such fuel cell systems, gases need to be both heated and cooled. Both processes can be combined without a net loss of heat by using heat exchangers. These were discussed briefly in Section 7.2.4 in the previous chapter.

A diagram of a fuel-processing system that would be needed for a natural-gas-powered phosphoric acid fuel cell is shown in Figure 8.5. This needs a fuel gas at about 220°C, with carbon monoxide levels down to about 0.5%. The system may look complicated enough, but even this has some simplifications. The following is an explanation of the process.

- The natural gas enters at around 20°C and is heated in a heat exchanger E, to a temperature suitable for desulphurisation. Steam, sufficient for both reforming and the water-gas shift reaction, is then added. The steam/methane mixture is then further heated before being passed to the steam reformer. Here it is heated to around 800/850°C by the burner and is converted to hydrogen and carbon monoxide, with some unreacted steam still present.
- The reformate gas then passes through the heat exchanger C again, losing heat to the incoming fuel gas. Further heat is lost to the incoming gas at E, and to the anode exhaust gas at F.
- The gas is now sufficiently cool for the first shift converter, where the majority of the CO is converted to CO_2. At D the gas is further cooled, giving up heat to the incoming steam, before passing to the low-temperature shift converter, where the final CO is converted. The final cooling is accomplished at B, where the incoming steam is heated.
- The hydrogen-rich fuel gas is then passed to the fuel cell. Here, most, but not all, of the hydrogen is converted into electrical energy. The remaining gas returns at heat exchanger F, still at about 220°C. Here it is preheated prior to reaching the burner.
- Also coming to the burner is air, which will have been preheated by the heat exchanger A, using energy from the burner exhaust gas.
- The steam arriving at about 120°C at heat exchanger B may have been created by heat from the fuel cell cooling system.
- The burner exhaust gas, which is still very hot, may be used to raise steam to drive any compressors needed to drive the process.

There are many other possible ways of configuring the gas flows and heat exchangers to get the required result, but they are unlikely to be simpler than this. As an example of another system, Figure 8.6 shows the schematic arrangement of the fuel processing within the Ballard 250-kW PEM fuel cell plant. Notice that we now have a selective oxidiser to

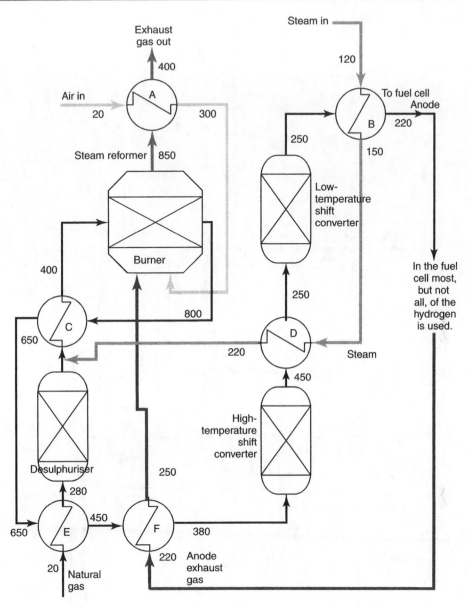

Figure 8.5 Diagram of a fuel-processing system for a phosphoric acid fuel cell. The numbers indicate approximate likely temperatures.

further remove carbon monoxide. Another key difference is that this system operates at pressure and the energy in the hot exhaust gases is used to drive the turbocompressors at the bottom of the diagram (see Chapter 9) rather than in heat exchangers. We will analyse a stationary fuel cell power system in more detail in Chapter 10.

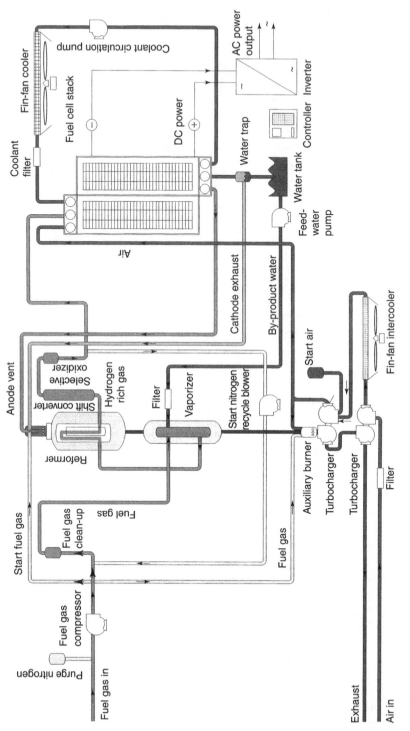

Figure 8.6 Process flow diagram for a Ballard 250-kW PEMFC plant. (Reproduced by kind permission of Alstom Ballard GmbH.)

8.5.3 Reformer and partial oxidation designs

Compact regenerative reformers with concentric annular catalyst beds

In all these systems the desulphuriser, shift reactors, and carbon monoxide clean-up systems are essentially packed-bed catalytic reactors of traditional design. As mentioned in the previous section, new catalyst materials are being developed for the shift reaction and other duties. It is likely that we will see traditional pelleted material replaced by coated monolith catalysts in the next few years. However, when it comes to the reformer reactors and catalysts for stationary plants, several novel features are being pursued, particularly in integrating some of the heat transfer duties. Figure 8.7 shows an example of a reformer designed by Haldor Topsoe for PAFC systems in which the heat for the reforming reaction is provided by combustion of the lean anode exhaust gas supplemented if necessary by fresh fuel gas. In this system, reformer fuel is combusted at a pressure of some 4.5 bar in a central burner located in the bottom of a pressure vessel. Feed gas is fed downwards through the first catalyst bed where it is heated to around 675°C by convection from the combustion products and the reformed product gas, both flowing countercurrent to the feed. On leaving the first bed of catalyst, the partially reformed gas is transferred through a set of tubes to the top of the second reforming section. The gas flows down through the catalyst, being heated typically to 830°C by convection from the co-currently flowing combustion products and also by radiation from the combustion tube. The combination of co-current and countercurrent heat transfer minimises metal

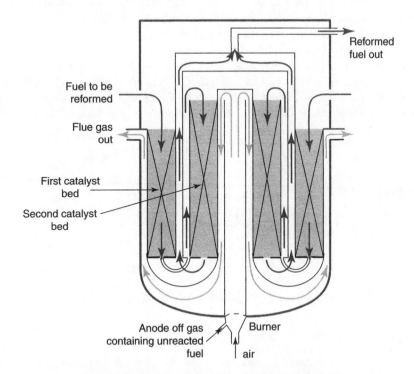

Figure 8.7 Haldor–Topsoe heat exchange reformer.

temperatures, an important consideration in the high-temperature reformer design. The advantages of such a reformer for fuel cell applications are small size and suitability for small-scale use, pressurised combustion of lean anode exhaust gas giving good process integration with the fuel cell, improved load following, and lower cost. Several companies have been developing reformers of this type including Haldor–Topsoe, International Fuel Cells (recently renamed UTC Fuel Cells), Ballard Generation Systems, Sanyo Electric, Osaka Gas, and ChevronTexaco.

Plate reformers and micro-channel reformers

In the plate reformer, a stack of alternate combustion and reforming chambers are separated by plates. The chambers are filled with suitable catalysts to promote the combustion and reforming reactions, respectively. Alternatively, either side of each plate can be coated with combustion catalyst and reforming catalyst. The heat from the combustion reaction is used to drive the reforming reaction. Plate reformers have the advantage that they can be very compact and that they offer a means of reducing heat transfer resistance to a minimum. The use of a combustion catalyst means that low heating value gases (e.g. anode exhaust gases) can be burnt without the need for a supplementary fuel. Plate reformers were first developed by IHI in the 1980s, in which the catalyst was in the form of spherical pellets located on either side of the heat exchanger surface (Hamada et al., 1997). More recently, researchers at Gastec designed a plate reformer in which the plates were coated with a ceramic-supported catalyst, and this technology has been acquired by the US fuel cell company Plug Power. Osaka Gas also built a plate reformer (Shinke et al., 2000), and several other companies hold patents on the technology.

The most advanced types of plate reformers use compact heat exchanger hardware in which the catalyst is coated directly onto the exchanger surfaces or shims in the form of a thin film a few microns thick (Goulding, 2001). The concept is shown in Figure 8.8. Such devices are being developed in the United States by Pacific Northwest National Laboratory who have demonstrated a 1-kW steam reformer, and Argonne National Laboratory who

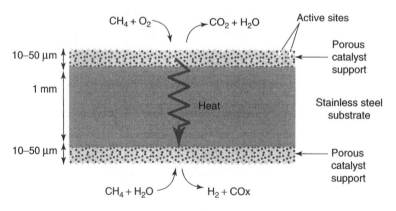

Figure 8.8 Plate reformer concept. Catalyst is coated as a thin film onto one or both sides of a heat exchanger material.

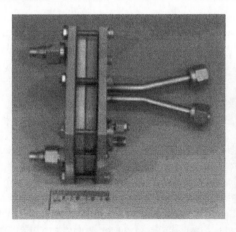

Figure 8.9 Experimental compact reformer reactors 1, the Pacific Northwest National Laboratory gasoline reformer. (Reproduced by kind permission of Pacific Northwest National Laboratory.)

Figure 8.10 Experimental compact reformer reactors 2, proof-of-concept Advantica natural gas reformer, a diffusion-bonded multi-channel reactor block ($3 \times 3 \times 10$ cm). (Reproduced by kind permission of Advantica Technologies Ltd.)

are developing a monolithic catalyst-based reformer for diesel. Several organisations have developed plate reformers for methanol, for example Idatech, Mitsubishi Electric, Innovatek Inc., NTT telecommunications laboratory, and Honeywell (Ogden, 2001). Examples of hardware are shown in Figures 8.9 and 8.10. Micro-channel reactor technology could be applied to other parts of a fuel processor such as fuel vaporisers and gas clean-up reactors. However, they suffer from two disadvantages: firstly, the plugging of the channels due to catalyst degradation and carbon laydown and secondly, the fact that the catalyst is incorporated into the reactor for life. In other words, the catalyst cannot be replaced when it becomes degraded.

Membrane reactor

Hydrogen is able to permeate selectively through palladium or palladium alloy membranes. This has led to the demonstration in the laboratory of membrane reformers where hydrogen is selectively removed from the reformer as it is produced. This hydrogen removal increases the methane conversion for a given operating temperature above that which is predicted thermodynamically, and the hydrogen so produced is very pure, making it very suitable for PEM fuel cell systems. Some large companies (Exxon, BP, Air Products, and Praxair) are developing membrane reformers for large-scale hydrogen production. It is likely that some of these can be scaled to meet the needs of fuel cell systems. Several developers are also known to be working on membrane reformers for fuel cell systems, examples being Praxair, Tokyo Gas, Wellman Defence, Aspen, and Idatech Inc. An example of the use of membrane technology is to be found in the Tokyo Gas Co. reformer, which has been built and tested for 1-kW PEMFC systems, and produces pure hydrogen from natural gas at a rate of $15\,\mathrm{Nm^3\,h^{-1}}$.

The systems being developed by IdaTech Inc. are perhaps the most advanced of the small-scale membrane reformer systems for fuel cell applications, and the company has recently announced early product sales. The characteristic of the Idatech fuel processor is that it combines an imperfect (but lower cost) membrane filter with a chemical purification system to generate relatively low-cost but high-quality hydrogen. The fuel processor combines the functions of a steam reformer, hydrogen purification, and heat generation into a single device producing 99.8% pure hydrogen with <3 ppm carbon monoxide and <25 ppm carbon dioxide.

A schematic of the Idatech membrane reformer is shown in Figure 8.11. In this case, methanol, or another vapourised liquid fuel and steam are preheated before being fed through a series of tubes containing the reforming catalyst. Steam reforming occurs on the catalyst and the product gas is fed into a planar reactor containing a hydrogen-selective membrane. Hydrogen selectively permeates through this and out as a product stream. Excess fuel from the catalyst bed is mixed with air and is passed to a burner that generates heat at the centre of the reformer tubes, thus providing heat for the reforming reaction. Hydrogen purification is accomplished using a two-stage purifier. Since the purification process is driven by a pressure gradient, the steam reforming reactions are conducted at elevated pressures, typically between 3 and 50 bar. The heat input necessary for steam raising and promoting the endothermic steam reforming reaction is derived solely from the combustion of a fuel gas stream. The fuel gas may consist of some or all the impurities (waste gases) rejected by the hydrogen purification module.

Idatech is now marketing a 2.5-kW packaged PEM power plant system, powered by a catalytic steam reformer, to serve the residential market. For this and its other fuel processor developments, Idatech has focused on using liquid fuels, including methanol,

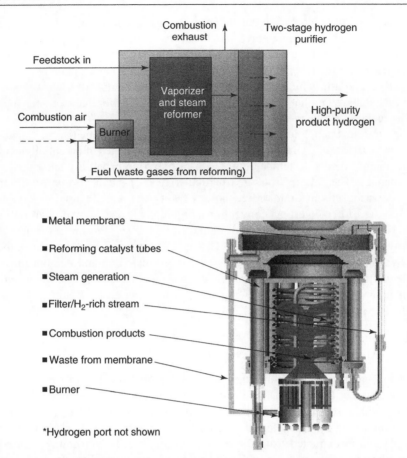

Figure 8.11 Membrane reactor fuel processor developed by IdaTech Inc. (a) Schematic of the membrane fuel processor and (b) arrangement of the reactor components. Reproduced by kind permission of ItaTech Inc.

propane, diesel, and ethanol. The reason is that liquid water and liquid feedstock can be pumped to pressure with low power requirements. The endothermic reaction heat is supplied by combustion of reformate waste gases at ambient pressure (Kalhammer et al., 1998).

Non-catalytic partial oxidation reactors

Non-catalytic partial oxidation is applied industrially by Texaco and Shell for the conversion of heavy oils to synthesis gas. In the case of the Shell, basic partial oxidation process, liquid fuel is fed to a reactor together with oxygen, and steam. A partial combustion takes place in the reactor, which yields a product at around 1150°C. It is this high temperature that poses a particular problem for conventional partial oxidation. The reactor has to be made of expensive materials to withstand the high temperatures, and the product gas needs to be cooled to enable unreacted carbon material to be separated from the gas stream. The high temperatures also mean that expensive materials of construction are required for the heat exchangers. In addition, the effluent from non-catalytic partial oxidation reactors invariably contains contaminants (including sulphur compounds), as well as carbon and ash, which need to be dealt with. The complexity, which these would add to a fuel cell system, has meant that simple partial oxidation has not been a preferred option for fuel cell applications.

One interesting application of non-catalytic reactors has been the so-called plasma reformer or 'plasmatron'. This type of reactor has the advantages of potentially being compact, operating at moderate temperatures with fast start-up capability and good response to load changes. The 'plasmatron' is a particular type of small plasma reformer developed by Massachusetts Institute of Technology (MIT) and now licensed to ArvinMeritor. It was designed for converting conventional liquid fuels into a hydrogen-rich gas for enhancing the performance of internal combustion engines. Plasma reformers are also being developed by Wangtec, the Idaho National Energy and Environment Laboratory, and by Syngen Inc. All these devices have yet to be scaled up and demonstrated in real fuel cell systems; however, the Syngen process does look promising for the generation of synthesis gas, which has commercial implications for both gas-to-liquids processing and for fuel cell systems.

Catalytic partial oxidation (CPO) reactors

CPO reactors can be very simple in design, requiring only one bed of catalyst into which the fuel and oxidant (usually air) are injected. Often steam is added as well, in which case some conventional reforming also occurs. As mentioned before, the combination of CPO and steam reforming is often referred to as *autothermal reforming*, because there is no net heat supplied to or extracted from the reactor. All the heat for reforming is provided by partial combustion of the fuel. In such reactors, two types of catalyst are sometimes used, one primarily for the CPO reaction, and the other to promote steam reforming, depending on the nature of the fuel and the application.

A well-known example of a CPO reactor is the Johnson Matthey HotSpot™ reactor (Edwards et al., 1998). A schematic representation of an early version of this technology is shown in Figure 8.12 and further on in Figure 8.15. It utilises a platinum/chromium oxide catalyst on a ceramic support. The novel feature of the reactor is the hot spot caused by point injection of the air-hydrocarbon mixture. This arrangement eliminates the need for pre-heating the fuel gas and air during operation, although for start-up on natural gas, the fuel should be preheated to around 500°C. Alternatively, the reactor can be started from ambient temperature, by introducing an initiating fuel such as methanol or a hydrogen-rich gas. Such initiating fuels are oxidised by air at ambient temperature over the catalysts, and this oxidation serves to raise the bed to the temperature needed for natural gas to react (typically over 450°C).

Several other developers have built CPO reactors for both mobile and stationary fuel cell applications. Shell has developed their CPO technology for reforming gasoline within the DaimlerChrysler/Ballard company Excellsis as well as in collaboration with UTC Fuel Cells for stationary applications. CPO of methanol proceeds with much lower heats of

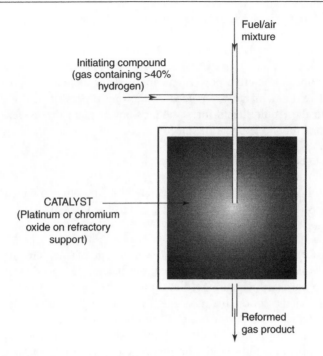

Figure 8.12 Simplified schematic of Johnson Matthey HotSpot reactor. Current versions of these devices have proprietary features not shown here. See also the photograph shown in Figure 8.15.

reaction compared with hydrocarbons and a very simple design of the CPO reactor can be used. A good example is that developed by Kumar et al. (1996) at Argonne National Laboratory. In this method, the methanol is simply mixed with water in liquid form and fed to an igniter, which is fed with a controlled flow of air so that only some of the methanol is burnt, and enough heat is produced to vaporise the water and methanol. The unburnt methanol and steam are fed to a reactor. In this reactor, the catalyst is supported on a honeycomb monolith material similar to that found in automobile exhaust catalysts. More recently, Schmidt (2001) has reported successful catalytic partial oxidation of iso-octane (a popular gasoline standard) on rhodium coated monoliths at 600°C.

These smaller scale reformers are also suitable for mobile applications, and so they are mentioned again in the following section.

8.6 Practical Fuel Processing – Mobile Applications

8.6.1 General issues

There are two types of mobile applications for fuel cells. The first is where the fuel cell provides the motive power for the vehicle, for example, in combination with electric traction for cars and buses. For such fuel cell vehicles, the PEM fuel cell (and for special applications the alkaline (electrolyte) fuel cell (AFC)) is the preferred option. The second application is where the fuel cell provides only auxiliary power. An example of this is the hotel load onboard ships or the provision of supplementary electric power for trucks for heating and air conditioning when stationary. Interestingly, BMW and others are interested in developing SOFCs as an auxiliary power source for automobiles. However, it is when the fuel cell provides the motive power, and the fuel for the stack is processed onboard, that considerable extra demands are placed on the fuel processor.

In addition to the requirements for stationary power plant discussed earlier, onboard fuel processors for mobile applications need

- to be compact (both in weight and volume),
- to be capable of starting up quickly,
- to be able to follow demand rapidly and operate efficiently over a wide operating range,
- to be capable of delivering low-CO content gas to the PEM stack,
- to emit very low levels of pollutants.

Over the past few years, research and development of fuel processing for mobile applications and small-scale stationary applications has mushroomed. Many organisations are developing proprietary technology, but almost all of them are based on the options outlined in the previous section, that is, steam reforming, CPO, or autothermal reforming. However, there is a vigorous debate being waged between vehicle manufacturers and fuel suppliers regarding the preferred fuel for FCVs. Methanol and gasoline are the front-runners in the short to medium term, with hydrogen preferred by many as the long-term option. Each fuel has its advantages and disadvantages, and it is likely that as FCVs do become commercial, the choice will be governed by both the type and duty of the vehicle, as well as the availability and distribution infrastructure for the fuel. In Chapter 11 we discuss some of the issues in analysing the fuelling options. In Chapter 6 we also discussed some of the safety issues pertaining to the use of methanol.

It should also be said that vehicles are by no means the only area of application for fuel processors. They are also being developed for the rapidly growing market in portable electronics equipment. Below we describe some of the processors that are being developed now for mobile processing of methanol and gasoline.

8.6.2 Methanol reforming for vehicles

Leading developers of methanol reforming for vehicles at present are Excellsis Fuel cell Engineers (DaimlerChrysler), General Motors, Honda, International Fuel Cells, Mitsubishi, Nissan, Toyota, and Johnson Matthey. Most of the organisations use steam reforming, although some are also working on partial oxidation. Two examples of practical methanol steam-reforming systems are given below.

DaimlerChrysler methanol processor

DaimlerChrysler developed a methanol processor for the NeCar 3 experimental vehicle. This was demonstrated in September 1997 as the world's first methanol-fuelled fuel cell car. It was used in conjunction with a Ballard 50-kW fuel cell stack. Characteristics of

Table 8.8 Characteristics of the methanol processor for NeCar 3 (Kalhammer et al., 1998)

Maximum unit size	50 kWe
Power density	1.1 kW$_e$ L^{-1} (reformer = 20 L, combustor = 5 L, CO selective oxidiser 20 L)
Specific power	0.44 kW$_e$ kg^{-1} (reformer = 34 kg, combustor = 20 kg, CO selective oxidiser = 40 kg)
Energy efficiency	not determined
Methanol conversion Efficiency	98–100%
Turn-down ratio	20 to 1
Transient response	<2 s

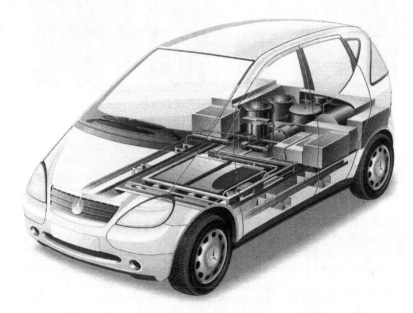

Figure 8.13 NeCar 3 packaging of stack and fuel processor.

the methanol processor are given in Table 8.8, and Figure 8.13 shows how these were packaged in the car.

Since the NeCar3 demonstration, DaimlerChrysler and Excellsis have been working with BASF to develop a more advanced catalytic reformer system for their vehicles. In November 2000, DaimlerChrysler launched the NeCar5, which, together with a Jeep Commander vehicle, represents state of the art in methanol fuel cell vehicles. In the past six years, the fuel cell drive system has been shrunk to such an extent that it presently requires no more space than a conventional drive system. The Necar 5, therefore, has the full complement of seats and interior space as a conventional gasoline-fuelled internal combustion engine car. The car is based on the A-Class Mercedes design and the methanol

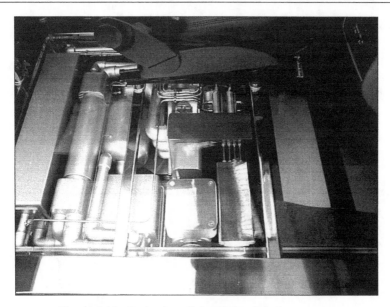

Figure 8.14 Packaging of the NeCar 5 fuel processor.

reformer is under the passenger compartment, as illustrated in Figure 8.14. The Necar 5 uses a Ballard 75 kW Mk 9 stack giving an impressive top speed of over 150 km h^{-1}.

General Motors/Opel methanol processor

General Motors European subsidiary Opel presented its first fuel cell powered vehicle, a concept car based on the Zafira, in September 1998. This uses a methanol reformer based on that developed by GM in a programme over many years working with the Los Alamos National Laboratory in the United States. The main features of the processor are given in Table 8.9.

Mercatox methanol processor

The EU-supported 'Mercatox' project led by the company Wellman CJB in the UK has led to the development of a complete process for the on-board steam reforming of methanol for vehicles. The basic premise was the use of compact, aluminium heat exchangers

Table 8.9 Features of the GM methanol fuel processor

Maximum unit size	30 kWe
Power density	0.5 kW$_e$ L^{-1}
Specific power	0.44 kW$_e$ kg^{-1}
Energy efficiency	82–85%
Methanol conversion	>99%

comprising corrugated plates coated on the one side with a methanol reforming catalyst, and on the opposite side with combustion catalyst for burning, amongst other things, fuel cell off gas. The principle is that of the plate reformer described in the previous section. A gas clean-up unit designed by Loughborough University also used the same basic construction with a selective oxidation catalyst coated on to the process side for converting carbon monoxide into carbon dioxide in the presence of hydrogen. The targets for the Mercatox project were a 50-kWe system, a volume of 49 L, mass of 50 kg, a warm-up time of less than 5 s, transient response time with storage support of less than 5 s, and without-storage support of less than 50 ms. The assessment of the manufacturing and vehicle design for the Mercatox project was carried out by Rover cars with modelling performed by Instituto Superior Technico (Dams, 2000).

Several organisations are developing catalytic partial oxidation or autothermal reactors for methanol processing, examples being Johnson Matthey (the HotSpot™ Fuel Processor), Argonne National Laboratory (US), Honda, DaimlerChrysler Shell, and Nuvera. The Johnson Matthey HotSpot™ fuel processor is shown below in Figure 8.15. Each one of the units is capable of producing in excess of 6000 L of hydrogen per hour. It is left as an exercise for the reader to show, using equation A2.8 from Appendix 2, and the fact that the density of hydrogen is $0.084 \, kg \, m^{-3}$ at normal temperature and pressure (NTP), that this is sufficient for a fuel cell of about 8 kW.

As mentioned before, the design for methanol CPO reactors can be very simple. There has been a push in recent years towards a multi-fuel processor, brought about by the requirements of the US Department of Energy to keep open all options for fuelling vehicles. Examples of multi-fuel processors are the integrated designs of NorthWest Power Systems, LLC – now known as Innovatek (Edlund and Pledger, 1998).

8.6.3 Micro-scale methanol reactors

The drive to make fuel processors more compact is now being fired by the prospects of using compact methanol fuel cells for consumer electronics devices. In Chapter 6 we

Figure 8.15 The Johnson Matthey HotSpot™ reactor. These can be made in different forms suitable for methanol, methane, or gasoline processing. (Picture by kind permission of Johnson Matthey plc.)

covered the direct methanol fuel cell (DMFC), and we saw that its performance is rather limited, and the efficiency not particularly high. They are also quite complex systems, and it could well be that small micro-reformers producing hydrogen for a PEMFC offer a better solution. We also saw in Chapter 6 that battery technology is reaching a limit in terms of storage capacity and fuel cells are seen as a means of extending the operating time of laptop computers and other devices.

Rapid progress in recent years in miniaturising chemical reactors has led some researchers to explore the fabrication of micro-scale fuel-processing reactors for fuel cells. Here we are considering systems that require a few watts or less of power, where the systems can be 'built on a chip' and where, in principle, the mass manufacturing methods of the semiconductor industry could be applied. Figure 8.16 shows a conceptual system, devised by researchers at the University of Bethlehem, PA (Pattekar et al., 2001).

Figure 8.16 shows a methanol/water vaporiser, followed by a catalytic steam reformer operating at about 250°C, in which the catalyst is a thin film of Cu/ZnO coated onto the silica reactor, and finally a membrane shift reactor consisting of a palladium diffusion layer mounted on top of a perforated copper-based shift catalyst. Built onto the chip are integrated resistive heaters for getting the reformer and vaporiser up to temperature, together with micro-scale sensors and control electronics. Whilst such systems are a long

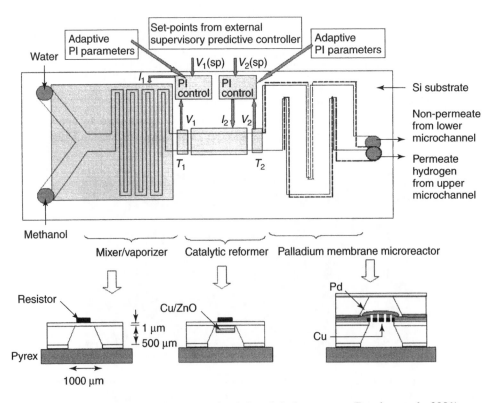

Figure 8.16 A conceptual micro-scale methanol fuel processor (Pattekar et al., 2001).

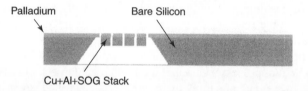

Figure 8.17 Palladium-based membrane shift reactor.

way from commercialisation, good progress is being made in demonstrating the feasibility of some aspects. For example, Pattekar (2001) has fabricated a micro-scale membrane shift reactor, shown schematically in Figure 8.17, using four layers of material: copper, aluminium, spin-on-glass (SOG), and palladium. Copper, aluminium, and SOG layers have a pattern of holes etched into them. They serve as a structural support for the main element of the membrane, the palladium film. Copper also acts as a catalyst in the shift reaction that is taking place in the same channel. The product gases from steam reforming of methanol gases flow in the channel etched in the silicon substrate. Carbon monoxide at a temperature of 250°C in the presence of the copper catalyst will get converted into more hydrogen gas, which in turn gets selectively separated through the palladium membrane. Separated hydrogen will pass through a channel etched in another silicon wafer bonded to the original silicon wafer on top of the palladium membrane.

Several research groups as well as consumer electronics companies are working on micro-scale reactor technology and it remains to be seen how successful these become. Many of the details of fabrication methods are proprietary and the example of the membrane shift reactor is just one example of a whole new area of development.

8.6.4 Gasoline reforming

Companies such as Arthur D. Little have been developing reformers aimed at utilising gasoline type hydrocarbons (Teagan et al., 1998). The company felt that the adoption of gasoline as a fuel for FCVs would be likely to find favour amongst oil companies, since the present distribution systems can be used. Indeed, Shell has demonstrated its own CPO technology on gasoline, and ExxonMobil in collaboration with GM has also been developing a gasoline fuel processor. Arthur D. Little spun out its reformer development into Epyx, which later teamed up with the Italian company De Nora to form the fuel cell company Nuvera. In the Nuvera fuel-processing system, the required heat of reaction for the reforming is provided by *in situ* oxidising a fraction of the feedstock in a combustion (POX) zone. A nickel-based catalyst bed following the POX zone is the key to achieving full fuel conversion for high efficiency. The POX section operates at relatively high temperatures (1100–1500°C), whereas the catalytic reforming operates in the temperature range 800 to 1000°C. The separation of the POX and catalytic zones allows a relatively pure gas to enter the reformer, permitting the system to accommodate a variety of fuels. Shift reactors (high-and low-temperature) convert the product gas from the reformer so that the exit concentration of CO is less than 1%. As described earlier, an additional CO-removal stage is therefore needed to achieve the CO levels necessary for a PEM fuel cell.

When designed for gasoline, the fuel processor also includes a compact desulphurisation bed integrated within the reactor vessel prior to the low-temperature shift.

Most of the development by Nuvera has been for stationary systems; however, Johnson Matthey has demonstrated its HotSpot reactor on reformulated gasoline (Ellis et al., 2001). They built a 10-kW fuel processor that met their technical targets, but they also addressed issues relating to mass manufacture; their work has identified areas that will require further work to enable gasoline reforming to become a commercial reality. These included the following:

- hydrogen storage for start-up and transients,
- an intrinsically safe after-burner design with internal temperature control and heat exchange that can cope with transients,
- effect of additives on fuels,
- better understanding of the issues relating to sulphur removal from fuels at source,
- improved sulphur trapping and regeneration strategies.

Johnson Matthey is now engaged in a commercialisation programme for their technology. The pace of development is now such that in April 2001, GM demonstrated their own gasoline fuel processor in a Chevrolet 2–10 pickup truck, billed as the world's first gasoline-fed fuel cell electric vehicle. As if to underline the synergy between stationary and mobile fuel processors, GM subsequently announced, in August 2001, a 5-kW stationary fuel cell system that could run on natural gas, methanol, or even gasoline. With the rapid developments being made in this area it remains to be seen which of the various fuel-processing systems will become economically viable in the future.

With all the fuel-processing options described for vehicular applications, the assumption is generally made that onboard storage and processing of fuel will be needed. The main reason for this is to ensure a sufficient driving range for the vehicles, since we have come to expect this for our present day cars, lorries, and buses. One way to side-step all the problems associated with onboard fuel processing is to make the fuel processing plant stationary, and to store the hydrogen produced, which can be loaded onto the mobile system as required. In fact, this may well be a preferred option for some applications, such as buses. However, as ever, solving one problem creates others, and the problems of storing hydrogen are quite severe. These are dealt with in Sections 8.9 and 8.10.

8.7 Electrolysers

8.7.1 Operation of electrolysers

Electrolysers use electricity to divide water (H_2O) into hydrogen and oxygen. They are thus the opposite of a fuel cell.

All the basic theory, and the reactions taking place at the electrodes, is the same for electrolysers as for fuel cells – except that the reactions go the other way. Different electrolytes can be used, just as for fuel cells. However, it is not very practical to use the high-temperature systems as steam would have to be supplied, which is not so convenient as liquid water. In practice, then, the only electrolytes in use are alkaline liquids and

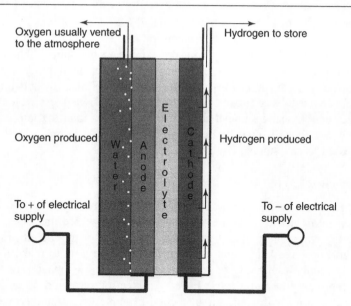

Figure 8.18 Simplified diagram of a PEM-based electrolyser. Note that in an electrolyser the positive electrode is properly called the anode, and the negative the cathode – unlike fuel cells.

solid proton exchange membranes. In the past, potassium hydroxide was by far the most common electrolyte used. However, the proton exchange membrane is now more usual in modern designs. The basic arrangement of a PEM-based electrolyser is shown in Figure 8.18. The basic structure of the electrodes and the electrolyte is the same as for the PEMFC considered in some detail in Chapter 4. The electrodes will usually be made somewhat differently, as the idea is to expel the product gas as quickly as possible, rather than to draw it in.

At the negative electrode[3], protons are removed from the electrolyte, electrons are provided by the external electrical supply and hydrogen is formed via the reaction

$$4H^+ + 4e^- \rightarrow 2H_2$$

At the positive electrode, the water is oxidised (electrons removed) and oxygen is made via the reaction

$$2H_2O \rightarrow O_2 + 4H^+ + 4e^-$$

These reactions are the opposite of those in Figure 1.3. One reason for the success of PEM-based electrolysers is that many of the problems of PEM fuel cells do not apply. Cooling is fairly trivial, as the water supplied to the cathode can be pumped around the

---

[3] The question of the names of electrodes was addressed in Chapter 1. Since this is an electrolysis cell the negative electrode should be called the cathode. It is the electrode the electrons flow into. This is somewhat muddling, as for a fuel cell, and thus in the rest of the book this electrode is the anode! Confusion is best avoided by calling it the negative electrode.

cell to remove heat. Another key problem of the PEMFC, that of water management, is also massively simplified, as the positive electrode must be flooded with water.

The hydrogen produced by such electrolysers will have a high purity. The only problem will be the presence of water vapour. This might be present in significant quantities, as water will diffuse through the electrolyte. The protons will also drag it through, in the same way as for fuel cells (see Figure 4.10). Electrolyser designers use various techniques to reduce this, such as using somewhat thicker electrolytes than that are common with fuel cells. However, it will often be necessary to have some drying of the product gas before it is passed to any storage equipment.

8.7.2 Applications of electrolysers

At first it would seem foolish to use electricity to make hydrogen for use in fuel cells to turn it back into electricity. However, there are cases where this does make sense. Usually these are where the fuel cell is mobile. The value of mobile electricity is much greater than grid power. A base station electrolyser can be used to make hydrogen that is then used in mobile fuel cells. A good example would be a boat yard that services a number of fuel cell powered electric boats. An electrolyser could be an ideal way of producing the necessary hydrogen. Another good example would be the owner of a fuel cell–powered car. They could have an electrolyser working away slowly and steadily during the night in their hall, lobby, or garage. It has been shown (Hammerli and Stuart, 2000) that such an arrangement would have a lower running cost than a gasoline powered car, even in the low tax regime of the United States. At low usage rates, the electrolyser is cheaper and simpler than using a small natural gas reformer. At what rate of usage the 'changeover' point comes when natural gas reformers become cheaper is something that will become clear in the years ahead. Unlike electrolysers, which are readily available, it is not possible (in 2003) to buy small ready to use natural gas-to-hydrogen reformers.

Electrolysers are also a good way of converting surplus electrical energy into chemical energy, and as such they are used in places where there are plentiful hydroelectric schemes that carry on producing power in the night. Some see a future where large numbers of photovoltaic panels and wind-driven generators will be connected to electrolysers and the hydrogen produced will be the major way in which energy is stored and transmitted.

At the time of going to print, the majority of electrolysers are used to make hydrogen for applications that are not connected with fuel cells. Hydrogen has many uses, for example, in chemical analysis equipment. For medium-scale users, the key advantages of electrolysers are as follows:

- The product hydrogen is very pure.
- It is produced as needed and does not have to be stored, and so is safer.
- Electricity is much easier and safer to supply than bottled hydrogen.
- The marginal cost is only a few cents per kilogram – much cheaper than gas supplied in high-pressure cylinders.

8.7.3 Electrolyser efficiency

The efficiency of an electrolyser is calculated in almost the same way as for a fuel cell. If V_c is the operating voltage for one cell of a fuel cell stack, then it was shown in Chapter 2,

Section 2.4, that the efficiency is given by the formula

$$\eta = \frac{V_c}{1.48}$$

In the case of an electrolyser the formula is just the inverse of this

$$\eta = \frac{1.48}{V_c}$$

The losses in electrolysers follow exactly the same pattern as for fuel cells as described in Chapter 3. Real values of V_c are around 1.6 to 2.0 V, depending on the current density. Because the problems of cooling and water management are so much more easily solved, the performance of electrolysers is routinely as good as for the best of fuel cells, with current densities of around $1.0 \, \text{A cm}^{-2}$ being normal.

An electrolyser can be operated very efficiently, with V_c about 1.6 V, if the current density is kept low. Low current density means a slow rate of production of hydrogen or a large electrolyser – hence, higher costs. There is always a balance to be struck between efficiency of production, and high rate of production per dollars worth of electrolyser. Typical operating efficiencies claimed by commercial makers of units, including the energy needed to compress the product gas, are around 60 to 70%.

8.7.4 Generating at high pressure

One of the interesting design questions in electrolysers is whether to run the electrolyser at high pressure. The gas can be generated and not released until a high pressure has built up inside the electrolyser. The product hydrogen can then be released *directly* into a high-pressure cylinder. Alternatively, the unit can operate at air pressure, or at a little higher pressure, and then the product gas is pumped to a high pressure.

It is interesting to compare these two approaches, as both are used. The energy cost for the pumping option is the work done to compress the gas. It is often considered that pumps work adiabatically,[4] indeed, a compressor for the air going into a fuel cell, such as we consider in the next chapter, will do so. However, in this case the situation will be approximately isothermal, as the stored product is in cylinders at normal ambient temperature. The isothermal operation is also promoted by the very high heat conductivity and heat capacity of hydrogen. The fact that hydrogen behaves quite differently from an 'ideal gas' at these kind of pressures also promotes isothermal operation. The standard 'text book' formula for the work done in isothermally compressing a gas from pressure P_1 to P_2 is

$$\Delta W = P_1 V_1 \ln\left(\frac{P_2}{P_1}\right) = nRT \ln\left(\frac{P_2}{P_1}\right) \qquad [8.12]$$

Here n is the number of moles and R is the universal gas constant. The temperature term T needs no subscript because the temperature is constant. The motor and pump driving

---

[4] No heat lost to the surroundings. In other words that gas is heated as it is compressed. The formulae for the work done compressing the gas are very different in this case. See equation 9.7.

the compressor will not be 100% efficient, and so the real figure will be higher than this. However, it will serve our purposes for the moment.

For the alternative approach – operation of an electrolyser at high pressure – the energy cost is the Nernst voltage rise, as discussed in Section 2.5. Here it was shown that the voltage rise resulting from an increase in hydrogen pressure only is

$$\Delta V = \frac{RT}{2F} \ln\left(\frac{P_2}{P_1}\right) \text{ Volts} \qquad [8.13]$$

This increase in voltage can be transposed to work and Joules as follows. The extra work done when a charge Q flows through the electrolyser cell is

$$\Delta W = \Delta V \times Q$$

We saw in Chapter 2 that the charge produced by one mole of hydrogen is $2F$, where F is the Faraday constant. Similarly, $2F$ coulombs are needed for each gram mole of hydrogen. The work done in producing n moles of hydrogen at high pressure is thus

$$\Delta W = \Delta V \times 2F \times n$$

Substituting equation 8.13 into this, we have

$$\Delta W = \frac{RT}{2F} \ln\left(\frac{P_2}{P_1}\right) \times 2F \times n = nRT \ln\left(\frac{P_2}{P_1}\right) \qquad [8.14]$$

This equation is *exactly* the same as equation 8.12. The fact that two such different methods of compressing a gas should require the same work is one of those 'coincidences' that makes science, engineering, and thermodynamics so interesting. Of course it is not really a coincidence. The Nernst equation that gave us equation 8.13 was derived using the same thermodynamic laws that gave us equation 8.12.

Science theory then does not give a clear-cut answer to the question of which is better – pumps or high-pressure electrolysers. In engineering terms though, it obviously makes sense to use one device for both generation and compression, if possible. Also, the pump system has more scope for inefficiencies and for making the real work noticeably greater than the formula would predict. A problem is that it is not that easy to design an electrolyser that operates at high pressure on the hydrogen side only, while venting the oxygen to the atmosphere at normal air pressure. However, some companies, for example, Proton Energy Systems in the United States, using various proprietary techniques, have succeeded in this. Figure 8.19 shows an electrolyser and a high-pressure hydrogen store used to provide hydrogen for a small fleet of vehicles in Munich, Germany.

Figure 8.19 Electrolyser with a high-pressure hydrogen store visible behind it at Munich airport. (Photograph reproduced by kind permission of Hamburgische Electricitäts Werke AG.)

8.7.5 Photo-electrolysis

It is not difficult to understand how electricity generated from sunlight, via photovoltaic cells, could be used to generate hydrogen by electrolysis. Combining the two processes, that is, photoelectric generation and electrolysis gives rise to a hybrid process known as photo-electrolysis. The idea is to develop a solid-state PV device that gathers sunlight and decomposes water that is in contact with the device. In contrast to the PV/electrolysis systems, external wiring or converters are not required, but a reasonably high voltage needs to be generated by the cells to enable the water to be decomposed. This means that several cells need to be joined together or 'cascaded'. Some success has been achieved with multiple layers of amorphous silicon cells, and the US Department of Energy is supporting the National Renewable Energy Laboratory and the Hawaii Natural Energy Institute in the development of these systems (Lakeman and Browning, 2001). The theoretical efficiency for tandem junction systems is 42%. In practice, a solar-to-hydrogen efficiency of 12.4% LHV using concentrated light has been achieved, and an outdoor test of amorphous silicon cells resulted in a solar-to-hydrogen efficiency of 7.8% LHV under natural sunlight.

8.8 Biological Production of Hydrogen

8.8.1 Introduction

Reference has already been made in this chapter to the generation of hydrogen from bio-fuels (Section 8.3). There we considered various bio-fuels and how they could be used in fuel cells, mainly by being reformed to a hydrogen-rich gas using steam reforming and other methods mentioned in Sections 8.4, 8.5, and 8.6. This section is NOT about such topics. This section is about how biological methods might be used to extract the hydrogen from *any* fuel – natural gas as well as bio-fuels. In other words, we will discuss the biological alternatives to the various reforming and shift-reaction processes we have been considering.

There is another alternative we consider in this section, which is the direct use of sunlight to generate hydrogen using microbes. This is a biological alternative to using a photovoltaic panel and an electrolyser, as considered in the previous section.

Unlike the previous fuel-processing systems, no commercial bio-fuel systems are available, although many systems for generating fuel gases from biomass (such as wood gasifiers) have been tried. Most biological hydrogen-generation systems are either currently under research or in the early stages of demonstration. However, there is renewed effort being made now in developing biological processes for generating hydrogen, both directly using photosynthesis and from organic materials. These generally use one of three types of metabolic processes such as

- photosynthesis using unicellular microorganisms that use either hydrogenase or nitrogenase reactions,
- fermentation[5] using bacteria to produce hydrogen anaerobically,
- various stepwise processes that use a combination of bacteria to predigest complex organic molecules to make less complex organic material that can then be transformed using hydrogen-producing organisms.

Kaplan (1998) points out that the success rate in developing these processes has not been very good, because

- the growth of organisms is inhibited by catabolites produced in the growth culture,
- hydrogen production is limited because growth slows down as hydrogen concentrations build up,
- the range of feedstocks the organisms can use is narrow,
- the rate of production of other gases is high or the rate of production of hydrogen is low.

8.8.2 Photosynthesis

Photosynthesis consists of two processes – the conversion of light energy to biochemical energy by a photochemical reaction and the reduction of atmospheric carbon dioxide to organic compounds such as sugars. Biochemical catalysts – enzymes – catalyse the carbon dioxide reduction. These processes usually take place in plants, but under certain conditions the pigments in some types of algae can also absorb solar energy. A few groups of algae and cyanobacteria (formerly known as blue-green algae) produce hydrogen rather than sugars via a photosynthesis route. These materials contain hydrogenase or nitrogenase enzymes, and it is these that have the ability to catalyse hydrogen formation.

It was Gaffron and Rubin (1942) who observed that the green alga, *Scenedesmus* produces hydrogen when exposed to light after being kept in the dark and under anaerobic conditions. Further work aimed at elucidating the mechanism of this process identified

---

[5] We do not clearly differentiate between fermentation and digestion as processes. Fermentation has a narrower definition, and so the broader term 'digestion' will normally be used for the processing of organic matter.

hydrogenase as the key enzyme, which appeared to work by reducing water to hydrogen, and at the same time oxidising an electron carrier (ferredoxin). The green algae, therefore, became known as 'water splitting' organisms, and the process is known as *biophotolysis*. Much work was done in the 1980s to develop a continuous flow reactor using green algae and Greenbaum (1998) reported hydrogen production by a *Chlamydomonas* mutant. High efficiencies of light to hydrogen (10–20%) were achieved. Unfortunately, the hydrogenase in green algae is very sensitive to oxygen, which can rapidly deactivate the enzyme's activity, and this has meant that green algae have not been seriously considered for large-scale development. However, recent research at the National Renewable Energy Laboratory (NREL) in the USA (Ghirardi et al., 2001) suggests that hydrogen production can be improved in C. *reinhardtii* by curbing oxygen activity by depleting the algae cells of sulphur and controlling sulphur re-addition. After allowing the algae culture to grow under normal conditions, the NREL team deprived it of both sulphur and oxygen, causing it to switch to an alternate metabolism that generates hydrogen. After several days of generating hydrogen, the algae culture was returned to normal conditions for a few days, allowing it to store up more energy. The process could be repeated many times.

Workers in Japan have found another mode of hydrogen formation in certain green algae. They found that in some cases carbon dioxide can be reduced to starch by photosynthesis in the presence of light and that this can be decomposed to hydrogen and alcohols under anaerobic conditions in the dark (Miura, 1986). Hydrogenases have also been purified and partly characterised in a few cyanobacteria and microalgae (Schulz, 1996) but the mechanisms for the metabolic processes remain unclear.

Nitrogenase is an enzyme responsible for catalysing the fixation of atmospheric nitrogen, and it was Benemann and Weare (1974) who first demonstrated that the nitrogen-fixing cyanobacterium *Anabaena Cylindrica* produces hydrogen and oxygen simultaneously in an argon atmosphere for several hours. Hydrogen production does occur under normal atmospheric conditions but at a much lower rate than nitrogen fixation. It has been shown that hydrogen production can be stimulated by starving the cyanobacteria of nitrogen.

Cyanobacteria have been screened for their hydrogen-producing ability by several researchers. Mitsui (1986) isolated a unicellular aerobic nitrogen fixer, *Synechococcus* sp. Miami BG043511, which is also an excellent hydrogen producer, with a conversion efficiency estimated at around 3.5% based on PAR (photosynthetically active radiation – light of energy 400 to 700 nm in wavelength) using an artificial light source. Certain other bacteria are also able to photosynthesise. These use either organic compounds or reduced sulphur compounds as the electron donors. Some use organic acids as electron donors, and since water reduction is not involved, the efficiency of light energy to hydrogen can be much higher than in the cyanobacteria. Efficiencies of 6 to 8% have been demonstrated using *Rhodobacter* sp in the laboratory (Miyake and Kawamura, 1987). A review of the biochemistry of hydrogenases in cyanobacteria has been published recently by Tamagnini et al. (2002).

Recently, it has been shown that certain photosynthetic bacteria can utilise carbon monoxide and water to produce carbon dioxide and hydrogen gas via the shift reaction (Wolfum, 2001). The implications of this are important since the shift reaction normally requires catalysts operating at high temperatures to ensure reasonable rates (see

Section 8.4.9). If a bio-reactor could host the shift reaction at ambient temperature, the product gas would contain little CO and further expensive gas cleanup could be eliminated, or at least greatly reduced. The work at NREL is at an early stage but preliminary data from a trickle-bed reactor using *Rubrivivax gelatinosus* CBS2 suggests that the rate of reaction is more rapid than the rate at which carbon monoxide can be supplied to the bacterial culture.

A laboratory reactor using a combined system of green algae and photosynthetic bacteria was also operated recently by Kansai Electric Power Co. Ltd. This used the green algae species *Chlamydomonas* MGA 161 and the photosynthetic bacterium, *Rhodovulum sulfidophilum* W-1S. *Chlamydomonas* MGA 161, when allowed to ferment, produces carbohydrates. Under anaerobic conditions and in the presence of argon gas, the photosynthetic bacterium *Rhodovulum sulfidophilum* W-1S then converts the organic carbohydrates to hydrogen.

All the photosynthesis schemes used so far have low efficiencies. However, the recent leaps made in our knowledge of genetic engineering may enable more efficient systems to be developed. One of the most efficient systems seen so far is that of R. *sphaeroides* at $250\,mLh^{-1}$ hydrogen generated per milligram of organism, with a photo energy conversion efficiency of 7% (Miyake and Kawamura, 1987). Before long it is likely that solar efficiencies of 10% will be achieved, close to the efficiencies of photovoltaic systems.

8.8.3 Hydrogen production by digestion processes

Hydrogen can be produced by microbial digestion of organic matter in the absence of light energy. Under relatively mild conditions of temperature and pressure many bacteria will readily produce hydrogen together with acetic acid and other low molecular weight organic acids. Unfortunately, the rates of reaction are usually low, and significant amounts of hydrogen are not produced because of two mitigating factors. Firstly, the inhibition of microbial hydrogenases (hydrogen-generating enzymes) may occur as hydrogen builds up and secondly, hydrogen may react with other organic species present or with carbon dioxide in the system, resulting in methane generation. As the partial pressure of hydrogen builds up, the forward reaction of organic matter to hydrogen becomes thermodynamically unfavourable. The challenge in using digestion processes is therefore to increase the rate of hydrogen production whilst preventing methane formation.

Simply increasing the temperature, as with other catalytic systems, is not an option as this can render organic enzymes and bacteria ineffective. Most digestion processes use batch reactors and work that has been carried out at the University of Glamorgan has shown that increased yields may be achieved by sparging a continuously operating reactor with an inert gas. The project is investigating methods of optimising hydrogen production from agricultural feedstocks (Lakeman and Browning, 2001).

Quantitative measurements for the evolution of hydrogen from various anaerobic bacterial systems have been made (Hart and Womack, 1967) but the values are generally very low. *Ecscherichica coli, aerobacter aerogenes, aerobacter cloacae,* and *pseudomonas* sp are all bacteria that yield measurable amounts of hydrogen. One of the best identified is *Clostridium butyricum* that has been measured to yield 35 mmol (784 mls) hydrogen per hour from 1 g of the microorganism at 37°C (Suzuki, 1983). More recently, the thermophilic *Thermotoga neapolitana* also shows considerable promise (Van Ooteghem et al., 2001). This bacterium has the potential to utilise a variety of organic wastes and to cost-effectively generate significant quantities of hydrogen.

The use of biological processes for hydrogen production is presently at the point of technical system development. Many questions still remain concerning the fundamental biochemical processes that are occurring. At the moment, the photosynthesis-algae-bacteria-system seems to be the best candidate for the first technical application. Present indications are that hydrogen production costs of 12 cents/kWh$_{H2}$ or less are achievable. On the basis of results achieved so far, current research programmes plan to be able to demonstrate technical feasibility within the next 2 years (Logan et al., 2002), with commercialisation being realized within the next 5 to 8 years. This will marry up well with the demand that FCVs in particular are expected to make for hydrogen.

8.9 Hydrogen Storage I – Storage as Hydrogen

8.9.1 Introduction to the problem

Up to now we have mainly considered the production of hydrogen from fossil fuels on an 'as needed' basis. However, there are times when it is more convenient and efficient to store the hydrogen fuel as hydrogen. This is particularly likely to be the case with low-power applications, which would not justify the cost of fuel-processing equipment. It can also be a reasonable way of storing electrical energy from sources such as wind-driven generators and hydroelectric power, whose production might well be out of line with consumption. Electrolysers, as outlined in Section 8.7, might be used to convert the electrical energy to hydrogen during times of high supply and low demand.

A small local store of hydrogen is also essential in the use of fuel cells for portable applications, unless the direct methanol fuel cell is being used.

As a result of its possible importance in the world energy scene as a general-purpose energy vector, a great deal of attention has been given to the very difficult problem of hydrogen storage. The difficulties arise because although hydrogen has one of the highest specific energies (energy per kilogram) – which is why it is the fuel of choice for space missions – its density is very low, and it has one of the lowest energy densities (energy per cubic metre). This means that to get a large mass of hydrogen into a small space very high pressures have to be used. A further problem is that, unlike other gaseous energy carriers, it is very difficult to liquefy. It cannot be simply compressed in the way that LPG or butane can. It has to be cooled down to about 22 K, and even in liquid form its density is quite low at 71 kg m^{-3}.

Although hydrogen can be stored as a compressed gas or a liquid, there are other methods that are being developed. Chemical methods can also be used. These are considered in the next section. The methods of storing hydrogen that will be described in this section are as follows:

- Compression in gas cylinders
- Storage as a cryogenic liquid
- Storage in a metal absorber – as a reversible metal hydride
- Storage in carbon nanofibres.

None of these methods are without considerable problems, and in each situation their advantages and disadvantages will play differently. However, before considering them in detail we must address the vitally important issue of safety in connection with storing and using hydrogen.

8.9.2 Safety

Hydrogen is a unique gaseous element, possessing the lowest molecular weight of any gas. It has the highest thermal conductivity, velocity of sound, mean molecular velocity, and the lowest viscosity and density of all gases. Such properties lead hydrogen to have a leak rate through small orifices faster than all other gases. Hydrogen leaks 2.8 times faster than methane and 3.3 times faster than air. In addition, hydrogen is a highly volatile and flammable gas, and in certain circumstances hydrogen and air mixtures can detonate. The implications for the design of fuel cell systems are obvious, and safety considerations must feature strongly (see Figure 8.20).

Hydrogen therefore needs to be handled with care. Systems need to be designed with the lowest possible chance of any leaks and should be monitored for such leaks regularly.

Figure 8.20   Icon of a myth. The 'Hindenburg disaster' of 6th May, 1937 put an end to the airship as a means of transport, and it has also been a major public relations disaster for hydrogen, since this was the lifting gas used. The accident led to the widely held myth that hydrogen is a particularly dangerous substance. Although the accident was tragic for those involved, the number of casualties was 37, quite low for an aircraft crash. About 2/3 of those on board survived. Many of those who died were burnt by the diesel fuel for the propulsion system, and in any case the fire did not start with the hydrogen, but with the skin of the airship, which was made of a highly flammable compound (Bain and VanVorst, 1999).

Table 8.10 Properties relevant to safety for hydrogen and two other commonly used gaseous fuels

	Hydrogen	Methane	Propane
Density, $kg\,m^{-3}$ at NTP	0.084	0.65	2.01
Ignition limits in air, volume% at NTP	4.0 to 77	4.4 to 16.5	1.7 to 10.9
Ignition temperature, °C	560	540	487
Min. ignition energy in air, MJ	0.02	0.3	0.26
Max. combustion rate in air, $m\,s^{-1}$	3.46	0.43	0.47
Detonation limits in air, volume%	18 to 59	6.3 to 14	1.1 to 1.3
Stoichiometric ratio in air	29.5	9.5	4.0

However, it should be made clear that, all things considered, hydrogen is no more dangerous, and in some respects it is rather less dangerous than other commonly used fuels. Table 8.10 gives the key properties relevant to safety of hydrogen and two other gaseous fuels widely used in homes, leisure, and business – methane and propane.

From this table the major problem with hydrogen appears to be the minimum ignition energy, apparently indicating that a fire could be started very easily. However, all these energies are in fact very low, lower than those encountered in most practical cases. A spark can ignite any of these fuels. Furthermore, against this must be set the much higher minimum concentration needed for detonation −18% by volume. The lower concentration limit for ignition is much the same as for methane, and a considerably lower concentration of propane is needed. The ignition temperature for hydrogen is also noticeably higher than for the other two fuels.

Another potential hazard arises from the rather greater range of concentrations needed to cause detonation. This means that care must be taken to prevent the build up of hydrogen in confined spaces. Fortunately, this is usually easily done; the high buoyancy, and high average molecular velocity ensures that hydrogen is the most rapidly dispersing of all gases.

A safety problem that can arise with hydrogen is that when it is burning the flame is virtually invisible. Fire-fighting teams will almost certainly have the necessary equipment to detect such fires, but this point should be borne in mind by non-specialists.

Considering these figures, hydrogen seems much the same as the other fuels from the point of view of potential danger. It is the much lower density that gives hydrogen a comparative advantage from a safety point of view. The density of methane is similar to air, which means it does not disperse quickly, but tends to mix with the air. Propane has a lower density than air, which tends to make it sink and collect at low points, such as in basements, in drains, and in the hulls of boats, where it can explode or set alight with devastating effects. Hydrogen on the other hand is so light that it rapidly disperses upwards. This means that the concentration levels necessary for ignition or detonation are very unlikely to be achieved.

Hydrogen, like all fuels, must be carefully handled. However, taking all things into account, it does not present any greater potential for danger than any other flammable liquids or gases in common use today. In some applications, for example boats, it has many safety advantages over what is generally used at the moment. In the sections that follow, we consider different storage methods. Some of these bring special safety hazards of their own, which we consider in turn.

8.9.3 The storage of hydrogen as a compressed gas

Storing hydrogen gas in pressurised cylinders is the most technically straightforward method and the most widely used one for small amounts of the gas. Hydrogen is stored in this way at thousands of industrial, research, and teaching establishments, and in most locations local companies can readily supply such cylinders in a wide range of sizes. However, in these applications the hydrogen is nearly always a chemical reagent in some analytical or production process. When we consider using and storing hydrogen in this way as an energy vector, then the situation appears less satisfactory.

Two systems of pressurised storage are compared in Table 8.11. The first is a standard steel alloy cylinder at 200 bar, of the type commonly seen in laboratories. The second is for larger scale hydrogen storage on a bus, as described by Zieger (1994). This tank is constructed with a 6-mm-thick aluminium inner liner, around which is wrapped a composite of aramide fibre and epoxy resin. This material has a high ductility, which gives it good burst behaviour, in that it rips apart rather than disintegrating into many pieces. The burst pressure is 1200 bar, though the maximum pressure used is 300 bar.[6]

The larger-scale storage system is, as expected, a great deal more efficient. However, this is slightly misleading. These large tanks have to be held in the vehicle, and the weight needed to do this should be taken into account. In the bus described by Zieger (1994), which used hydrogen to drive an internal combustion engine, 13 of these tanks were mounted in the roof space. The total mass of the tanks and the bus structure reinforcements is 2550 kg, or 196 kg per tank. This brings down the 'storage efficiency' of the system to 1.6%, not so very different from the steel cylinder. Another point is that in both systems we have ignored the weight of the connecting valves, and of any pressure reducing regulators. For the 2-L steel cylinder system, this would typically add about 2.15 kg to the mass of the system, and reduce the storage efficiency to 0.7% (Kahrom, 1999).

Table 8.11 Comparative data for two cylinders used to store hydrogen at high pressure. The first is a conventional steel cylinder and the second a larger composite tank for use on a hydrogen-powered bus

	2 L steel, 200 bar	147 L composite, 300 bar
Mass of empty cylinder	3.0 kg	100 kg
Mass of hydrogen stored	0.036 kg	3.1 kg
Storage efficiency (% mass H_2)	1.2%	3.1%
Specific energy	0.47 kWh kg^{-1}	1.2 kWh kg^{-1}
Volume of tank (approx.)	2.2 L (0.0022 m^3)	220 L (0.22 m^3)
Mass of H_2 per litre	0.016 kg L^{-1}	0.014 kg L^{-1}

---

[6] It should be noted that at present composite cylinders are about three times the cost of steel cylinders of the same capacity.

The reason for the low mass of hydrogen stored, even at such very high pressures, is of course its low density. The density of hydrogen gas at NTP is $0.084\,\mathrm{kg\,m^{-3}}$, compared to air, which is about $1.2\,\mathrm{kg\,m^{-3}}$. Usually less than 2% of the storage system mass is actually hydrogen itself.

Pressurised hydrogen gas is mainly used in fairly small quantities and has to be transported with great care. For small-scale users of fuel cells the figures quoted by Kahrom (1999), which take into account all costs, including depreciation of cylinders, administration, cost of pressure reducing valves and so on, estimate the cost of hydrogen as about \$2.2 per gram. Using the figures given in Appendix 2, Section A2.4, we can see that this corresponds to about \$56 per kilowatthour, or about \$125 per kilowatthour for the electricity from a fuel cell of efficiency 45%. This is absurdly expensive when compared to mains electricity, but is considerably cheaper than the cost of electricity from primary batteries.

The metal that the pressure vessel is made from needs very careful selection. Hydrogen is a very small molecule, of high velocity, and so it is capable of diffusing into materials that are impermeable to other gases. This is compounded by the fact that a very small fraction of the hydrogen gas molecules may dissociate on the surface of the material. Diffusion of atomic hydrogen into the material may then occur, which can affect the mechanical performance of materials in many ways. Gaseous hydrogen can build up in internal blisters in the material, which can lead to crack promotion (hydrogen-induced cracking). In carbonaceous metals such as steel, the hydrogen can react with carbon forming entrapped CH_4 bubbles. The gas pressure in the internal voids can generate an internal stress high enough to fissure, crack, or blister the steel. The phenomenon is well known and is termed *hydrogen embrittlement*. Certain chromium-rich steels and Cr-Mo alloys have been found that are resistant to hydrogen embrittlement. Composite reinforced plastic materials are also used for larger tanks, as has been outlined above.

In addition to the problem of very high mass, there are considerable safety problems associated with storing hydrogen at high pressure. A leak from such a cylinder would generate very large forces as the gas is propelled out. It is possible for such cylinders to become essentially jet-propelled torpedoes and to inflict considerable damage. Furthermore, vessel fracture would most likely be accompanied by autoignition of the released hydrogen and air mixture, with an ensuing fire lasting until the contents of the ruptured or accidentally opened vessel are consumed (Hord, 1978). Nevertheless, this method is widely and safely used, provided the safety problems, especially those associated with the high pressure, are avoided by correctly following the due procedures. In vehicles, for example, pressure relief valves or rupture discs (see next section) are fitted, which will safely vent gas in the event of a fire, for example. Similarly, pressure regulators attached to hydrogen cylinders are fitted with flame traps to prevent ignition of the hydrogen. The main advantages of storing hydrogen as a compressed gas are as follows:

- Simplicity
- Indefinite storage time
- No purity limits on the hydrogen.

It is most widely used in places where the demand for hydrogen is variable and not so high. It is also used for buses, both for fuel cells and internal combustion engines. It is well suited to storing the hydrogen from electrolysers that are run at times of excess electricity supply. One such system is shown in Figure 8.19.

8.9.4 Storage of hydrogen as a liquid

The storage of hydrogen as a liquid (commonly called LH_2), at about 22 K, is currently the only widely used method of storing large quantities of hydrogen. A gas cooled to the liquid state in this way is known as a cryogenic liquid. Large quantities of cryogenic hydrogen are currently used in processes such as petroleum refining and ammonia production. Another notable user is NASA, which has huge $3200\,m^3$ ($850\,000$ US gallon) tanks to ensure a continuous supply for the space programme. Figure 8.21 shows a tank for storing cryogenic hydrogen in Hamburg, Germany. Two fuel cell systems, of power about 200 kW, can also be seen to the right of the picture.

The hydrogen container is a large, strongly reinforced vacuum (or Dewar) flask. The liquid hydrogen will slowly evaporate, and the pressure in the container is usually maintained below 3 bar, though some larger tanks may use higher pressures. If the rate of evaporation exceeds the demand, then the tank is occasionally vented to make sure the pressure does not rise too high. A spring-loaded valve will release and close again when the pressure falls. The small amounts of hydrogen involved are usually released to the atmosphere, though in very large systems it may be vented out through a flare stack and burnt. As a back-up safety feature, a rupture disc is usually also fitted. This consists of a ring covered with a membrane of controlled thickness, so that it will withstand a certain pressure. When a safety limit is reached, the membrane bursts, releasing the gas. However, the gas will continue to be released until the disc is replaced. This will not be done till all the gas is released and the fault is rectified.

Figure 8.21 LH2 tank on the left. To the right of the picture are two fuel cell systems, each providing about 200 kW of electrical power and 200 kW of heat to the houses and offices nearby. (Photograph reproduced by kind permission of Hamburgische Electricitäts Werke AG.)

Table 8.12   Details of a cryogenic hydrogen container suitable for cars

Mass of empty container	51.5 kg
Mass of hydrogen stored	8.5 kg
Storage efficiency (% mass H_2)	14.2%
Specific energy	5.57 kWh kg^{-1}
Volume of tank (approx.)	0.2 m^3
Mass of H_2 per litre	0.0425 kg L^{-1}

When the LH_2 tank is being filled, and when fuel is being withdrawn, it is most important that air is not allowed into the system, otherwise an explosive mixture could form. The tank should be purged with nitrogen before filling.

Although usually used to store large quantities of hydrogen, considerable work has gone into the design and development of LH_2 tanks for cars, though this has not been directly connected with fuel cells. BMW, among other automobile companies, has invested heavily in hydrogen power internal combustion engines, and these have nearly all used LH_2 as the fuel. Such tanks have been through very thorough safety trials. The tank used in their hydrogen-powered cars is cylindrical in shape, and is of the normal double wall, vacuum or Dewar flask type of construction. The walls are about 3 cm thick and consist of 70 layers of aluminium foil interlaced with fibreglass matting. The maximum operating pressure is 5 bar. The tank stores 120 L of cryogenic hydrogen. The density of LH_2 is very low, about 71 kg m^{-3}, so 120 L is only 8.5 kg (Reister and Strobl, 1992). The key figures are shown in Table 8.12.

The hydrogen fuel feed systems used for car engines cannot normally be applied unaltered to fuel cells. One notable difference is that in LH_2 power engines the hydrogen is often fed to the engine still in the liquid state. If it is a gas, then being at a low temperature is an advantage, as it allows a greater mass of fuel/air mixture into the engine. For fuel cells, the hydrogen will obviously need to be a gas, and preheated as well. However, this is not a very difficult technical problem, as there is plenty of scope for using waste heat from the cell via heat exchangers.

One of the problems associated with cryogenic hydrogen is that the liquefaction process is very energy-intensive. Several stages are involved. The gas is first compressed and then cooled to about 78 K using liquid nitrogen. The high pressure is then used to further cool the hydrogen by expanding it through a turbine. (See Section 8.8.) An additional process is needed to convert the H_2 from the isomer where the nuclear spins of both atoms are parallel (*ortho*-hydrogen) to that where they are anti-parallel (*para*-hydrogen). This process is exothermic, and if allowed to take place naturally would cause boil-off of the liquid. According to figures provided by a major hydrogen producer, and given by Eliasson and Bossel (2002), the energy required to liquefy the gas under the *very best of circumstances* is about 25% of the specific enthalpy or heating value of the hydrogen. For modern plants, this means liquefying over 1000 kg h^{-1}. For plants working at about 100 kg h^{-1}, hardly a small rate, the proportion of the energy lost rises to about 45%. In overall terms then, this method is a highly inefficient way of storing and transporting energy.

Figure 8.22 Refuelling with LH2. Note that there are two pipes. The system is fully sealed to prevent the inlet of air. One pipe carries in the LH2, the other draws off the hydrogen gas that is formed by the slowly evaporating liquid. (Photograph reproduced by kind permission of Hamburgische Electricitäts Werke AG.)

In addition to the regular safety problems with hydrogen, there are a number of specific difficulties concerned with cryogenic hydrogen. Frostbite is a hazard of concern. Human skin can easily become frozen or torn if it comes into contact with cryogenic surfaces. All pipes containing the fluid must be insulated, as must any parts in good thermal contact with these pipes. Insulation is also necessary to prevent the surrounding air from condensing on the pipes, as an explosion hazard can develop if liquid air drips onto nearby combustibles. Asphalt, for example, can ignite in the presence of liquid air. (Concrete paving is used around static installations.) Generally though, the hazards of hydrogen are somewhat less with LH2 than with pressurised gas. One reason is that if there is a failure of the container, the fuel tends to remain in place and vent to the atmosphere more slowly. Certainly, LH2 tanks have been approved for use in cars in Europe. Figure 8.22 shows a van being filled with liquid hydrogen in a scheme for company vehicles operating in Hamburg, Germany. Although this vehicle uses an internal combustion engine, it shows that an infrastructure based on LH2 is possible and could be applied to fuel cells.

8.9.5 Reversible metal hydride hydrogen stores

The reader might well question the inclusion of this method in this section, rather than with the chemical methods that follow. However, although the method is chemical in its operation, that is not in any way apparent to the user. No reformers or reactors are needed to make the systems work. They work exactly like a hydrogen 'sponge' or 'absorber'. For this reason it is included here.

Certain metals, particularly mixtures (alloys) of titanium, iron, manganese, nickel, chromium, and others, can react with hydrogen to form a metal hydride in a very easily controlled reversible reaction. The general equation is

$$M + H_2 \leftrightarrow MH_2 \qquad [8.15]$$

One example of such an alloy is titanium iron hydride, the last entry in Table 8.16 of the following section. In terms of *mass* this is not a very promising material; it is the *volumetric* measure that is the advantage of these materials. It requires one of the lowest volumes to store 1 kg in Table 8.16; certainly it is one of the lowest practical materials. It actually holds more hydrogen per unit volume than pure liquid hydrogen.[7]

To the right, the reaction of 8.15 is mildly exothermic. To release the hydrogen then, small amounts of heat must be supplied. However, metal alloys can be chosen for the hydrides so that the reaction can take place over a wide range of temperatures and pressures. In particular, it is possible to choose alloys suitable for operating at around atmospheric pressure and room temperature.

The system then works as follows: hydrogen is supplied at a little above atmospheric pressure to the metal alloy, inside a container. Reaction 8.15 proceeds to the right, and the metal hydride is formed. This is mildly exothermic, and in large systems some cooling will need to be supplied, but normal air cooling is often sufficient. This stage will take a few minutes, depending on the size of the system, and if the container is cooled. It will take place at approximately constant pressure.

Once all the metal has reacted with the hydrogen, then the pressure will begin to rise. This is the sign to disconnect the hydrogen supply. The vessel, now containing the metal hydride, will then be sealed. Note that the hydrogen is only stored at modest pressure, typically up to 2 bar.

When the hydrogen is needed, the vessel is connected to, for example, the fuel cell. Reaction 8.15 then proceeds to the left and hydrogen is released. If the pressure rises above atmospheric pressure, the reaction will slow down or stop. The reaction is now endothermic, so energy must be supplied. This is supplied by the surroundings – the vessel will cool slightly as the hydrogen is given off. It can be warmed slightly to increase the rate of supply, using, for example, warm water or the air from the fuel cell cooling system.

Once the reaction is completed, and all the hydrogen has been released, then the whole procedure can be repeated.

Usually several hundred charge/discharge cycles can be completed. However, rather like rechargeable batteries, these systems can be abused. For example, if the system is filled at high pressure, the charging reaction will proceed too fast and the material will get too hot and will be damaged. Also, the system is damaged by impurities in the hydrogen, so high purity hydrogen should be used.

The raw hydride material, as listed in Table 8.16, cannot of course be used by itself. It has to be contained in a vessel. Although the hydrogen is not stored at pressure, the container must be able to withstand reasonably high pressure, as it is likely to be filled from a high-pressure supply, and allowance must be made for human error. For

---

[7] This may at first sight seem impossible. However, contrary to what a basic understanding of kinetic theory might suggest, the molecules in liquid hydrogen are still very widely spaced apart. This can be seen from its density, which is only $71 \, kg \, m^{-3}$. When bonded to the metal the molecules are actually closer together, though the material is obviously very much denser, at $5470 \, kg \, m^{-3}$.

example, the unit shown in Figure 8.23 will be fully charged at a pressure of 3 bar, but the container can withstand 30 bar. The container will also need valves and connectors. Even taking all these into account, impressive practical devices can be built. Table 8.13 gives details of the very small 20-SL holder for applications such as portable electronics equipment, manufactured by GfE Metalle und Materialien GMBH of Germany, and shown in Figure 8.23. The volumetric measure, mass of hydrogen per litre, is nearly as good as for LH$_2$, and the gravimetric measure is not a great deal worse than for compressed gas, and very much the same as for a small compressed cylinder.

One of the main advantages of this method is its safety. The hydrogen is not stored at a significant pressure, and so cannot rapidly and dangerously discharge. Indeed, if the valve is damaged, or there is a leak in the system, the temperature of the container will fall, which will inhibit the release of the gas. The low pressure greatly simplifies the design of the fuel supply system. It thus has great promise for a very wide range of applications

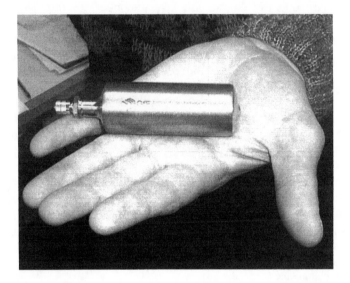

Figure 8.23 Small metal hydride hydrogen store for fuel cells used with small portable electronic equipment.

Table 8.13 Details of a very small metal hydride-hydrogen container suitable for portable electronics equipment

Mass of empty container	0.26 kg
Mass of hydrogen stored	0.0017 kg
Storage efficiency (% mass H$_2$)	0.65%
Specific energy	0.26 kWh kg^{-1}
Volume of tank (approx.)	0.06 L
Mass of H$_2$ per litre	0.028 kg L^{-1}

where small quantities of hydrogen are stored. It is also particularly suited to applications where weight is not a problem, but space is. A good example is fuel cell – powered boats, where weight near the bottom of the boat is an advantage, and is often artificially added, but space is at a premium.

The disadvantages are particularly noticeable where larger quantities of hydrogen are to be stored, for example, in vehicles. The specific energy is poor. Also, the problem of heating during filling and cooling during release of hydrogen becomes more acute. Large systems have been tried for vehicles, and a typical refill time is about 1 h for an approximately 5-kg tank. The other major disadvantage is that usually very high purity hydrogen must be used; otherwise the metals become contaminated as they react irreversibly with the impurities.

8.9.6 Carbon nanofibres

In 1998 a paper was published on the absorption of hydrogen in carbon nanofibres (Chambers et al., 1998). The authors presented results suggesting that these materials could absorb in excess of 67% of hydrogen. This amazingly large figure took the academic world by storm, as it offered the prospect of levels of hydrogen storage in material at ambient temperatures and pressures, which would ensure hydrogen a certain future as a fuel for vehicles. Using the new graphitic storage material, it was predicted that a fuel cell car with hydrogen stored in carbon nanofibres would run for 5000 miles on one charge of fuel! Before long several groups around the world were carrying out research on hydrogen storage in carbon nanofibres. The initial euphoria became tempered as various groups tried to repeat the work and found that they could not – only relatively small amounts of hydrogen were absorbed. Nevertheless, the US Department of Energy set a benchmark for hydrogen storage in FCVs of $62 \, kg \, H_2/m^3$ and 6.5% weight (i.e. the percentage weight of stored hydrogen to the total system weight). This would provide enough hydrogen in a car for a journey of some 300 miles. The DOE target continues to stimulate a lot of research in this area but so far no carbon nanoscale material has reproducibly come near this figure.

So what are these new *nanoscale* fibrous materials? The story of carbon nanofibres really started with the discovery of fullerenes[8] in the early 1980s (Kroto et al., 1985) when it was found that new molecular structures could be synthesised at the nanoscale, which have fascinating and useful properties. By nanoscale we mean of the dimension of nanometres (1 nm is 10^{-9} m). So we are talking about the sorts of units you need to describe the sizes of individual atoms/molecules.[9] Carbon nanomaterials can be made by a variety of processes and can be examined under the electron microscope when their structure becomes apparent.

Amorphous and graphitic carbon have been used as an absorbant for some materials for many years. 'Activated carbon' has been available as a storage and filter medium

---

[8] Interestingly, in theory a molecule of Buckminster fullerene (C60) can be hydrogenated up to $C_{60}H_{60}$, which corresponds to almost 7.7 wt.% hydrogen absorbed. In the United States, MER Corporation has demonstrated absorptions up to 6.7% on fullerenes but the absorption is very slow and so far no practical use has been made of this.

[9] These distances are tiny. To use an oft-quoted comparison, a human hair is between 100,000 and 200,000 nanometres thick, while a typical virus can be just 100 nm wide. Atoms themselves are typically between one-tenth and half of a nanometre wide. Scientists also measure small distances in Angstrom units. 1 Angstrom is 10^{-10} m.

industrially for decades. However, activated carbon is really ineffective for storing hydrogen because the hydrogen molecules only weakly absorb on the carbon surface at room temperatures, despite such materials having high surface areas $(2000\,m^2\,g^{-1})$.[10] On the other hand, carbon whose filaments are at the nanometre scale would be expected to have a larger surface area and porosity, and therefore be able to store much more hydrogen than normal graphite. The question is whether carbon nanofibres can live up to their promise Molecular simulations and quantum-mechanical calculations tell us that the very high storage densities that were first reported are inconsistent with theory. Later work (Park et al., 1999 and Yang, 2000) has suggested that some of the high absorption figures were a result of the presence of water vapour in the samples, which tended to expand the graphitic layers in the carbon structure so that multiple layers of hydrogen atoms would absorb. Meticulous care is needed in measuring hydrogen uptake on very small samples of materials.

Without going into too much detail, we should point out that there are really three types of carbon nanofibres that are now being investigated for hydrogen storage: graphitic nanofibres, single-walled carbon nanotubes, and multi-walled carbon nanotubes. The essential structures of these materials are shown in Figure 8.24.

Graphitic nanofibres are made by decomposing hydrocarbons or carbon monoxide over suitable catalysts. The catalyst influences the arrangement of the graphitic sheets

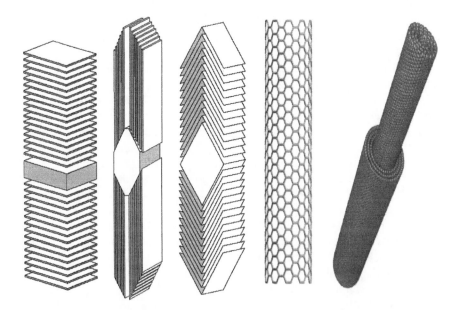

Figure 8.24 Schematic representation of different types of carbon nanofibres: (a) graphitic platelets, (b) graphitic ribbons, (c) graphitic herringbone, (d) single-walled nanotube and (e) multi-walled nanotube.

---

[10] Physical absorption on activated carbon is increased at low temperatures. Super activated carbons have been developed which can store up to 7 wt.% hydrogen at pressures of 40 bar and temperatures of 165 K.

that are produced, and many different types of structures have been observed, with the most common being described as platelets, ribbons, and a herringbone arrangement (see Figure 8.24). Graphitic nanofibres vary from around 50 nm to 1000 nm in length and 5 to 100 nm in diameter (Chambers et al., 1998). There is a wide range of reported hydrogen absorption values on graphitic nanofibres (Atkinson et al., 2001), and a consensus has yet to be reached on the practicality of using these materials in real gas storage systems.

Carbon nanotubes, either single walled (typical diameter 1–2 nm), or multi-walled (diameter 5–50 nm), are an alternative to the graphitic nanofibre systems. They are special because of their small dimensions, each tube possessing a smooth and highly ordered graphitic surface. Single-walled nanotubes (SWNT) (Figure 8.24d) can be thought of as a one-dimensional Buckminster fullerene, or as a tube made up of hexagonal graphite plates linked together. Carbon nanotubes were first identified by Iijkima (1991) who prepared them by accident, using an electric arc drawn between two carbon electrodes. Nanotubes are still produced by electric arcs, but more recently other methods have also been employed, such as laser ablation, and chemical vapour deposition (Ajayan, 2001). Unfortunately, there is a cost penalty associated with using such high-tech methods of preparation.

As with graphitic nanofibres, there is yet to be a consensus reached on the efficacy of carbon nanotubes for hydrogen storage. Experimental results are often difficult to interpret, and issues of purity of materials and subtleties in experimental techniques (gravimetric vs. volumetric methods) make it difficult to draw firm conclusions. The experimentalist is often trying to measure very small changes in weight or volume, for example, Dillon et al. (1997) reported absorptions of ca 0.01 wt% H_2 on SWNT, which had a purity of only 0.01%, from which he extrapolated that 5 to 10 wt% could be absorbed on pure materials. Before long, Dillon and some others were able to show absorption capacities of a few percent were indeed measurable on some purer materials. However, these findings were later contradicted by Hirscher et al. (2001), and even more recently by Dillons group (Dillon, 2001) who showed that absorptions of only ca. 1 to 2 wt% H+ could reproducibly be absorbed on impure SWNTs.

Some success has been reported using alkali-doped multi-walled carbon nanotubes (MWCT). Values of 20% uptake have been reported at 20 to 400°C at ambient pressure by Chen et al. (1999). Yet again, lower figures (2.5 wt%) were obtained when other researchers tried to repeat the experiments. Most recent work has, however, started to demonstrate a reproducibility and consistency amongst the experiments (Dillon, 2001). This is giving grounds for hope that carbon nanofibres will feature in real hydrogen storage systems in the future. Certainly the fabrication of nanofibres is an interesting and fascinating area of research. It is somewhat controversial and the jury is still out on the practical feasibility of using these materials for cost-effective hydrogen storage. The variability in experimentally measured hydrogen storage values is illustrated in Table 8.14, which shows values obtained with different carbon nanotubes, compared with graphitic nanofibres and other methods of hydrogen storage.

8.9.7 Storage methods compared

Table 8.15 shows the range of gravimetric and volumetric measures for the three systems described above that are available now. Obviously these figures cannot be used in isolation – they do not include cost, for example. Safety aspects do not appear in this table

Table 8.14 Summary of experimentally reported hydrogen storage capacities in carbon nanofibres (adapted from Atkinson (2001) and Cheng (2001))

Material	Gravimetric hydrogen storage amount (wt%)	Storage temp. (K)	Storage Pressure (Mpa)
GNFs (tubular)	11.26	298	11.35
GNFs (herring bone)	67.55	298	11.35
GNFs (platelet)	53.68	298	11.35
GNFs	6.5	300	12
GNFs	~10	300	8, 12
CNFs	~10	300	10.1
CNFs	~5	300	10.1
SWNTs (low purity)	5–10 (prediction)	273	0.04
SWNTs (high purity)	3.5~4.5	298	0.04
SWNTs (high purity)	8.25	80	7.18
SWNTs (50% purity)	4.2	300	10.1
SWNTs	0.1	300~520	0.1
SWNT (50–70% purity electrochemical)	2	—	—
Li–MWNTs	20.0	473–673	0.1
K–MWNTs	14.0	<313	0.1
Li–MWNTs	~2.5	473–673	0.1
K–MWNTs	~1.8	313	0.1
Li/K–GNTs(SWNT)	~10	300	8, 12
MWNTs	~5	300	10
MWNTs (electrochemical)	<1	—	—
Nanostructured graphite	7.4	300	1.0
FeTi hydride	<2	>263	2.5
NiMg-hydride	<4	>523	2.5
Cryoadsorption	~5	~77	2.0
Iso-octane/gasoline	17.3	>233	0.1

GNF = graphitic nanofibre, SWNT = single-walled nanotube, MWNT = multi-walled nanotube.

Table 8.15 Data for comparing methods of storing hydrogen fuel

Method	Gravimetric storage efficiency, % mass hydrogen	Volumetric mass (in kg) of hydrogen per litre
Pressurised gas	0.7–3.0	0.015
Reversible metal hydride	0.65	0.028
Cryogenic liquid	14.2	0.040

either. The liquid hydrogen system is plainly not suitable for very small-scale applications. Another factor is that two of the systems (pressurised and liquid) are also methods of hydrogen transport and supply, whereas the reversible metal hydride system is only a store and must be refilled from a *local* source of hydrogen.

8.10 Hydrogen Storage II – Chemical Methods

8.10.1 Introduction

None of the methods for storing hydrogen outlined in Section 8.9 is entirely satisfactory. Other approaches that are being developed rely on the use of chemical 'hydrogen carriers'. These could also be described as 'man-made fuels'. There are many compounds that can be manufactured that hold, for their mass, quite large quantities of hydrogen. To be useful these compounds must pass three tests:

1. It must be possible to very easily make these compounds give up their hydrogen – otherwise there is no advantage over using a reformed fuel in one of the ways already described in this chapter.
2. The manufacturing process must be simple and use little energy – in other words the energy and financial costs of putting the hydrogen into the compound must be low.
3. They must be safe to handle.

A large number of chemicals that show promise have been suggested or tried. Some of these, together with their key properties, are listed in Table 8.16. Many of them do not warrant a great deal of consideration, as they easily fail one or more of the three tests mentioned above. Hydrazine is a good example. It passes the first test very well, and it has been used in demonstration fuel cells with some success, and was mentioned in connection with 'dissolved fuel' alkaline fuel cells in Chapter 5. However, hydrazine is both highly toxic and very energy intensive to manufacture, and so fails the second and third tests. Nevertheless, several of the compounds of Table 8.16 are being considered for practical applications, and will be described in more detail. (Note that the metal hydride 'hydrogen absorbers' at the bottom of the table have already been considered, in Section 8.9.5.)

8.10.2 Methanol

The important properties of methanol have already been outlined in Chapter 6. Also, reformers for extracting the hydrogen from methanol, designed for mobile applications, have been described already in Section 8.6. Methanol is the 'man-made' carrier of hydrogen that is attracting the most interest among fuel cell developers.

As we have seen, methanol can be reformed to hydrogen by steam reforming, according to the following reaction:

$$CH_3OH + H_2O \rightarrow CO_2 + 3H_2$$

Table 8.16 Potential hydrogen storage materials. The 'volume to store 1 kg' of H_2 figure excludes all the extra equipment needed to hold or process the compound, so it is not a practical figure. For example, all the simple and complex hydrides need large quantities of water, from which some of the hydrogen is also released

Name	Formula	Percent hydrogen	Density kg L^{-1}	Vol. (L) to store 1 kg H_2	Notes
Liquids					
Liquid H_2	H_2	100	0.07	14	Cold, $-252°C$
Ammonia	NH_3	17.76	0.67	8.5	Toxic, 100 ppm
Liquid methane	CH_4	25.13	0.415	9.6	Cold $-175°C$
Methanol	CH_3OH	12.5	0.79	10	
Ethanol	C_2H_5OH	13.0	0.79	9.7	
Hydrazine	N_2H_4	12.58	1.011	7.8	Highly toxic
30% sodium borohydride solution	$NaBH_4 + H_2O$	6.3	1.06	15	Expensive, but works well
Simple hydrides					
Lithium hydride	LiH	12.68	0.82	6.5	Caustic
Beryllium hydride	BeH_2	18.28	0.67	8.2	Very toxic
Diborane	B_2H_6	21.86	0.417	11	Toxic
Sodium hydride	NaH	4.3	0.92	25.9	Caustic, but cheap
Calcium hydride	CaH_2	5.0	1.9	11	
Aluminium hydride	AlH_3	10.8	1.3	7.1	
Silane	SiH_4	12.55	0.68	12	Toxic 0.1 ppm
Potassium hydride	KH	2.51	1.47	27.1	Caustic
Titanium hydride	TiH_2	4.40	3.9	5.8	
Complex hydrides					
Lithium borohydride	$LiBH_4$	18.51	0.666	8.1	Mildly toxic
Aluminium borohydride	$Al(BH_4)_3$	16.91	0.545	11	
Sodium borohydride	$NaBH_4$	10.58	1.0	9.5	
Palladium hydride	Pd_2H	0.471	10.78	20	
Titanium iron hydride	$TiFeH_2$	1.87	5.47	9.8	See Section 8.9.5

The equipment is much more straightforward, though the process is not so efficient, if the partial oxidation route is used, for which the reaction is

$$2CH_3OH + O_2 \rightarrow 2CO_2 + 4H_2$$

The former would yield 0.188 kg of hydrogen for each kilogram of methanol, the latter 0.125 kg of hydrogen for each kilogram of methanol. We have also seen in Section 8.6 that *autothermal* reformers use a combination of both these reactions, and this attractive alternative would provide a yield somewhere between these two figures. However, whatever reformer is used, full utilisation is not possible – it never is with gas mixtures containing carbon dioxide, as there must still be some hydrogen in the exit gas. Also, in

Table 8.17   Speculative data for a hydrogen source,
storing 40 L (32 kg) of methanol

Mass of reformer and tank	64 kg
Mass of hydrogen stored[a]	4.4 kg
Storage efficiency (% mass H2)	6.9%
Specific energy	$5.5 \, kWh \, kg^{-1}$
Volume of tank + reformer	$0.08 \, m^3$
Mass of H_2 per litre	$0.055 \, kg \, L^{-1}$

[a]Assuming 75% conversion of available H_2 to usable H_2.

the case of steam reforming, some of the product hydrogen is needed to provide energy
for the reforming reaction.

If we assume that the hydrogen utilisation can be 75%, then we can obtain 0.14 kg
of hydrogen for each kilogram of methanol. As we have seen, a particularly pertinent
application of methanol-derived hydrogen is the case of motor vehicles. We can speculate
that a 40-L tank of methanol might be used, with a reformer of about the same size and
weight as the tank. Such a system should be possible in the reasonably near term, and
would give the figures shown in Table 8.17:

Similar tables could be drawn up for a POX reformer. The yield of hydrogen is less,
but the reformer is smaller and lighter. If the storage efficiency (13.9%) and mass of
H_2 per litre (0.055) figures are compared with those for the three systems described in
the previous section, and shown in Table 8.15, it will immediately become clear why
methanol systems are looked on with such favour, and why they are receiving a great
deal of attention for systems of power above about 10 W right through to tens of kilowatts.

We should note that *ethanol*, according to the figures of Table 8.16, should be just as
promising as methanol as a hydrogen carrier. In Chapter 6 we compared methanol with
ethanol. In this context its main disadvantage is that the equivalent reformation reactions
of equations given above for methanol do not proceed nearly so readily, making the
reformer markedly larger, more expensive, less efficient, and more difficult to control.
Ethanol is also usually somewhat more expensive. All these disadvantages more than
counter its very slight higher hydrogen content.

8.10.3 Alkali metal hydrides

An alternative to the reversible metal hydrides (Section 8.9.5) are alkali metal hydrides,
which react with water to release hydrogen and produce a metal hydroxide. Some of these
are shown in Table 8.16. Bossel (1999) has described a system using calcium hydride that
reacts with water to produce calcium hydroxide and release hydrogen:

$$CaH_2 + 2H_2O \rightarrow Ca(OH)_2 + 2H_2 \qquad [8.16]$$

It could be said that the hydrogen is being released from the water by the hydride.

Another method that is used commercially, under the trade name 'Powerballs', is
based on sodium hydride. These are supplied in the form of polyethylene-coated spheres

of about 3 cm diameter. They are stored underwater, and cut in half when required to produce hydrogen. An integral unit holds the water, product sodium hydroxide, and a microprocessor-controlled cutting mechanism that operates to ensure a continuous supply of hydrogen. In this case the reaction is

$$NaH + H_2O \rightarrow NaOH + H_2 \qquad [8.17]$$

This is a very simple way of producing hydrogen, and its energy density and specific energy can be as good or better than the other methods we have considered so far. Sodium is an abundant element, and so sodium hydride is not expensive. The main problems with these methods are as follows:

• The need to dispose of a corrosive and unpleasant mixture of hydroxide and water. In theory, this could be recycled to produce fresh hydride, but the logistics of this would be difficult.
• The fact that the hydroxide tends to attract and bind water molecules, which means that the volumes of water required tend to be considerably greater than equations 8.16 and 8.17 would imply.
• The energy required to manufacture and transport the hydride is greater than that released in the fuel cell.

A further point is that the method does not stand very good comparison with metal air batteries. If the user is prepared to use quantities of water, and is prepared to dispose of water/metal hydroxide mixtures, then systems such as the aluminium/air or magnesium/air battery are preferable. With a salt-water electrolyte, an aluminium/air battery can operate at 0.8 V at quite a high current density, producing three electrons for each aluminium atom. The electrode system is much cheaper and simpler than a fuel cell.

Nevertheless, the method compares quite well with the other systems in several respects. The figures in Table 8.18 below are calculated for a self-contained system capable of producing 1 kg of hydrogen, using the sodium hydride system. The equipment for containing the water and gas, and the cutting and control mechanism is assumed to weigh 5 kg. There is three times as much water as equation 8.14 would imply is needed.

The storage efficiency compares well with other systems. This method may well have some niche applications where the disposal of the hydroxide is not a problem, though these are liable to be limited.

Table 8.18 Figures for a self-contained system producing 1 kg of hydrogen using water and sodium hydride

Mass of container and all materials	45 kg
Mass of hydrogen stored	1.0 kg
Storage efficiency (% mass H2)	2.2%
Specific energy	0.87 kWh kg^{-1}
Volume of tank (approx.)	50 L
Mass of H$_2$ per litre	0.020 kg L^{-1}

Another method of stabilising alkali metal hydrides is as a slurry in an organic liquid. This has been investigated by Thermo Technologies in the US (McClaine, 2000). The principle is that a light metal hydride such as lithium hydride is slurried in an organic dispersant such as light mineral oil. When hydrogen is required the slurry is mixed with water. Using an oil to stabilise the hydride has the advantage that more reactive hydrides containing a greater proportion of hydrogen may be considered. In theory a 60% by weight slurry of lithium hydride in mineral oil has a storage capacity of 13 wt%. In practice this could drop as low as 6.4 wt% if it is assumed that the water for the reaction is carried. Thermo Technologies had proposed that the hydride could be recycled by reduction with carbon from coal or biomass according to the reactions

$$LiOH + C = Li + CO + H_2$$
$$CO + H_2O = CO_2 + H_2$$

8.10.4 Sodium borohydride

A good deal of interest has recently been shown in the use of sodium tetrahydridoborate, or sodium borohydride as it is usually called, as a chemical hydrogen carrier. This reacts with water to form hydrogen according to the reaction

$$NaBH_4 + 2H_2O \rightarrow 4H_2 + NaBO_2 \qquad \Delta H = -218\,kJ\,mol^{-1} \qquad [8.18]$$

This reaction does not normally proceed spontaneously, and solutions of $NaBH_4$ in water are quite stable. Some form of catalyst is usually needed. The result is one of the great advantages of this system – it is highly controllable. Millennium Cell Corp. in the United States has been actively promoting this system and has built demonstration vehicles running on both fuel cells and internal combustion engines using hydrogen made in this way. Companies in Europe, notably NovArs GmbH in Germany (Koschany, 2001), have also made smaller demonstrators. Notable features of equation 8.18 are as follows:

- It is exothermic, at the rate of 54.5 kJ per mole of hydrogen.
- Hydrogen is the only gas produced, it is not diluted with carbon dioxide
- If the system is warm, then water vapour will be mixed with the hydrogen, which is highly desirable for PEM fuel cell systems.

Although rather overlooked in recent years, $NaBH_4$ has been known as a viable hydrogen generator since 1943. The compound was discovered by the Nobel laureate Herbert C. Brown, and the story is full of interest and charm, but is well told by Prof. Brown himself elsewhere (Brown, 1992). It will suffice to say that shortly before the end of the 1939–1945 war, plans were well advanced to bulk manufacture the compound for use in hydrogen generators by the US Army Signals Corps, when peace rendered this unnecessary. However, in the following years many other uses of sodium borohydride, notably in the paper-processing industries, were discovered, and it is produced at the rate of about 5000 tonnes per year,[11] mostly using Brown's method, by Morton International (merged with Rohm and Haas in 1999).

If mixed with a suitable catalyst, $NaBH_4$ can be used in solid form, and water is added to make hydrogen. The disadvantage of this method is that the material to be transported is a flammable solid, which spontaneously gives off H_2 gas if it comes into contact with water. This is obviously a safety hazard. It is possible to purchase sodium borohydride mixed with 7% cobalt chloride for this purpose. However, this is not the most practical way to use the compound.

Current work centres on the use of solutions. This has several advantages. Firstly, the hydrogen source becomes a single liquid – no separate water supply is needed. Secondly, this liquid is not flammable, and only mildly corrosive, unlike the solid form. The hydrogen releasing reaction of equation 8.18 is made to happen by bringing the solution into contact with a suitable catalyst. Removing the catalyst stops the reaction. The gas generation is thus very easily controlled – a major advantage in fuel cell applications.

The maximum practical solution strength used is about 30%. Higher concentrations are possible, but take too long to prepare, and are subject to loss of solid at lower temperatures. The solution is made alkaline by the addition of about 3% sodium hydroxide – otherwise the hydrogen evolution occurs spontaneously. The 30% solution is quite thick, and so weaker solutions are sometimes used, even though their effectiveness as a hydrogen carrier is worse.

Such solutions are stable for long periods, though hydrogen evolution does occur slowly. The 'half-life' of such solutions has empirically been shown to follow the equation

$$\log_{10}(t_{\frac{1}{2}}) = \text{pH} - (0.034T - 1.92)$$

In this equation the half-life $t_{1/2}$ is in minutes, and the temperature T is in Kelvin. A solution of the type described has a half-life of about two years at 20°C. So it is best if the solutions are used fairly promptly after being made up, but *rapid* loss of hydrogen is not a problem. One litre of a 30% solution will give 67 g of hydrogen, which equates to about 800 NL. This is a very good volumetric storage efficiency.

Generators using these solutions can take several forms. The principle is that to generate hydrogen, the solution is brought into contact with a suitable catalyst, and that generation ceases when the solution is removed from the catalyst. Suitable catalysts include platinum and ruthenium, but other less-expensive materials are effective, including iron oxide. Fuel cell electrodes make very good reactors for this type of generator.

A practical sodium borohydride system is shown in Figure 8.25. The solution is pumped over the reactor, releasing hydrogen. The motor driving the pump is turned on and off by a simple controller that senses the pressure of the hydrogen, and which turns it on when more is required. The solution is forced through the reactor, and so fresh solution is continually brought in contact with the catalyst. The rate of production is simply controlled by the duty cycle of the pump. The reaction takes place at room temperature, and the

[11] *Kirk-Othmer Encyclopedia of Chemical Technology*, J. Wiley and Sons.

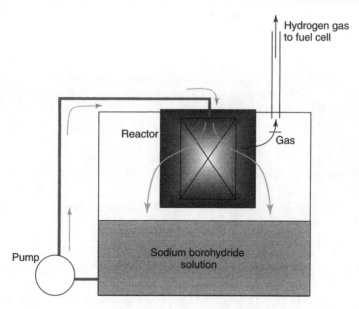

Figure 8.25 Example reactor for releasing hydrogen from a solution of sodium borohydride in water, stabilised with sodium hydroxide. The rate of production of hydrogen is easily controlled by controlling the rate at which the solution is pumped over the reactor.

whole system is extremely simple when compared to any of the other generators that have been outlined in this section.

However, when the solution is weak, the reaction rate will be much slower, and the system will behave differently. It is likely that it will not be practical to obtain highly efficient solution usage. Also, the solution cannot be renewed at the user's convenience – it must be completely replaced when the $NaBH_4$ has been used up, and at no other time.

Another method that is used is that of the Hydrogen on Demand system promoted by Millennium Cell, and promoted at their website. This uses a single-pass catalyst, rather than the recirculation system of Figure 8.25. A major advantage of this is that the tank of fresh solution can be topped up at any time. A disadvantage is that two tanks are needed – the second one for the spent solution that has passed over the catalyst.

In terms of the general figure of merit 'volume required to produce 1 kg of hydrogen', the 30% $NaBH_4$ solution is the worst of the liquid carriers shown in Table 8.16. However, it is competitive, and is only very slightly worse than pure liquid hydrogen. However, it has many advantages over the other technologies. Liquid hydrogen has huge practical difficulties associated with the fact that it has to be kept at such very low temperatures, which makes storage, transport, and filling a major problem. Superficially ammonia looks attractive, but we have seen in the previous section that it has major practical difficulties. Methanol requires a fairly substantial chemical plant to extract the hydrogen, which adds considerably to the cost, volume, and weight of any system. Hydrazine is included in Table 8.16 for completeness – but it is so highly toxic and carcinogenic that its use in any but the most extreme of circumstances is ruled out. In comparison, we note the following points about the $NaBH_4$ solution:

- It is arguably the safest of all the liquids to transport.
- Apart from cryogenic hydrogen, it is the only liquid that gives pure hydrogen as the product. *This is very important*, as it means it is the only one where the product gas can be 100% utilized within the fuel cell.
- The reactor needed to release the hydrogen requires no energy and can operate at ambient temperature and pressure.
- The rate of production of hydrogen can be simply controlled.
- The reactor needed to promote the hydrogen production reaction is very simple – far simpler than that needed for any of the other liquids.
- If desired, the product hydrogen gas can contain large quantities of water vapour, which is highly desirable for PEM fuel cells.

In order to compare a complete system, and produce comparative figures for gravimetric and volumetric storage efficiency, we need to speculate on what a complete hydrogen-generation system would be like. Systems have been built where the mass of the unit is about the same as the mass of the solution stored, and about twice the volume of solution held. So, a system that holds 1 L of solution has a volume of about 2 L, and weighs about 2 kg. Such a system would yield the figures shown in Table 8.19.

These figures are very competitive with all other systems. So what are the disadvantages?

There are three main problems, the second two being related. The first is the problem of disposing of the borate solution. This is not unduly difficult, as it is not a hazardous substance. However, the other disadvantages are far more severe. The first is the cost. Sodium borohydride is an expensive compound. By simple calculation and reference to catalogues it can be shown that the cost of producing hydrogen this way is about $630 per kilogram. This is over 100 times more expensive than using an electrolyser driven by grid-supplied electricity (Larminie, 2002). At this sort of cost the system is not at all viable.

Linked to this problem of cost is the energy required to manufacture sodium borohydride. Using current methods this far exceeds the requirements of compounds such as methanol. Currently sodium borohydride is made from borax ($NaO·2B_2O_3·10H_2O$), a naturally occurring mineral with many uses that is mined in large quantities, and

Table 8.19 Speculative data for a hydrogen source, storing 1.0 L of 30% $NaBH_4$, 3% NaOH and 67% H_2O solution

Mass of reformer, tank, and solution	2.0 kg
Mass of hydrogen stored	0.067 kg
Storage efficiency (% mass H_2)	3.35%
Specific energy	1.34 kWh kg^{-1}
Volume of system (approx.)	2.0 L
Mass of H_2 per litre	0.036 kg L^{-1}

methanol. The aim is that the sodium metaborate produced by the hydrolysis reaction of equation 8.18 is recycled back to sodium borohydride. The table below shows the molar enthalpies of the formation of the key compounds. It can be seen that such recycling will be a formidable challenge, requiring at least $788\,kJ\,mol^{-1}$. However, the 'prize' is *four* moles of hydrogen, so that is at least $197\,kJ\,mol^{-1}$, which is not quite so daunting. Nevertheless, there are many problems to be overcome before such recycling is viable.

	$NaBH_4$	$NaBO_2$
Molar enthalpy of formation	$-189\,kJ\,mol^{-1}$	$-977\,kJ\,mol^{-1}$

The companies, such as Millennium Cell Inc., who are hoping to commercialise this process are working hard on this problem of production cost – finance and energy. If they succeed, there will be a useful hydrogen carrier, but until the costs come down by a factor of at least 10 this method is only suitable for special niche applications.

8.10.5 Ammonia

Ammonia is a colourless gas with a pungent choking smell that is easy to recognise. It is highly toxic. The molecular formula is NH_3, which immediately indicates its potential as a hydrogen carrier. It has many uses in the chemical industry, the most important being in the manufacture of fertiliser, which accounts for about 80% of the use of ammonia. It is also used in the manufacture of explosives. Ammonia is produced in huge quantities. The annual production is estimated at about 100 million tonnes, of which a little over 16 million is produced in the United States.[12]

Ammonia liquefies at $-33°C$, not an unduly low temperature, and can be kept in liquid form at normal temperature under its own vapour pressure of about 8 bar – not an unduly high pressure. Bulk ammonia is normally transported and stored in this form. However, it also readily dissolves in water – in fact it is the most water-soluble of all gases. The solution (ammonium hydroxide) is strongly alkaline, and is sometimes known as 'ammonia water' or 'ammonia liquor'. Some workers have built hydrogen generators using this as the form of ammonia supplied, but this negates the main advantage of ammonia, which is its high 'hydrogen density', as well as adding complexity to the process.

Liquid ammonia is one of the most compact ways of storing hydrogen. In terms of volume needed to store 1 kg of hydrogen, it is better than almost all competing materials – see Table 8.16. Counter-intuitively it is approximately 1.7 times as effective as liquid hydrogen. (This is because, even in liquid form, hydrogen molecules are very widely spaced, and LH_2 has a very low density.)

Table 8.16 shows ammonia to be the best liquid carrier, in terms of space to store 1 kg of hydrogen – apart from hydrazine, which is so toxic and carcinogenic that it is definitely not a candidate for regular use. However, the margin between the leading candidates is not very large. The figures ignore the large size of the container that would be needed, especially in the case of ammonia, liquid methane, and LH2, though not for the key rival compound, methanol.

---

[12] Information provided by the Louisiana Ammonia Association, www.lammonia.com.

Two other features of ammonia lie behind the interest in using it as a hydrogen carrier. The first is that large stockpiles are usually available because of the seasonal nature of fertiliser use. The second is that ammonia prices are sometimes somewhat depressed because of an excess of supply over demand. However, when the details of the manufacture of ammonia, and its conversion back to hydrogen, are considered, it becomes much less attractive.

Using ammonia as a hydrogen carrier involves the manufacture of the compound from natural gas and atmospheric nitrogen, the compression of the product gas into liquid form, and then, at the point of use, the dissociation of the ammonia back into nitrogen and hydrogen.

The production of ammonia involves the steam reformation of methane (natural gas), as outlined in Section 8.4. The reaction has to take place at high temperature, and the resulting hydrogen has to be compressed to very high pressure (typically 100 bar) to react with nitrogen in the Haber process. According to the Louisiana Ammonia Producers Association, who make about 40% of the ammonia produced in the United States, the efficiency of this process is about 60%. By this they mean that 60% of the gas used goes to provide hydrogen, and 40% is used to provide energy for the process. This must be considered a 'best case' figure, since there will no doubt be considerable use of electrical energy to drive pumps and compressors that is not considered here. The process is inherently very similar to methanol production – hydrogen is made from fuel, and is then reacted with another gas. In this case it is nitrogen instead of carbon dioxide. The process efficiencies and costs are probably similar.

Since the manufacture of ammonia requires the steam reforming of natural gas to make hydrogen, it is self-evident that for stationary systems the use of ammonia is pointless and inefficient. Anywhere that ammonia can be delivered to, so can natural gas. It is therefore only for mobile/portable systems that ammonia could be considered.

The recovery of hydrogen from ammonia involves the simple dissociation reaction:

$$NH_3 \rightarrow \tfrac{1}{2}N_2 + \tfrac{3}{2}H_2 \qquad \Delta H = +46.4 \, kJ \, mol^{-1} \qquad [8.19]$$

For this reaction to occur at a useful rate, the ammonia has to be heated to between 600 and 800°C, and passed over a catalyst. Higher temperatures of about 900°C are needed if the output from the converter is to have remnant ammonia levels down to the ppm level. On the other hand the catalysts need not be expensive: iron, copper, cobalt, and nickel are among many materials that work well. Systems doing this have been described in the literature, Kaye et al. (1998) and Faleschini et al. (2000), for example – the latter paper has a good review of the catalysts that can be used.

The reaction is endothermic as shown. However, this is not the only energy input required. The liquid ammonia absorbs large amounts of energy as it vaporises into a gas, which is why it is still quite extensively used as a refrigerant.

$$NH_3(l) \rightarrow NH_3 \, (g) \qquad \Delta H = +23.3 \, kJ \, mol^{-1} \qquad [8.20]$$

Once a gas at normal temperature, it then has to be heated, because the dissociation reaction only takes place satisfactorily at temperatures of around 800 to 900°C. For simplicity we will assume an 800°C temperature rise. The molar specific heat of ammonia is 36.4 J mol^{-1} kg^{-1}. So

$$\Delta H = 800 \times 36.4 = 29.1 \text{ kJ mol}^{-1}$$

This process results in the production of 1.5 moles of hydrogen, for which the molar enthalpy of formation (higher heating value (HHV)) is −285.84 kJ mol^{-1}. The best possible efficiency for this stage of the process is thus

$$\frac{(285.84 \times 1.5) - (23.3 + 29.1 + 46.4)}{285.84 \times 1.5} = 0.77 = 77\%$$

This should be considered an upper limit of efficiency, as we have not considered the fact that the reformation process will involve heat losses to the surroundings. However, systems should be able to get quite close to this figure, since there is scope for using heat recovery, as the product gases would need to be cooled to about 80°C before entering the fuel cell. The vaporisation might also take place at below ambient temperature, allowing some heat to be taken from the surroundings.

Since the product gas is a mixture of hydrogen and nitrogen, it cannot be 100% used in the fuel cell stack. The gas that flows out of the cell anodes must still contain some hydrogen. (This is shown in Section 7.2.2.) The system would have to be designed so that sufficient hydrogen gas was left in the anode off gas to provide energy in a burner to heat the ammonia prior to entering the reactor. It would be possible to make such a system – it is inherently simpler than a methane steam reformer for example. However, ammonia is only worthy of consideration for mobile systems, and so the complexity should be compared with systems such as an autothermal or POX reformer for methane. One major problem is how would the ammonia reformer be started? A considerable amount of hydrogen fuel would need to be stored to provide the initial heat to make the system start.

The corrosive nature of ammonia and ammonium hydroxide is another major problem. Water is bound to be present in a fuel cell. Any traces of ammonia left in the hydrogen and nitrogen product gas stream will dissolve in this water, and thus form an alkali (ammonium hydroxide) inside the cell. In small quantities, in an alkaline electrolyte fuel cell, this is tolerable. However, in the PEM fuel cell, which is the most developed and most likely to be used in mobile applications, it would be fatal. This point is admitted by some proponents of ammonia, and is used by them as an advantage for alkaline fuel cells (Kordesch et al., 1999). Hydrogen from other hydrogen carriers such as methanol and methane also contains poisons, notably carbon monoxide. However, these can be removed, and do not permanently harm the cell; they just temporarily degrade performance. Ammonia on the other hand, will do permanent damage, and this damage will steadily get worse and worse.

Ammonia as a hydrogen carrier can well be compared to methanol. If it were, the following points would be made:

- The production methods and costs are similar.
- The product hydrogen per litre of carrier is slightly better.
- Ammonia is far harder to store, handle, and transport.

- Ammonia is more dangerous and toxic.
- The process of extracting the hydrogen is more complex, there is no equivalent to the POX or the autothermal reformers.
- The reformer operates at very high temperatures, making integration into small fuel cell systems much more difficult than for methanol.
- The product gas is difficult to use with any type of fuel cell other than alkaline.

The conclusion must be that the use of ammonia as a hydrogen carrier is going to be confined to only the most unusual circumstances.

8.10.6 Storage methods compared

We have looked at a range of hydrogen storage methods. In Section 8.9, we looked at fairly simply 'hydrogen in, hydrogen out' systems. In Section 8.10, we looked at some more complex systems involving the use of hydrogen-rich chemicals that can be used as carriers.

None of the methods is without major problems. The main conclusion is that wherever possible hydrogen should NOT be stored at all. Rather, it should be made from natural gas (or other fuel) using a reformer near the fuel cell as and when it is needed, using the methods described in Sections 8.4, 8.5, and 8.6. However, in the case of mobile systems this will not always be possible. Table 8.20 compares the systems that are currently feasible in relation to gravimetric and volumetric effectiveness. Together with the summary comments, this should enable the designer to choose the least difficult alternative. It is worth noting that the method with the worst figures (storage in high-pressure cylinders) is actually the most widely used. This is because it is so simple and straightforward. The figures also show why methanol is such a promising candidate for the future.

Table 8.20 Data for comparing methods of storing hydrogen fuel. The figures include the associated equipment, for example, tanks for liquid hydrogen, or reformers for methanol

Method	Gravimetric storage efficiency, % mass H_2	Volumetric mass (in kg) of H_2 per litre	Comments
High pressure in cylinders	0.7–3.0	0.015	'Cheap and cheerful' widely used
Metal hydride	0.65	0.028	Suitable for small systems
Cryogenic liquid	14.2	0.040	Widely used for bulk storage.
Methanol	6.9	0.055	Low-cost chemical. Potentially useful in a wide range of systems
Sodium hydride pellets	2.2	0.02	Problem of disposing of spent solution.
$NaBH_4$ solution in water	3.35	0.036	Very expensive to run.

References

Ajayan P.M. and Zhou O.A. (2001) "Applications of carbon nanotubes", in M.S. Dresselhaus, G. Dresselhaus, Ph. Avouris (eds) *Carbon Nanotubes, Topics in Applied Physics*, Vol. 80, Springer-Verlag, Berlin, pp. 391–425.

Andreeva D., Idakiev V., Tabakova T., Ilieva L., Falara P., Bourlinos A., and Travlos A. (2002) "Low-temperature water-gas shift reaction over Au/CeO_2 catalysts", *Catalysis Today*, **72**, 61–67.

Atkinson K., Roth S., Hirscher M., and Grunwald W. (2001) "Carbon nanostructures: an efficient hydrogen storage medium for fuel cells?" *Fuel Cells Bulletin*, **38**, 9–12.

Bain A. and VanVorst W.D. (1999) "The Hindenburg tragedy revisited: the fatal flaw found", *International Journal of Hydrogen Energy*, **24**(3), 399–403.

Baker B.S. (1965) *Hydrocarbon Fuel Cell Technology*, Academic Press, New York and London.

Bennemann J.R. and Weare N.M. (1974) *Science*, **184**, 174–175.

Bossel U.G. (1999) "Portable fuel cell battery charger with integrated hydrogen generator. *Proceedings of the European Fuel Cell Forum Portable Fuel Cells Conference*, Lucerne, pp. 79–84.

Brown H.C. (1992) "The Discovery of New Continents of Chemistry", lecture given in 1992, available at www.chem.purdue.edu/hcbrown/Lecture.htm.

Brown D.J., Brown R.A., Cooke B.H., Haynes D.A., Taylor M.R., and Blyth Z. (1996) "*A Study Assessing Electricity Generation Integrating Coal Gasifiers and Fuel Cells*", ETSU report F/02/00026/REP, AEA Technology, Harwell, UK.

Chambers A., Park C., Baker R.T.K., and Rodriguez N.M. (1998) "Hydrogen storage in graphite nanofibres", *Journal of Physical Chemistry B*, **102**, 4253–4256.

Chen P., Wu X., Lin J., and Tan K.L. (1999) "High H-2 uptake by alkali-doped carbon nanotubes under ambient pressure and moderate temperatures", *Science*, **285**, 91–93.

Cheng H., Pez G., and Cooper A.C. (2001) "Mechanism of Hydrogen Sorption in Single-walled Carbon Nanotubes", *J. Am. Chem. Soc.*, **123**, 5845–5846.

Dams R.A.J., Moore S.C., and Hayter P.R. (2000) "Compact fast-response methanol fuel processing system for PEMFC electric vehicles", *2000 Fuel Cell Seminar Abstracts*, Palm Springs, Calif., pp. 234–237.

Dicks A.L. (1996) "Hydrogen generation from natural gas for the fuel cell systems of tomorrow," *Journal of Power Sources*, **61**, 113–124.

Dillon A.C., Jones K.M., Bekkedahl T.A., Kiang C.H., Bethune D.S., and Heben M.J. (1997) "Storage of hydrogen in single-walled carbon nanotubes", *Nature*, **386**, 377–379.

Dillon A.C., Gennett T., Alleman J.Z.L., Jones K.M., Parilla P.A., and Hebden M.J. (2001) "Carbon nanotube materials for hydrogen storage." *Proceedings of the 2001 U.S. DOE Hydrogen Program* (http://www.eren.doe.gov/hydrogen/pdfs/30535am.pdf).

Dvorak K., van Nusselrooy P.F.M.T., Vasalos I.A., Berger R.J., Olsbye U., Verykios X.E., Roberts M.P., Hildebrandt U., and Hageman R. (1998) "*Catalytic Partial Oxidation of Methane to Synthesis Gas*", report of EU Contract JOF3-CT95-0026.

Edlund D.J. and Pledger W.A. (1998) "Pure hydrogen production from a multi-fuel processor", *Proceedings of the US Fuel Cell Seminar*, Palm Springs, Calif., pp. 16–19.

Edwards N., Ellis S.R., Frost J.C., Golunski S.E., van Keulen A.N.J., Lindewald N.G., and Reinkingh J.G. (1998) "Onboard hydrogen generation for transport applications: the HotSpot™ methanol processor", *Journal of Power Sources*, **71**(1–2), 123–128.

Eliasson B. and Bossel U. (2002) "The future of the hydrogen economy, bright or bleak?" *The Fuel Cell World EFCF Conference Proceedings*, pp. 367–382.

Ellis S.R., Golunski S.E., and Petch M.I. (2001) "*Hotspot Processor for Reformulated Gasoline*", ETSU report no. F/02/00143/REP. DTI/Pub URN 01/958.

Farrauto R. (2001) "Catalytic generation of hydrogen for the solid polymer membrane fuel cell", *Abstracts of Papers of the American Chemical Society*, **222** (FUEL Part 1), 89.

Faleschini G., Hacker V., Muhr M., Kordesch K., and Aronsson R. (2000) "*Ammonia for High Density Hydrogen Storage*", published on the world wide web at www.electricauto.com/HighDensity_STOR.htm.

Freni S. and Maggio G. (1997) *International Journal of Energy Research*, **21**, 253–264.

Gaffron H. and Rubin J. (1942) *Journal of General Physiology*, **26**, 219–240.

Ghirardi M.L., Kosourov S., and Siebert M. (2001) "Cyclic photobiological algal H2-production." *Proceedings of the 2001 DOE Hydrogen Program Review*, NREL/CP-570-30535.

Gorte R., Kim H., and Vohs J.M. (2002) "Novel SOFC anodes for the direct electrochemical oxidation of hydrocarbon", *Journal of Power Sources*, **106**(1–2), 10–15.

Goulding P.S., Judd R.W., and Dicks A.L. (2001) 8UK Patent Application GB 2353801.

Greenbaum E. et al. (1998) "CO_2 fixation and photoevolution of H^{-2} and O^{-2} in a mutant of chlamydomonas lacking photosystem-I", *Nature*, **376**(6539), 438–441.

Hamada K, Mizusawa M., and Koga K. (1997) "*Plate Reformer*" US Patent no. 5,609,834.

Hammerli M. and Stuart A. (2000) "Build-up of a hydrogen infrastructure." *Proceedings of the Fuel Cell 2000 Conference*, Lucerne, pp. 339–348.

Hart A.B. and Womack G.J. (1967) *Fuel Cells – Theory and Applications*, Chapman and Hall Ltd, London.

Hilaire S., Wang X., Luo T., Gorte R.J., and Wagner J. (2001) "A comparative study of water-gas-shift reaction over ceria supported metallic catalysts", *Applied Catalysis A (General)*, **215**(1–2), 271–278.

Hirscher M., Becher M., Haluska M., Dettlaff-Weglikowska U., Quintel A., Duesberg G.S., Choi Y.-M., Downes P., Hulman M., Roth S., Stepanek I., and Bernier P. (2001) "Hydrogen storage in sonicated carbon materials", *Applied Physics A*, **72**(2), 129–132.

*Hord J. (1978) "Is hydrogen a safe fuel?" *International Journal of Hydrogen Energy*, **3**, 157–176.

Iijkima S. (1991) *Nature*, **354**(56), 391.

Irvine J.T.S. and Sauvet A. (2001) "Improved oxidation of hydrocarbons with new electrodes in high temperature fuel cells", *Fuel Cells*, **1**(34), 205–210.

Joensen F. and Rostrup-Nielsen J.R. (2002) "Conversion of hydrocarbons and alcohols for fuel cells", *Journal of Power Sources*, **105**(2), 195–201.

Kahrom H. (1999) "Clean hydrogen for portable fuel cells." *Proceedings of the European Fuel Cell Forum Portable Fuel Cells Conference*, Lucerne, pp. 159–170.

Kalhammer F.R., Prokopius P.R., Roan V., and Voecks G.E. (1998) "*Status and Prospects of Fuel Cells as Automobile Engines*", report prepared for the State of California Air Resources Board.

Kalk T.F., Mahlendorf F., and Roes J. (2000) "Cracking of hydrocarbons to produce hydrogen for PEM fuel cells", *Fuel Cell Seminar Abstracts*, Courtesy Associates Inc., Portland, Oregan, pp. 317–320.

Kaplan S. and Moore M.D. (1998) "*Process for the Production of Hydrogen Using Photosynthetic Proteobacteria*", U.S. Patent 5,804,424.

Kaye I.W. and Bloomfield D.P. (1998) "Portable ammonia powered fuel cell", *Conference of the Power Sources*, Cherry Hill, pp. 759–766.

Kordesch K., Hacker V., Gsellmann J., Cifrain M., Faleschini G., Enzinger P., Ortner M., Muhr M., and Aronsson R. (1999) "Alkaline fuel cell applications." *Proceedings of the 3rd International Fuel Cell Conference*, Nagoya, Japan.

Koschany P. (2001) "Hydrogen sources integrated in fuel cells." *The Fuel Cell Home EFCF Conference Proceedings*, pp. 293–298.

Kramer G.J., Wieldraaijer W., Biesheuvel P.M., and Kuipers H.P.C.E. (2001) "The determining factor for catalysts selectivity in shell's catalytic partial oxidation process", *American Chemical Society, Fuel Chemistry Division Preprints*, **46**(2), 659–660.

Kroto H.W., Heath J.R., O' Brien S.C., Curl S.C., and Smalley R.E. (1985) *Nature*, **381**(162), 391.

Kumar R., Ahmed S., and Krumpelt M. (1996) "The low temperature partial oxidation reforming of fuels for transportation fuel cell systems." *Proceedings of the US Fuel Cell Seminar*, Kissimee, Florida, November, pp. 17–20.

Lakeman J.B. and Browning D.J. (2001) "*Global Status of Hydrogen Research*," ETSU report F/03/00239/REP, U.K.

Langnickel U. (1999) "Fuel cell using digester gas", *European Fuel Cell News*, **6**(2), 4.

Larminie J. (2002) "Sodium borohydride: is this the answer for fuelling small fuel cells?" *The Fuel Cell World EFCF Conference Proceedings*, pp. 60–67.

Ledjeff-Hey K., Kalk Th., Mahlendorf F., Niemzig O., and Roes J. (1999) "Hydrogen by cracking of propane." *Proceedings of the International Conference on Portable Fuel Cells*, Lucerne, 21–24, June (ISBN 3-905592-3-7) pp. 193–203.

Liebhafsky H.A. and Cairns E.J. (1968) *Fuel Cells and Fuel Batteries*, John Wiley & Sons, New York.

Logan B.E., Oh S.E., Kim O.S., and Van Ginkel S. (2002) "Biological hydrogen production measured in batch anaerobic respirometers", *Environmental Science and Technology*, **36**(11), 2530–2535.

McClaine A.W., Breault R.W., Larsen C., Konduri R., Rolfe J., Becker F., and Miskolczy G. (2000) "Hydrogen transmission/storate with metal hydride organic slurry and advanced chemical hydride/hydrogen for PEMFC vehicles." *Proceedings of the 2000 U.S. DOE Hydrogen Program Review* NREL/CP-570-28890.

Mitsui A. et al. (1986) *Nature*, **323**, 720–722.

Miura Y. et al. (1986) *Agricultural and Biological Chemistry*, **50**, 2837–2844.

Miyake J. and Kawamura S. (1987) *International Journal of Hydrogen Energy*, **39**, 147–149.

Ogden J.A. (2001) "Review of Small Stationary Reformers for Hydrogen Production," International Energy Agency Report No. IEA/H2/TR-02/002, available from US Department of Energy at http://www.eren.doe.gov/hydrogen/iea/iea_publications.html.

Okada O., Tabata T., Takami S., Hirao K., Masuda M., Futjita H., Iwasa N., and Okhama T. (1992) "Development of an advanced reforming system for fuel cells." *International Gas Research Conference*, Orlando Florida, USA, Nov, 15–19.

Park C., Anderson P.E., Chambers A., Tan C.D., Hidalgo R., and Rodriguez N.M. (1999) "Further studies of the interaction of hydrogen with graphite nanofibres", *Journal of Physical Chemistry, B.*, **103**, 10572–10581.

Pattekar A.V., Kothare M.V., Karnik S.V., and Hatalis M.K. (2001) "A Microreactor for in-situ Hydrogen Production by Catalytic Methanol Reforming," *Proceedings of the 5th International Conference on Microreaction Technology held in Strasbourg*, France, May 27–30, http://www.lehigh.edu/~inmifuel/Microreactor_IMRET5.pdf

Perry Murray E., Tsai T., and Barnett S.A. (1999) "A direct methane fuel cell with ceria-based anode", *Nature*, **400**, 649–651.

*Reister D. and Strobl W. (1992) "Current development and outlook for the hydrogen fuelled car", *Hydrogen Energy Progress IX*, 1202–1215.

Rostrup-Nielsen J.R. (1993) *Catalysis Today*, **18**, 305–324.

Rostrup-Nielsen J.R. (1984) in J.R. Anderson and M. Boudart (eds.) "Catalytic Steam Reforming" *Catalytic Science and Engineering*. Vol. 5, Springer-Verlag, Copenhagen.

Schmidt L.D., Klein E.J., O'Connor R.P., and Tummala S. (2001) "Production of hydrogen in millisecond reactors: combination of partial oxidation and water-gas shift", *Abstracts of Papers of the American Chemical Society*, **222** (FUEL Part 1) 111.

Schulz R. (1996) *Journal of Marine Biotechnology*, **4**, 16–22.

Shinke N., Higashiguchi S., and Hirai K. (2000) *Fuel Cell Seminar Abstracts*, Courtesy Associates Inc., Portland, Oregan, pp. 292–295.

Smith J.G. (1966) Chapter 6, in *"An introduction to Fuel Cells"*, K.R. Williams (ed.) (1996) Elsevier Publishing, London, pp. 214–246.

Suzuki S. and Karube I. (1983) *Applied Biochemistry and Bioengineering*, **4**, 281.

Swartz S.L. (2002) "Nanoscale water gas shift catalysts." presented at the *DOE Fuel Cells Review Meeting Golden*, Colorado, 10 May. DOE Contract Number: DE- FC02- 98EE50529.

Tamagnini P., Axelsson R., Lindberg P., Oxelfelt F., Wunschiers R., and Lindblad P. (2002) "Hydrogenases and hydrogen metabolism of cyanobacteria", *Microbiology and Molecular Biology Reviews*, **66**(1), 1–20.

Teagan W.P., Bentley J., and Barnett B. (1998) "Cost implications of fuel cells for transport applications: fuel processing options", *Journal of Power Sources*, **71**, 80–85.

Twigg M. (1989) *Catalyst Handbook*, 2nd ed., Wolfe, London.

Van Hook J.P. (1980) *Catalysis Review/Science and Engineering*, **21**(1).

Van Ooteghem S.A., Beer S.K., and Yue P.C. (2001) "Hydrogen production by the thermophilic bacterium Thermotoga Neapolitana." *Proceedings of the 2001 DOE Hydrogen Program Review*, NREL/CP-570-30535.

Warren D. et al. (1986) "Performance evaluation for the 40 kW fuel cell power plant utilizing a generic landfill gas feedstock", KTI corporation report.

Wolfrum E.J. and Watt A.S. (2001) "Bioreactor design studies for a novel hydrogen-producing bacterium." *Proceedings of the 2001 DOE Hydrogen Program Review*, NREL/CP-570-30535.

Yang R.T. (2000) "Hydrogen storage by alkali-doped carbon nanotubes – revisited", *Carbon*, **38**, 623–641.

*Zieger J. (1994) "HYPASSE – hydrogen powered automobiles using seasonal and weekly surplus of electricity", *Hydrogen Energy Progress X*, 1427–1437.

*These papers are reprinted in Norbeck J.M. et al. (1996) *Hydrogen Fuel for Surface Transportation*, pub. Society of Automotive Engineers.

9

Compressors, Turbines, Ejectors, Fans, Blowers, and Pumps

9.1 Introduction

Air has to be moved around fuel cell systems for cooling and to provide oxygen to the cathode. Fuel gas often has to be pumped around the anode side of the fuel cell too. To do this, pumps, fans, compressors, and blowers have to be used. In addition, the energy of exhaust gases from a fuel cell can sometimes be harnessed using a turbine, making use of what would otherwise go to waste. The technology for such equipment is very mature, having already been developed for other applications. The fuel cell system designer will be choosing devices principally designed for other markets and products. The needs of fuel cells vary widely, as does their size and application, and so we need to look at a wide range of 'gas moving devices'. These would have been primarily designed for a wide field of applications, from turbochargers in diesel engines to aerators for fish tanks!

It is hoped that the reader will find much useful information in this chapter. It has been put here after the descriptions of and discussions about the main types of fuel cell, as the content is relevant to all of them. The topics dealt with are as follows:

- The very important, and quite complex, subject of *compressors*, their different types and performance. This is covered in Sections 9.2 to 9.7 and includes the derivation of two very useful equations. The first gives the temperature rise of a gas as it is compressed (equation 9.4). The second gives the power needed to run an air compressor (equation 9.7).
- *Turbines*, which can sometimes be used to harness the energy of exhaust gases, are briefly covered in Sections 9.8 and 9.9.
- *Ejectors*, a very simple type of pump that can often be used to circulate hydrogen gas if it comes from a high pressure store, or for recycling anode gases, are discussed in Section 9.10.

Fuel Cell Systems Explained, Second Edition James Larminie and Andrew Dicks
© 2003 John Wiley & Sons, Ltd   ISBN: 0-470-84857-X

- *Fans and blowers*, used for cooling and for cathode gas supply in small fuel cells, are covered in Section 9.11.
- *Membrane or diaphragm pumps*, used to pump reactant air and hydrogen through small (200 W) to medium (3 kW) proton exchange membrane (PEM) fuel cells, are discussed briefly in Section 9.12.

9.2 Compressors – Types Used

The types of compressors used in fuel cell systems are the same as those used in other engines, especially diesel engines. The four main types are illustrated in Figure 9.1.

The simplest to understand is the Roots compressor shown in Figure 9.1a. It is easy to envisage how this will pump and compress any gas when the two elements rotate, pushing the gas through. The Roots compressor is quite cheap to produce, and it works over a wide range of flow rates. However, it only works at reasonable efficiencies when producing a small pressure difference. Until recently it had been regarded as the rather lower quality and cheaper type of compressor. However, of late the Eaton Corporation in the United States has revived the Roots compressor. Their device has the refinements that the rotors have been twisted by 60° to form a partial helix and that the counterrotating rotors have three lobes, instead of the two shown in Figure 9.1a. These improvements, while increasing the cost, also greatly reduce the pressure variations and hence the noise,

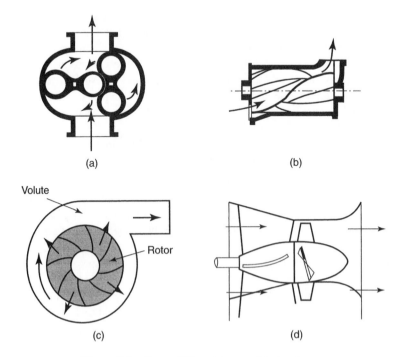

Figure 9.1   Some different types of compressors.

and also increase the efficiency, though they are still only suitable for a small pressure change, boosting by a factor of up to 1.8. They are used as superchargers for petrol engines and are fitted as standard on certain types of Ford cars (among others), and have a strong after-sales accessory market.

It is not so easy to see how the Lysholm or screw compressor of Figure 9.1b works. There are two screws, as can be seen in the photograph of the model of Figure 9.2b, which counter rotate, driving the gas up the region between the two screws and compressing it at the same time. It can be thought of as a refinement of the 'Archimedes Screw' that has been used for pumping water since ancient times. There are two variations of the screw compressor. In the first, an external motor drives only one rotor, and the second rotor is turned by the first. For this to work the rotors are in contact, and so must be lubricated with oil. This type of compressor is widely used to provide compressed air for pneumatic tools and other industrial application needs. The small amounts of oil that are inevitably carried out with the air do not matter. In the second type, the two rotors are connected by a synchronising gear – a separate pair of cogs provides the driving link from one rotor to the other. The counter rotating screws do not come into contact, though of course for a good efficiency they will run very close to each other. This type of compressor gives an oil-free output and is the sort we need in a fuel cell system, though it also has other applications, for example, as a compressor for refrigeration systems.

Among the advantages of the Lysholm compressor are that it can be designed to provide a wide range of compression ratios – the exit pressure can be up to eight times the input pressure. (This is done by changing the length and pitch of the screws.) Another advantage is that they work at a good efficiency over a wide range of flow rates. However, they are expensive to manufacture – the rotors are clearly precision items. The axles that they turn on are subject to both lateral and axial loading, which makes the bearings more complex and expensive, and the synchronising gears add further complication and expense.

The most common type of compressor is the centrifugal or radial type illustrated in Figure 9.1c. The gas is drawn in at the centre and flung out at high speed to the outer volute. Here the kinetic energy is 'converted' to a pressure increase. The radial compressor is used on the vast majority of engine turbocharging systems. The precise shape of the rotating vane has been developed after much research and experience. An example rotor is shown in Figure 9.2. Although the shape may be complex, it can usually be cast as one piece. This type of compressor is thus low cost, well developed, and available to suit a wide range of flow rates. The efficiency of the centrifugal compressor is as good as any other type, but it must be operated within well-defined flow rates and pressure changes to obtain these efficiencies. Indeed, it cannot operate at all at low flow rates, as is explained in Section 9.6. A further problem is that the rotor must rotate at very high speed −80 000 rpm being typical. (For comparison, screw and Roots compressors rotate at up to about 15 000 rpm.) This means that care must be taken with the lubrication of the bearings.

The axial flow compressor, as in Figure 9.1d, works by driving the gas using a large number of blades rotating at high speed. It is, in essence, the inverse of the turbine commonly used in gas and steam thermal power systems. As with these steam or gas turbines, there must be a gap as small as possible between the ends of the blades and the housing. This means they are expensive to manufacture. The efficiency is good, but only

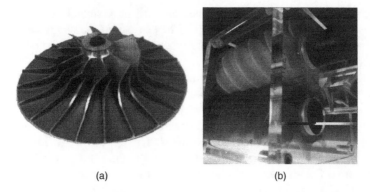

(a)   (b)

Figure 9.2 (a) Typical centrifugal air compressor rotor and (b) model of twin screw Lysholm compressor.

over a fairly narrow range of flow rates. Experience with diesel systems suggests that the axial flow compressor will only be considered for systems above a few megawatts, operating at between half and full power at all times (Watson and Janota, 1982).

A type of compressor that is used in some industrial air compressors is the rotating vane compressor. This claims advantages in terms of cost over the screw compressor. However, they cannot be used for fuel cells, since the tips of the rotating vanes *must* run over a film of oil, and so, even after filtering, there will always be some oil in the output gas, which is not generally acceptable for fuel cells.

Whatever type of compressor is being used, the symbol used in process flow diagrams is as shown in Figure 9.3.

9.3 Compressor Efficiency

As with fuel cells, compressor efficiency is not easy to define. Whenever a gas is compressed, work is done on the gas, and so its temperature will rise, unless the compression

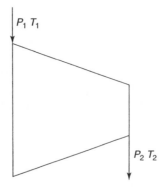

$P_1\ T_1$

$P_2\ T_2$

Figure 9.3 Symbol used for compressors.

is done very slowly or there is a lot of cooling. In a reversible process, which is also adiabatic (no heat loss), it can readily be shown (using advanced high school or early university physics) that if the pressure changes from P_1 to P_2, then the temperature will change from T_1 to T_2', where

$$\frac{T_2'}{T_1} = \left(\frac{P_2}{P_1}\right)^{\frac{\gamma-1}{\gamma}}$$

[9.1]

This formula gives the constant entropy, or isentropic temperature change, as is indicated by the $'$ on T_2'. γ is the ratio of the specific heat capacities of the gas, C_P/C_V.

In practice, the temperature change will be higher than this. Some of the motion of the moving blades and vanes serves only to raise the temperature of the gas. Also, some of the gas might 'churn' around the compressor, doing little but getting hotter. We call the actual new temperature T_2.

To derive the efficiency, we use the ratio between the following two quantities:

- The actual work done to raise the pressure from P_1 to P_2
- The work that would have been done if the process had been reversible or isentropic – the isentropic work.

To find these two figures we make some assumptions that are generally valid:

- The heat flow from the compressor is negligible
- The kinetic energy of the gas as it flows into and out of the compressor is negligible, or at least the change is negligible
- The gas is a 'perfect gas', and so the specific heat at constant pressure, C_P, is constant.

With these assumptions, the work done is simply the change in enthalpy of the gas:

$$W = c_p(T_2 - T_1)m$$

where m is the mass of gas compressed. The isentropic work done is

$$W' = c_p(T_2' - T_1)m$$

The isentropic efficiency is the ratio of these two quantities. Note that the name 'isentropic' does not mean we are saying the process *is* isentropic, rather we are *comparing it with* an isentropic process.

$$\eta = \frac{\text{isentropic work}}{\text{real work}} = \frac{c_p(T_2' - T_1)m}{c_p(T_2 - T_1)m}$$

and so

$$\eta_c = \frac{T_2' - T_1}{T_2 - T_1}$$

[9.2]

If we substitute equation 9.1 for the isentropic second temperature, we get

$$\eta_c = \frac{T_1}{(T_2 - T_1)} \left(\left(\frac{P_2}{P_1} \right)^{\frac{\gamma-1}{\gamma}} - 1 \right) \qquad [9.3]$$

In this derivation we have ignored the difference between the static and the stagnation temperatures. This is important when the gas velocity is high. However, in the fuel cell itself the kinetic energy of the gas is not important. Even in diesel engines, whose flow rates after the compressor are very unsteady and often rapid, the equations given here are usually used.

It is useful to rearrange the equation to give the change in temperature:

$$\Delta T = T_2 - T_1 = \frac{T_1}{\eta_c} \left(\left(\frac{P_2}{P_1} \right)^{\frac{\gamma-1}{\gamma}} - 1 \right) \qquad [9.4]$$

This definition of efficiency does not consider the work done on the shaft driving the compressor. To bring this in we should also consider the mechanical efficiency η_m, which takes into account the friction in the bearings, or between the rotors and the outer casing (if any). In the case of centrifugal and axial compressors, this is very high, almost certainly over 98%. We can then say that

$$\eta_T = \eta_m \cdot \eta_c \qquad [9.5]$$

However, it is the isentropic efficiency η_c that is the most useful, because it tells us how much the temperature rises.

The rise in temperature can be quite high. For example, using air at 20°C (293 K) for which $\gamma = 1.4$, a doubling of the pressure, and using a typical value for η_c of 0.6, equation 9.4 becomes

$$\Delta T = \frac{293}{0.6}(2^{0.286} - 1)$$
$$= 170\,K$$

For some fuel cells, this rise in temperature is useful as it preheats the reactants. On the other hand, for low-temperature fuel cells it means that the compressed gas needs cooling. Such coolers between a compressor and the user of the gas are called *intercoolers*. We need to know the temperature rise in any case, and equation 9.4 also allows us to derive a formula for compressor power.

9.4 Compressor Power

The power needed to drive a compressor can be readily found from the change in temperature. If we take unit time, then clearly

$$\text{Power} = \dot{W} = c_p \cdot \Delta T \cdot \dot{m}$$

where $\dot{m}$ is the rate of flow of gas in $\mathrm{kg\,s^{-1}}$. ΔT is given by equation 9.4, and so we have

$$\mathrm{Power} = c_p \frac{T_1}{\eta_c} \left(\left(\frac{P_2}{P_1} \right)^{\frac{\gamma-1}{\gamma}} - 1 \right) \cdot \dot{m} \qquad [9.6]$$

This is obviously a very useful general equation. In the case of an air compressor, which is a feature of many fuel cell systems, we use the values for air of

$$c_p = 1004\,\mathrm{J\,kg^{-1}\,K^{-1}}$$

and

$$\gamma = 1.4$$

$$\mathrm{So\ power} = 1004 \frac{T_1}{\eta_c} \left(\left(\frac{P_2}{P_1} \right)^{0.286} - 1 \right) \dot{m}\,\mathrm{W} \qquad [9.7]$$

The isentropic efficiency η_c can be found, or at least estimated, from charts such as those described in the following section. To find the power needed from the motor or turbine driving the compressor, the figure found from equation 9.7 should be divided by the mechanical efficiency, η_m. For a good estimate, 0.9 can be used for this. However, when specifying a compressor from a supplier, the figure from equation 9.7 is best used unaltered.

9.5 Compressor Performance Charts

The efficiency and performance of a compressor will depend on many factors that include the following:

- Inlet pressure, P_1
- Outlet pressure, P_2
- Gas flowrate, $\dot{m}$
- Inlet temperature, T_1
- Compressor rotor speed, N
- Gas density, ρ
- Gas viscosity, μ.

To try and tabulate or draw on some kind of map, the compressor performance with all these variables would clearly be hopeless. It is necessary to eliminate or group together these variables. This is often done in the following way:

- The inlet and outlet pressures are combined into one variable, the pressure ratio P_2/P_1.
- Since, for any gas, density $\rho = P/RT$, and P and T are being considered, the density can be ignored. It is considered via the inclusion of pressure and temperature.

- It turns out that the viscosity of the gas, bearing in mind the limited range of gases normally used, can be ignored.

Further simplification is done by a process of dimensional analysis, which can be found in texts on turbines and turbochargers (Watson and Janota, 1982). The result is to group together variables in 'non-dimensional' groups. They are not really non-dimensional, but that is because various constants, with dimensions, have been eliminated from the analysis.

$$\text{The two groups are } \frac{\dot{m}\sqrt{T_1}}{P_1} \text{ and } \frac{N}{\sqrt{T_1}}$$

These are called the *mass flow factor* and the *rotational speed factor*, respectively. Other texts call them the 'non-dimensional' mass flow and rotational speed. Charts are then plotted of the efficiency, for different pressure ratios and mass flow factors. Lines are plotted on these charts of constant rotational speed factor. The chart for a typical screw compressor (Lysholm) is shown in Figure 9.4. The lines of constant efficiency are like the contours of a map. Instead of indicating hills, they indicate areas of higher operating efficiency.

The unit generally used for P_1 is the bar, and for temperature we use Kelvin. We can relate the mass flow factor to the power of a fuel cell reasonably simply. If we assume typical fuel cell operating conditions, (i.e. the air stoichiometry = 2, and the average fuel

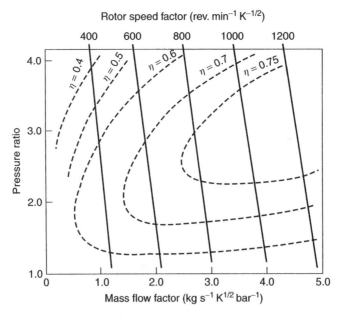

Figure 9.4 Performance chart for a typical Lysholm or screw compressor. A mass flow factor of 5 corresponds to the air needs of a fuel cell of power of about 250 kW.

cell voltage $= 0.6$ V), then, using equation A2.4, the flow rate of the air for a 250-kW fuel cell is

$$\frac{3.57 \times 10^{-7} \times 2 \times 250\,000}{0.6} = 0.3 \, \text{kg s}^{-1}$$

If we then assume standard conditions for the air (i.e. $P_1 = 1$ bar, $T = 298$ K), the mass flow factor is

$$\frac{0.3 \times \sqrt{298}}{1.0} = 5.18 \approx 5 \, \text{kg s}^{-1} \, \text{K}^{\frac{1}{2}} \, \text{bar}^{-1}$$

So, the horizontal x-axis of Figure 9.4 corresponds to the airflow needs of fuel cells of power approximately 0 to 250 kW. Similarly, if the rotor speed factor is 1000, this will correspond to a speed of about 17 000 rpm. (Note, these units have to run fast, the centrifugal compressors considered in the next section have to turn even faster!).

The use of these 'non-dimensional' quantities is standard practice in textbooks on compressors and turbines. However, it has to be said that in many manufacturers' data sheets they are not used. Standard condition figures are applied ($P = 1.0$ bar, $T = 298$ K), and the mass flow factor is replaced with mass flow rate, or even volume flow rate, and the rotational speed factor is replaced with speed in rev min^{-1}. Generally speaking, such charts will give satisfactory results – except in the case of multistage compressors. When gas has been compressed through the first stage, its temperature and pressure would obviously have changed markedly, and so the mass flow factor will be quite different, even though the actual mass flow rate is unchanged.

Worked Example

A fuel cell stack with an output power of 100 kW operated at a pressure of 3 bar. Air is fed to the stack using the Lysholm compressor whose chart is shown in Figure 9.4. The inlet air is at 1.0 bar and 20°C. The fuel cell is operated at an air stoichiometry of 2, and the average cell voltage in the stack is 0.65 V, corresponding to an efficiency of 52% (referred to as lower heating value (LHV)). Use the chart and other equations to find

- the required rotational speed of the air compressor,
- the efficiency of the compressor,
- the temperature of the air as it leaves the compressor,
- the power of the electric motor needed to drive the compressor.

Solution. First we find the mass flow rate of air using equation A2.4

$$\dot{m} = \frac{3.57 \times 10^{-7} \times 2 \times 100\,000}{0.65} = 0.11 \, \text{kg s}^{-1}$$

This is then converted to the mass flow factor.

$$\text{Mass flow factor} = \frac{0.11 \times \sqrt{293}}{1.0} = 1.9 \, \text{kg s}^{-1} \, \text{K}^{\frac{1}{2}} \, \text{bar}^{-1}$$

We now find the speed and the efficiency of the compressor using the chart. Find the intercept of a horizontal line drawn from pressure ratio = 3 and a vertical line starting from the x-axis at mass flow factor = 1.9. This will be very close to the 600 rotor speed factor line, and also the $\eta = 0.7$ 'efficiency contour'. So we can say that the rotor speed is

$$600 \times \sqrt{293} = 10\,300\,\text{rpm}.$$

The efficiency of the compressor is 0.7 or 70%. We can use this, and the mass flow rate, to find the temperature rise and the compressor power. The temperature rise is found using equation 9.4.

$$\Delta T = \frac{293}{0.7}(3^{0.286} - 1) = 155\,\text{K}$$

Since the entry temperature is 20°C, this means that the exit temperature is 175°C. (So we note that if this is a PEM fuel cell it will need cooling. On the other hand, if it is a phosphoric acid fuel cell the preheating would have already been done.)

To find the compressor power we use equation 9.7

$$\text{Power,}\ \dot{W} = 1004 \times \frac{293}{0.7}(3^{0.286} - 1) \times 0.11 = 17.1\,\text{kW}$$

This is the power of the compressor ignoring any mechanical losses in the bearings and drive shafts. The electric motor will not be 100% efficient either, so we could estimate that the power of the motor should be about 20 kW.

- Note that this 20 kW of electrical power will have to come from the 100-kW fuel cell, consuming 20% of its output. This parasitic load is a major problem when running systems at pressure, and its importance for PEM fuel cells is discussed in Chapter 4.
- Note also that in this worked example we have assumed that we are working with air of fairly standard (i.e. low) water content. It was pointed out in Chapter 4 that the inlet air to PEM fuel cells is sometimes humidified. This has the effect of altering both the specific heat capacity and γ, the ratio of the specific heat capacities, and this can alter the performance of the compressor. However, if the inlet air needs to be humidified, this is usually done after compression, when the air is hotter.

9.6 Performance Charts for Centrifugal Compressors

Centrifugal compressors are very common, being inherently simple and reliable. However, they do have special problems that merit careful consideration. These special features can be elucidated from their performance charts, an example being shown in Figure 9.5. All centrifugal compressors have performance charts that are similar in form. The two main points to note are as follows:

- There are regions of high efficiency, but they are very narrow. The constant efficiency 'contours' are very close together when moving across the chart at a constant pressure ratio.

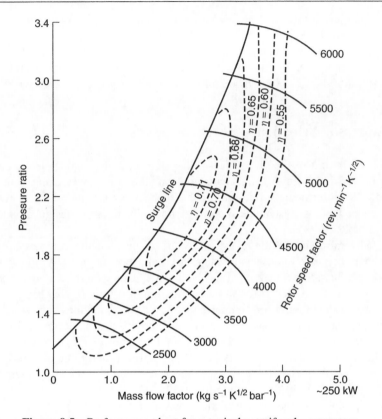

Figure 9.5 Performance chart for a typical centrifugal compressor.

- There is a distinct 'surge line'. To the left of this line the compressor is unstable and should not be used. The centrifugal compressor works by accelerating gas out from the centre. If it does not have any gas 'to work on', then it does not work, and the pressurised gas upstream of the pump will flow back, only to be pumped through again. The pressure thus becomes unstable and the gas gets hotter. The centrifugal pump must not be operated in this 'low flow rate' region. If the flow rate has to be low, the pressure ratio must also be reduced. (This is a very simple explanation of 'surge' in centrifugal compressors; for a fuller explanation, a book on turbochargers or compressors is recommended, e.g. Watson and Janota, 1982.)

A consequence of these two points, and the general shape of the performance chart, is that if a system using centrifugal compressors is to be operated at optimum efficiency, the pressure should *not* be constant. As the gas flow rate increases and decreases with varying power, the pressure should also rise and fall, following the maximum efficiency regions of Figure 9.5. In most applications of centrifugal compressors this is not a problem. For example, when turbocharging an internal combustion engine, at lower engine speeds, less pressure ratio, and hence supercharging, is expected. Similarly, with fuel cells, less pressure boost is needed at lower powers. However, both the reactant air and fuel must be at a similar pressure, otherwise the stack may be damaged by pressure differentials, and the control needed to keep both pressures changing, yet the same as each other, can be difficult. Also, if a fuel-processing system is involved, this can further complicate matters.

Note that a rotor speed factor of 4500, the value in the optimum operating region, corresponds to $4500 \times \sqrt{298} \approx 78\,000$ rpm. These are very high-speed devices.

9.7 Compressor Selection – Practical Issues

It would be tempting to suppose the fuel cell system designer could select a compressor simply using the mass flow rate, calculated from the cell power using equation A2.4 in Appendix 2. Unfortunately this is unlikely to be the case, unless he is working on an unlimited budget! A system designer will normally have to design around what is actually available.

For pressure ratios in the range 1.4 to around 3, most of the compressors available are designed principally for use with internal combustion engines. A good example is the Eaton Supercharger, as mentioned in Section 9.2. This would be a good choice for a pressure ratio around 1.35 to 1.7 (or an outlet pressure of 0.35 to 0.7 bar gauge). However, such units are produced only for comparatively large petrol engines. The smallest of the range from this particular company is shown in Figure 9.6. It is designed for use in the flow range 50–100 L s^{-1}, corresponding to a fuel cell power range of about 50 kW to 150 kW.[1] In terms of fuel cells, this is quite a high power, and remember that this is for the *smallest* available fuel cell. We can see then that obtaining such a compressor for fuel cells of power less than 50 kW is going to be difficult.

The same problem applies to the higher-pressure ratios, of above 1.6 up to 3, which corresponds to 0.6 to 2 bar gauge, or about 9 to 30 psig. In this pressure range, the

Figure 9.6 Eaton Supercharger. This unit is about 25 cm long. It will boost pressure by between 35 and 70 kPa. (Reproduced by kind permission of Eaton Corporation.)

---

[1] This was calculated using the equation A2.4 derived in Appendix 2 with typical values of stoichiometry ($\lambda = 2$) and average cell voltage ($V_c = 0.6$).

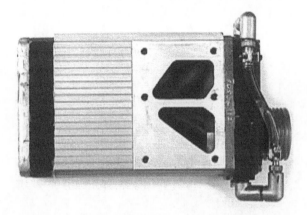

Figure 9.7   A Lysholm or screw compressor. The air enters a hole, not visible, on the left. It exits via the holes on the visible face. A pulley on the right drives the screws.

screw compressor or Lysholm would be the first choice for efficiency and flexibility, as is shown by its chart in Figure 9.4. Such screw compressors are known commercially in the motor trade as 'WhippleChargers', and the majority are manufactured by a company called *Autorotor* in Sweden. A typical unit is shown in Figure 9.7. The smallest unit in this range weighs just 5 kg, and its dimensions are $18 \times 9 \times 13$ cm, so it is quite a small unit. However, it is designed for flow rates up to about $0.12 \, \text{kg s}^{-1}$, which corresponds to about 100 kW for typical fuel cell operating conditions ($\lambda = 2$, cell voltage $= 0.6$). It can be seen from Figure 9.4 that when such units are operated at low flow rates, their efficiency falls off. So we can see, again, that we have a problem if we are looking for a compressor for small fuel cells. However, for fuel cells of 50 kW or more, and especially above 100 kW, there is a wide range of products available, where the designer can choose from tried and tested units that are produced in quantity and so are readily available at reasonable cost.

The units shown in Figures 9.6 and 9.7 are designed for use on internal combustion engines. They both have pulleys for connection to the crankshaft of the engine. In fuel cell systems the compressor will have to be driven by an electric motor. Typically, this will be as large or larger than the compressor, but motors are covered in more detail in the next chapter.

9.8 Turbines

Fuel cells usually output warm or hot gases. Fuel reformers do likewise. This energy can be harnessed with a turbine. The turbine will frequently be used to turn a compressor to compress the incoming air or fuel gas. In a few cases, there may be excess power, and in such cases a generator can be fitted to the same shaft.

There are only two types of turbines worth considering. The first is the centripetal or radial, which is essentially the inverse of the centrifugal compressor considered above.

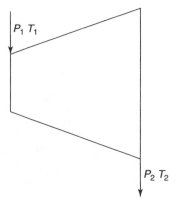

Figure 9.8   Symbol for a turbine.

This is the preferred choice unless the powers concerned are over about 500 kW, when the axial turbine used as standard in gas and steam turbine power generation sets may be considered. Whatever type of turbine is used, the symbol is as shown in Figure 9.8.

The efficiency of the turbine is treated in a similar way to that for compressors, with the same assumptions. If the turbine worked isentropically, then the outlet temperature would fall to T_2' where, as with the compressor

$$\frac{T_2'}{T_1} = \left(\frac{P_2}{P_1}\right)^{\frac{\gamma-1}{\gamma}}$$

However, because some of the energy will not be output to the turbine shaft, but will stay with the gas, the actual output temperature will be higher than this. The actual work done will therefore be *less* than the isentropic work. (For the compressor it was *more*). We thus define the isentropic efficiency as

$$\eta_c = \frac{\text{actual work done}}{\text{isentropic work}}$$

Making the same assumptions about ideal gases and so on that we did for compressors, this becomes

$$\eta_c = \frac{T_1 - T_2}{T_1 - T_2'}$$

$$= \frac{T_1 - T_2}{T_1\left(1 - \left(\frac{P_2}{P_1}\right)^{\frac{\gamma-1}{\gamma}}\right)} \qquad [9.8]$$

This is usually used to find the temperature change. So we can arrange it to

$$\Delta T = T_2 - T_1 = \eta_c \cdot T_1 \left(\left(\frac{P_2}{P_1} \right)^{\frac{\gamma-1}{\gamma}} - 1 \right) \qquad [9.9]$$

Note that because $P_2 < P_1$ this will always be negative. It is useful to use this equation to derive a formula for the power available from the turbine. Applying the same reasoning and simplifying assumptions that we did in Section 9.4, we have

$$\text{Power} = \dot{W} = c_p \cdot \Delta T \cdot \dot{m}$$

So

$$\text{Power}, \dot{W} = c_p \eta_c T_1 \left(\left(\frac{P_2}{P_1} \right)^{\frac{\gamma-1}{\gamma}} - 1 \right) \dot{m} \qquad [9.10]$$

To get the power available to drive an external load, we should multiply this by η_m, the mechanical efficiency. As with the compressor, this should be about 0.98 or better.

The representation of turbine performance using charts is done along the same lines as that for compressors. The performance map is drawn in a similar way, except that the vertical axis is now P_1/P_2, instead of P_2/P_1. The form of the map has similarities, though the shape and direction of the rotational speed factor lines are completely different. An example is shown in Figure 9.9. For any given speed the mass flow rate rises as the pressure drop increases, as we would expect, but tends towards a maximum value. This maximum value is called the *choking limit*. Naturally enough, this depends largely on the diameter of the turbine housing.

Notice that if you pick on a constant mass flow (e.g. 1.5) and move up through the map, the rotor speed factor increases as the pressure ratio increases. This again is what would be expected.

Worked Example

What power would be available from the exit gases of the 100-kW fuel cell system given in Section 9.5?

Solution. The mass of the cathode exit gas would have been increased by the presence of water produced in the cells, but since this is the result of replacing O_2 with $2H_2O$, the mass change will be very small, hydrogen being so light.[2] So we will approximate $\dot{m}$, the mass flow rate as still $0.11 \, \text{kg s}^{-1}$. We can only estimate the exit temperature as, say 90°C (363 K), a typical operating temperature for this type of fuel cell. The entry pressure is

---

[2] The addition of water to the air will raise the specific heat capacity c_p and lower γ, but the effect of this will not be great. We are only *estimating* the power.

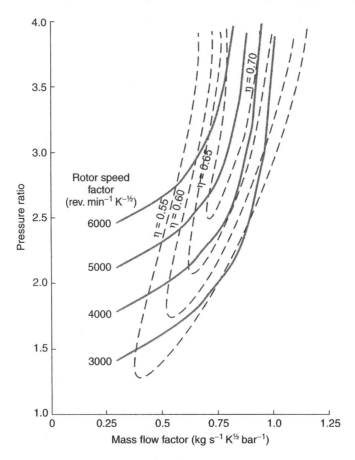

Figure 9.9 Performance chart for a typical small radial turbine.

3.0 bar. The exit pressure must be a little less than this, say 2.8 bar. So we can calculate the mass flow factor as

$$\text{Mass flow factor } \frac{0.11 \times \sqrt{363}}{2.8} = 0.75 \, \text{kg s}^{-1} \, \text{K}^{\frac{1}{2}} \text{bar}^{-1}$$

Using the turbine performance chart in Figure 9.9 we can find the efficiency and speed of the turbine. If we find the intercept up from 0.75 on the x-axis and 2.8 on the pressure ratio axis, we see that it is very close to the rotor speed factor = 5000 line, and in the $\eta = 0.70$ efficiency region. (More or less optimum operating conditions in fact!) So we can conclude that

$$\text{Rotor speed} = 5000 \times \sqrt{363} \approx 95\,000 \, \text{rpm}$$

Note that this is very fast! Such turbines are high-speed devices. They could NOT be used to drive a screw compressor, though such speeds are suitable for driving centrifugal compressors, as is discussed in Section 9.9 below.

Since we know that the efficiency is about 0.7, we can use equation 9.10 to find the power. The exit gas is not normal air; it has extra water vapour, and less oxygen, and so the specific heat capacity will have changed, as will γ, the ratio of the specific heat capacities. In engines, standard values used are $1150\,\mathrm{J\,kg^{-1}\,K^{-1}}$ for c_p and 1.33 for γ (e.g. Stone, 1999). In this case the change in gas composition is not so great, so we will use $1100\,\mathrm{J\,kg^{-1}\,K^{-1}}$ for c_p and $\gamma = 1.38$. The constant $(\gamma - 1/\gamma)$ thus becomes 0.275 T_1 is 363 K, so equation 9.10 becomes

$$\text{Power available} = 1100 \times 0.7 \times 363 \left(\frac{1}{2.8}^{0.275} - 1 \right) \times 0.11 \approx -7.6\,\mathrm{kW}.$$

The minus sign indicates that power is given out. This is a useful addition to the 100-kW electrical power output of the fuel cell, but note that it provides less than half of the power needed to drive the compressor, as calculated with the worked example of Section 9.5. Furthermore, this example is the best possible result; turbine efficiencies will usually be somewhat lower than the 0.7 we obtained here. As can be seen from the chart, much of the operating region is at greatly lower efficiencies.

9.9 Turbochargers

The turbine of the worked example in Section 9.8 might be used to drive a compressor. When this is done the turbine and compressor are mounted side by side on the same shaft. With the surrounding housing this makes a very compact and simple unit. They are usually called *turbochargers* because their main application is the supercharging of engines using a turbine, driven by the exhaust gases. Turbocompressors might be a better name. The symbol for such a unit is shown in Figure 9.10.

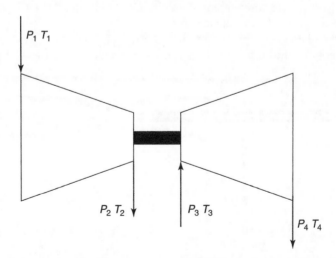

Figure 9.10   Representation of a linked turbine and compressor or 'turbocharger'.

The technology of these units is very mature, and they are available in a range of sizes, but only for the larger fuel cells, 50 kW or more. They are used on the majority of modern diesel engines and so are mass-produced. Compared to many other parts of a fuel cell system, they are readily available at low cost.

9.10 Ejector Circulators

The ejector is the simplest of all types of pumps. It has no moving parts. They are widely used in hydrogen fuel cells where the hydrogen is stored at pressure. In some Siemens Westinghouse SOFCs, they are used to recirculate the fuel gas. They harness the stored mechanical energy in the gas to circulate the fuel around the cell.

A simple ejector is shown in Figure 9.11. A gas or liquid passes through the narrow pipe A into the venturi B. It acquires a high velocity at B and hence produces suction in pipe C. The fluid coming through at A thus entrains with it the fluid from C and sends it out at D. Clearly, for this to work, the fluid from A must be at a higher pressure than that in C/B/D, otherwise it would not eject from pipe A at high velocity.

The fluid entering at A does not have to be the same as that in pipe C/B/D. The most common use of the ejector is in steam systems, with steam being the fluid passing through the narrow pipe and jet A. So, ejectors are used to pump air, for example, to maintain vacuum in the condensers of steam turbines. They are also used to pump water into boilers, and since the steam mixes with the pumped water it also preheats it. The steam is also used to circulate lower pressure steam in steam heating systems. In this application it is closest to its normal use in fuel cell systems.

In fuel cell systems in which, for example, the hydrogen fuel is circulated by an ejector as in Figure 9.11, the gas is supplied at high pressure at A, and in the ejector the energy of the expanding gas draws the gas from the fuel cell at C and sends it on through D back to the fuel cell. The pressure difference generated will be enough to drive the gas through the cell as well as through any humidifying equipment.

In terms of converting the mechanical energy stored in the compressed gas into kinetic energy of the circulating gas, the ejector circulator is very inefficient. However, since the mechanical energy of the stored gas is not at all convenient to harness, it is essentially

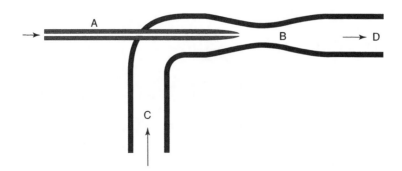

Figure 9.11   Diagram of a simple ejector circulation pump.

'free', and so as long as the circulation is sufficient, the inefficiency does not matter. The internal diameter of Pipe A, and the mixing region B, and C/D, have to be chosen bearing in mind the pressure differences and the flow rates. Tables and graphs for these can be found in chemical engineering reference books such as Perry (1984).

9.11 Fans and Blowers

For straightforward cooling purposes, fans and blowers are available in a huge range of sizes. The normal axial fan, such as we see cooling all types of electronic equipment, is an excellent device for moving air, but only against very small pressures. A small fan, such as is commonly used in electronic equipment, might move air at the rate of about $0.1\,\mathrm{kg\,s^{-1}}$, which is equivalent to $85\,\mathrm{L\,s^{-1}}$ at standard conditions. However, this flow rate will drop to zero if the back pressure even rises to $50\,\mathrm{Pa}$. These pressures are so low that they are often converted to centimetres of water and this helps us visualise how low they are – in this case just $0.5\,\mathrm{cm}$ of water! This is a typical maximum back pressure for such a fan. The result is that they can only be used when the air movement is in a very open area, such as equipment cabinets, and in a few very open designs of PEM fuel cell.

For a somewhat greater pressure, the centrifugal fan, such as we often see being used to blow the air in heating units, would be used. These draw air in and throw it out sideways. They are not totally different from the centrifugal compressors already considered, but turn at much slower speeds (by a factor of several hundred), have much longer blades, and a much more open construction. There are two main types, the 'backward curved' and the 'forward curved' centrifugal blowers. These names relate to the shape of the blades, as is shown in Figure 9.12 opposite. The forward curved blade gives a higher throughput of air, but with a lower maximum back pressure – around $3\,\mathrm{cm}$ of water being a typical value. This type of blower is used to drive the cooling air through small- to medium-sized PEM fuel cells. The backward curved impeller is more suitable in cases where a higher pressure is needed, but we accept a lower flow rate. However, in this context 'higher pressure' means 3 to $10\,\mathrm{cm}$ of water – still virtually nothing! This type of blower is used

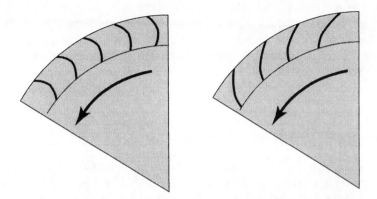

Figure 9.12 Forward curved and backward curved blades on a centrifugal blower.

for feeding air to furnaces, and might well be suitable for blowing the reactant air to small- and medium-sized PEM fuel cells. However, sometimes the pressure requirements might be greater, and a pump such as that described in the following section could be used.

With these fans and blowers the concept of efficiency is a difficult one. The input power is simply the electrical power used to run the motor. The purpose of the blower could be said to move the air, and so the output power is the rate of change of kinetic energy of the air. Using this measure, fans and blowers usually have very low efficiencies, which fall even lower as the back pressure against the airflow rises. However, this is not really a very helpful measure, since we do not actually want kinetic energy in the air. We usually want something else, for example, the removal of heat. Let us take as an example a small 120-mm axial fan such as is often used to cool electronic equipment. Such a fan might move air at $0.084 \, kg \, s^{-1}$ and consume 15 W of electric power. If the air it blows rises in temperature by just $10°C$, then the rate of removal of heat energy will be

$$\text{Power} = C_p \times \Delta T \times \dot{m} = 1004 \times 10 \times 0.084 = 843 \, \text{W}$$

That is, 843 W of heat removed for just 15 W of electrical power. Is that efficient? It may be. But it is not, if more of that heat could be removed by natural convection and radiation, and we are using a larger fan than is necessary. The point is that the efficiency of the heat removal system depends more on the design of the airflow path, and how much temperature the air can pick up as it goes through, than on the performance of the motor. Rather than the efficiency of such a cooling system, it is more useful to define *effectiveness* as

$$\text{Cooling system effectiveness} = \frac{\text{rate of heat removal}}{\text{electrical power consumed}}$$

In this case, that would give a figure of $843/15 = 56$. In the example system given in Section 4.8, page 101, a centrifugal blower, probably forward curved, consumes 70 W of electrical power to dispose of 1.84 kW of heat. The effectiveness of this system is thus 26, somewhat lower than in the previous case. This is because the rate of movement of air with a centrifugal blower is less than that with an axial fan, and is the price that has to be paid for a more compact fuel cell, where it is more difficult to get the air through. On the other hand, the more compact systems should be able to achieve a greater temperature change in the air as it goes through. Generally, fuel cell system designers should obtain effectiveness figures of between 20 and 30 fairly readily. There is always a balance in cooling systems between flow rate of air and electrical power consumed. Higher flow rates improve heat transfer, but at the expense of fan power consumption.

9.12 Membrane/Diaphragm Pumps

Small- to medium-sized PEM fuel cells are liable to be an important market for fuel cells in portable power systems. A problem with these cells is the pumps for circulating the reactant air, and the hydrogen fuel, if not dead-ended. In compact designs of, say, 200 W to 2 kW, the back pressure of the reactant air is liable to be around 10 kPa, equivalent to about 1 m of water, or 0.1 bar. This is too great for the blowers and fans we have just been considering. On the other hand the flow rates are too low for the commercial blowers and compressors we considered in the earlier sections of this chapter. Another type of pump is needed. The main features of these pumps should include the following:

- Low cost
- Silent
- Reliable when operated continuously for long periods
- Able to operate against pressures of about 10 kPa (1m water)
- Available in a suitable range of sizes, that is, about 2.5×10^{-4} to 2.5×10^{-3} kg s^{-1} (or 12 to 120 SLM)
- Efficient, low power consumption

The types of pumps already existing that satisfy a good deal of these requirements are made for applications such as gas sampling equipment, small-scale chemical processing, and fish tank aerators. They are either vane or diaphragm pumps. It is the fish tank aerator pumps that 'hit' most of the requirements mentioned above. These are mass produced, and so are of low cost. They have to be silent, and are meant for continuous operation day after day for several years. The larger versions are also designed for operating against a metre or two of water, and so offer exactly the pressure range required. These pumps are diaphragm pumps, as in Figure 9.13. There are many types of diaphragm (or membrane) pumps, since they have been used since ancient times. The diaphragm is moved up and down, shifting the air through the system by way of the two valves in a fairly obvious mechanical way. The use of the soft rubber makes for very quiet[3] and long-term reliable operation, though it does limit the operating pressure, but in this regard it is still suitable for our 10–20 kPa needs. Unfortunately, it is not possible to use this type of pump directly 'as sold', since the actuators they are usually supplied with are only suitable for alternating current (AC) operation, and not particularly efficient. However, the basic pump module, as in Figure 9.13, can be purchased very readily for a few dollars. The mass-produced units will shift air at only about 5×10^{-4} kg s^{-1}, but they can be readily operated in parallel. It is a fairly straightforward task to add a direct current (DC) operated actuator or motor to the pump, or to several pumps. Furthermore, the flow rate can be easily controlled by modulating the force applied by the actuator.

An example of a published system design that uses this approach is for a 300-W PEM fuel cell (Popelis et al., 1999). Using equation A2.4, and a stoichiometry of 2, and operation at about 0.6 V per cell, a 300-W cell would require a reactant air flow rate of 3.57×10^{-4} kg s^{-1}, which is equivalent to 18 SLM. The published details give a flow rate of 15–20 SLM.

This flow rate is delivered at a pressure of between 1.10 and 1.15 bar. The pump unit is manufactured by KNF Neuberger and is driven by a 12-VDC motor. According to the published data, this motor/pump combination consumes between 14 and 19 W. The parasitic power loss is thus about 6%. This represents about as low a figure as can

---

[3] In some systems the diaphragm pump is quite noisy. This is usually the result of an imperfect match between the motor and the actuator, or noise in the mechanism that converts the circular motor action into linear motion for the pump.

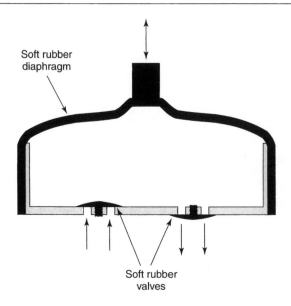

Figure 9.13 Diaphragm pump – low cost, quiet, and long-term reliable.

be expected for a small fuel cells system. For example, the 2-kW system presented in Section 4.9 used 200 W of electrical power to provide the reactant air, a 10% parasitic loss.

References

Perry R.H. and Green D.W. (1984) *Perry's Chemical Engineer's Handbook*, McGraw-Hill, New York, pp. 6–33.

Popelis I., Tsukada A., and Scherer G. (1999) "12 Volt 300 Watt PEFC power pack." Proceedings of the European Fuel Cell Forum Portable Fuel Cells Conference, Lucerne, pp. 147–155.

Stone R. (1999) *Introduction to Internal Combustion Engines*, 3rd edition, Macmillan, New York.

Watson N. and Janota M.S. (1982) *Turbocharging the Internal Combustion Engine*, Macmillan, New York.

10

Delivering Fuel Cell Power

10.1 Introduction

We have covered the details of fuel cell performance, how the different types work, how they can be fuelled, and how the gases can be blown and compressed. In this final chapter, we think about the basic purpose of a fuel cell – to generate electrical power. In this chapter we have brought together those areas of power electronics and electrical engineering that are of particular relevance to fuel cell systems, and hopefully explained them in a way that engineers from all specialities can understand. Fuel cells bring a few special problems, but by and large standard equipment and methods used in other electrical power systems can be employed. In other words, unlike many aspects of fuel cell systems, the electrical problems can generally be solved using more-or-less standard technologies. There are four technologies that we particularly need to address.

1. *Regulation*. The electrical output power of a fuel cell will often not be at a suitable voltage, and certainly that voltage will not be constant. We have seen in Chapter 3 that a graph of voltage against current for a fuel cell is by no means flat. Increasing the current causes the voltage to fall in all electrical power generators, but in fuel cells the fall is much greater. Voltage regulators, DC/DC converters, and chopper circuits are used to control and shift the fuel cell voltage to a fixed value, which can be higher or lower than the operating voltage of the fuel cell. These circuits are covered in Section 10.2.
2. *Inverters*. Fuel cells generate their electricity as direct current, DC. In many cases, especially for small systems, this is an advantage. However, in larger systems that connect to the mains grid, this DC must be converted to alternating current, AC. The inverters that do this are described and discussed in Section 10.3. Another important topic discussed in this section is that of the legislative framework that applies to generators connected to the mains.
3. *Electric motors*. A major area of application for fuel cells is transport, where the electrical power is delivered to an electric motor. In addition, motors are often needed

Fuel Cell Systems Explained, Second Edition James Larminie and Andrew Dicks
© 2003 John Wiley & Sons, Ltd   ISBN: 0-470-84857-X

to drive compressors and pumps within a fuel cell system. Most fuel cell systems of power greater than about 1 kW will have at least three electric motors. The advent of cheaper power electronic devices and controllers has led to the development of new types of motor with very high efficiencies. The competing claims of these different motor types are often hard to evaluate. In Section 10.4, we give an overview of these modern electric motors and their controllers, with a particular view to their application within fuel cell systems. This includes an indication of the size and mass a motor of any given power is likely to have.

4. *Hybrid systems*. In terms of cost, per watt of power fuel cells are expensive, and likely to remain so for some time to come. In many applications, for example, transport and communications equipment, the average power needed is much lower than the peak power. In these cases, lower-cost systems can be designed using a battery (or capacitor) with a fuel cell in a hybrid system. Basically, the fuel cell runs at the average power, with the battery or capacitor providing the peak power. These hybrid systems are described in Section 10.5.

10.2 DC Regulation and Voltage Conversion

10.2.1 Switching devices

The voltage from all sources of electrical power varies with time, temperature, and many other factors, especially current. However, fuel cells are particularly badly regulated. In Chapter 3, Figures 3.1 and 3.2, we see that the voltage from a cell falls rapidly with rising current density. Figure 10.1 summarises some data from a real 250-kW fuel cell used to drive a bus (Spiegel et al., 1999). The voltage varies from about 400 to over 750 V, and we also see that the voltage can have different values at the same current. This is because, as well as current, the voltage also depends on temperature, air pressure, and on whether the compressor has got to speed, among other factors.

Most electronic and electrical equipment require a fairly constant voltage. This can be achieved by dropping the voltage down to a fixed value below the operating range of the fuel cell, or boosting it up to a fixed value. This is done using 'switching' or 'chopping' circuits, which are described below. These circuits, as well as the inverters and motor controllers to be described in later sections, use *electronic switches*.

As far as the user is concerned, the particular type of electronic switch used does not matter greatly, but we should briefly describe the main types used, so that the reader has some understanding of their advantages and disadvantages. Table 10.1 shows the main characteristics of the most commonly used types.

The metal oxide semiconductor field effect transistor (MOSFET) is turned on by applying a voltage, usually between 5 and 10 V, to the gate. When 'on', the resistance between the drain (d) and source (s) is very low. The power required to ensure a very low resistance is small, as the current into the gate is low. However, the gate does have a considerable capacitance, so special drive circuits are usually used. The current path behaves like a resistor, whose ON value is $R_{DS_{ON}}$. The value of $R_{DS_{ON}}$ for a MOSFET used in voltage-regulation circuits can be as low as about 0.01 ohms. However, such low values are only possible with devices that can switch low voltages, in the region of 50 V. Devices that

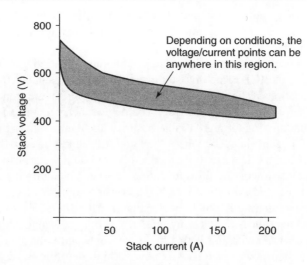

Figure 10.1 Graph summarising some data from a real 250-kW fuel cell used to power a bus. (Derived from data in Spiegel et al. 1999).

Table 10.1 Key data for the main types of electronic switches used in modern power electronic equipment.

Type	Thyristor	MOSFET	IGBT
Symbol			
Max. voltage (V)	4500	1000	1700
Max. current (A)	4000	50	600
Switching time (μs)	10–25	0.3–0.5	1–4

Note: insulated gas bipolar transistor, IGBT.

can switch higher voltages have values of $R_{DS_{ON}}$ of about 0.1 ohms, which causes higher losses. MOSFETs are widely used in low-voltage systems of power less than about 1-kW.

The insulated gate bipolar transistor, IGBT, is essentially an integrated circuit combining a conventional bipolar transistor and a MOSFET, and has the advantages of both. They require a fairly low voltage, with negligible current at the gate, to turn on. The main current flow is from the collector to the emitter, and this path has the characteristics of a *p-n* junction. This means that the voltage does not rise much above 0.6 V at all currents within the rating of the device. This makes it the preferred choice for systems in which the current is greater than about 50 A. They can also be made to withstand higher voltages. The longer switching times compared to the MOSFET, as given in Table 10.1, are a disadvantage in lower power systems. However, the IGBT is now almost universally the electronic switch of choice in systems from 1 kW up to several hundred kW, with the 'upper' limit rising each year.

The thyristor has been the electronic switch most commonly used in power electronics. Unlike the MOSFET and IGBT, the thyristor can only be used as an electronic switch – it has no other applications. The transition from the blocking to the conducting state is triggered by a pulse of current into the gate. The device then remains in the conducting state until the current flowing through it falls to zero. This feature makes them particularly useful in circuits for rectifying AC, where they are still widely used. However, various variants of the thyristor, particularly the gate-turn-off or GTO thyristor can be switched off even while a current is flowing by the application of a negative current pulse to the gate.

Despite the fact that the switching is achieved by just a pulse of current, the energy needed to effect the switching is much greater than for the MOSFET or the IGBT. Furthermore, the switching times are markedly longer. The only advantage of the thyristor (in its various forms) for DC switching is that higher currents and voltages can be switched. However, the maximum power of IGBTs is now so high that this is very unlikely to be an issue in fuel cell systems that are usually below 1 MW in power.

Ultimately, the component used for the electronic switch is not of great importance. As a result, the circuit symbol used is often the 'device independent' symbol shown in Figure 10.2. In use, it is essential that the switch moves as quickly as possible from the conducting to the blocking state, or vice versa. No energy is dissipated in the switch while it is in open circuit, and only very little energy is lost when it is fully on; it is while the transition is taking place that the product of voltage and current is non-zero and that power is lost.

10.2.2 Switching regulators

The 'step-down' or 'buck' switching regulator (or chopper) is shown in Figure 10.3. The essential components are an electronic switch with an associated drive circuit, a diode, and an inductor. In Figure 10.3a the switch is on, and the current flows through the inductor and the load. The inductor produces a back electromotive force (emf), making the current gradually rise. The switch is then turned off. The stored energy in the inductor keeps the current flowing through the load using the diode, as in Figure 10.3b. The different currents flowing during each part of this on–off cycle are shown in Figure 10.4. The voltage across the load can be further smoothed using capacitors if needed.

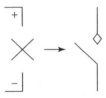

Figure 10.2 Symbol used for an electronically operated switch.

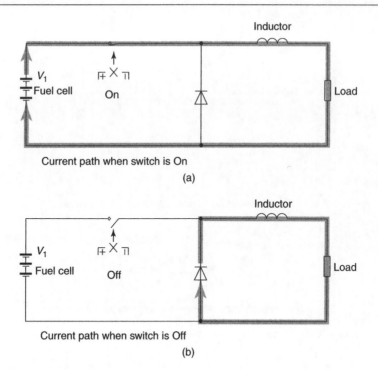

Current path when switch is On

(a)

Current path when switch is Off

(b)

Figure 10.3 Circuit diagram showing the operation of a switch mode step-down regulator.

If V_1 is the supply voltage, and the 'on' and 'off' times for the electronic switch are t_{ON} and t_{OFF}, then it can be shown that the output voltage V_2 is given by

$$V_2 = \frac{t_{ON}}{t_{ON} + t_{OFF}} V_1 \qquad [10.1]$$

It is also clear that the ripple depends on the frequency – at higher frequency, the ripple is less. However, each turn-on and turn-off involves the loss of some energy, so the frequency should not be too high. A control circuit is needed to adjust t_{ON} to achieve the desired output voltage – such circuits are readily available from many manufacturers.

The main energy losses in the step-down chopper circuit are

- switching losses in the electronic switch,
- power lost in the switch while it is turned on ($0.6 \times I$ for an IGBT or $R_{DS_{ON}} \times I^2$ for a MOSFET),
- power lost because of the resistance of the inductor,
- losses in the diode, $0.6 \times I$.

In practice all these can be made very low. The efficiency of such a step-down chopper circuit should be over 90%. In higher voltage systems, of about 100 V or more, efficiencies as high as 98% are possible.

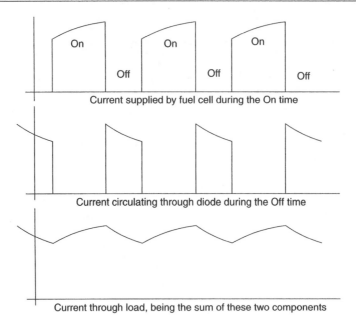

Figure 10.4   Currents in the step-down switch mode regulator circuit.

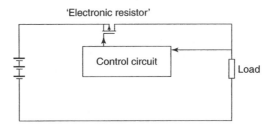

Figure 10.5   Linear regulator circuit.

We should, at this point, briefly mention the 'linear' regulator circuit. The principle is shown in Figure 10.5. A transistor is used again, but this time it is not switched fully on or fully off. Rather, the gate voltage is adjusted so that its resistance is at the correct value to drop the voltage to the desired value. This resistance will vary continuously depending on the load current and the supply voltage. This type of circuit is widely used in electronic systems, but should *never* be used with fuel cells. Simply converting the surplus voltage into heat does reduce the voltage, but wastes energy. Fuel cells will always be used where efficiency is of paramount importance, and linear regulators have no place in such systems.

Because fuel cells are essentially low-voltage devices, it is often desirable to step up or boost the voltage. This can also be done, quite simply and efficiently, using switching circuits. The circuit of Figure 10.6 is the basis usually used.

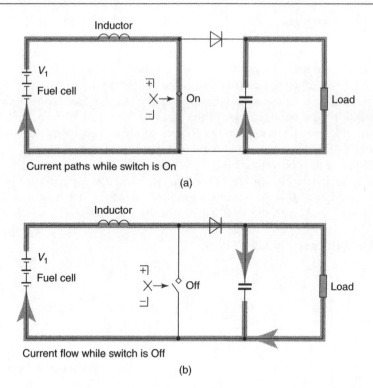

Figure 10.6   Circuit diagram to show the operation of a switch mode boost regulator.

We begin our explanation by assuming that some charge is in the capacitor. In Figure 10.6a the switch is on, and an electric current is building up in the inductor. The load is supplied by the discharging of the capacitor. The diode prevents the charge from the capacitor from flowing back through the switch. In Figure 10.6b the switch is off. The inductor voltage rises sharply because the current is falling. As soon as the voltage rises above that of the capacitor (plus about 0.6 V for the diode), the current will flow through the diode, charge up the capacitor, and flow through the load. This will continue as long as there is still energy in the inductor. The switch is then closed again, as in Figure 10.6a, and the inductor re-energised while the capacitor supplies the load.

Higher voltages are achieved by having the switch off for a short time. It can be shown that for an ideal converter with no losses

$$V_2 = \frac{t_{ON} + t_{OFF}}{t_{OFF}} V_1 \qquad [10.2]$$

In practice, however, the output voltage is somewhat less than this. As with the step-down (buck) switcher, control circuits for such boost or step-up switching regulators are readily available from many manufacturers.

The losses in this circuit come from the same sources as for the step-down regulator. However, because the currents through the inductor and switch are higher than the output current, the losses are higher. Also, all the charge passes through the diode this time, and so is subject to the 0.6-V drop and hence energy loss. The result is that the efficiency of these boost regulators is somewhat less than that for the buck. Nevertheless, over 80% efficiency should normally be obtained, and in systems in which the initial voltage is higher (over 100 V), efficiencies of 95% or more are possible.

A third possibility is to use a buck-boost regulator. In this case the final output is set somewhere within the operating range of the fuel cell. While such circuits are technically possible, their efficiency tends to be rather poor, certainly no better than the boost chopper, and often worse. The consequence is that this is not a good approach.

An exception to the above possibility is in cases in which a small variation in output voltage can be tolerated, and an up-chopper circuit is used at *higher currents only*. This is illustrated in Figure 10.7. At lower currents the voltage is not regulated. The circuit of Figure 10.6 is used, with the switch permanently off. However, the converter starts operating when the fuel cell voltage falls below a set value. Since the voltage shift is quite small, the efficiency would be higher.

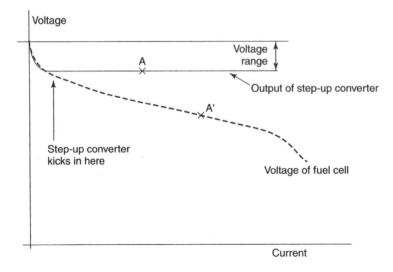

Figure 10.7 Graph of voltage against current for a fuel cell with a step-up chopper circuit that regulates to a voltage a little less than the maximum stack voltage.

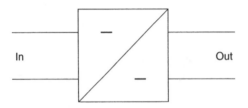

Figure 10.8 Typical fuel cell, DC–DC converter and inverter arrangement.

It should be pointed out that, of course, the current 'out' from a step-up converter is less than the current 'in'. In Figure 10.7, if the fuel cell is operating at point A′, the system output will be at point A – a higher voltage but a lower current. Also, the system is not entirely 'loss free' while the converter is not working. The current would all flow through the inductor and the diode, resulting in some loss of energy.

These step-up and step-down switching or chopper circuits are called *DC – DC converters*. The symbol usually used in diagrams is shown in Figure 10.8. Complete units, ready-made and ruggedly packaged, are available in a very wide range of powers, and with input and output voltages. In the cases in which the requirements cannot be met by an off-the-shelf unit, we have seen that the units are essentially quite simple and not hard to design. Standard control integrated circuits can nearly always be used to provide the switching signals.

10.3 Inverters

10.3.1 Single phase

Fuel cells will often be used in combined heat and power (CHP) systems in both homes and businesses. In these systems the fuel cell will need to connect to the AC mains grid. The output may also be converted to AC in some grid-independent systems.

In small domestic systems, the electricity will be converted to a single AC voltage. In larger industrial systems, the fuel cell will be linked to a three-phase supply, as discussed in Section 10.3.2.

The arrangement of the key components of a single-phase inverter is shown in Figure 10.9. There are four electronic switches, labelled A, B, C, and D, connected in what is called an *H-bridge*. Across each switch is a diode, whose purpose will become clear later. A resistor and an inductor represent the load through which the AC is to be driven.

The basic operation of the inverter is quite simple. First switches A and D are turned on and a current flows to the right through the load. These two switches are then turned off; at this point we see the need for the diodes. The load will probably have some inductance, and so the current will not be able to stop immediately, but will continue to flow in the same direction, through the diodes across switches B and C, back into the supply. The switches B and C are then turned on, and a current flows in the opposite direction, to the left. When these switches turn off, the current 'free wheels' on through the diodes in parallel with switches A and D.

The resulting current waveform is shown in Figure 10.10. In some cases, though increasingly few, this waveform will be adequate. The fact that it is very far from a sine wave will be a problem in almost all circumstances in which there is a connection to the mains grid.

The difference between a pure sine wave and any other waveform is expressed using the idea of 'harmonics'. These are sinusoidal oscillations of voltage or current whose frequency, f_v, is a whole number multiple of the fundamental oscillation frequency. It can be shown that *any* periodic waveform of *any* shape can be represented by the addition of harmonics to a fundamental sine wave. The process of finding these harmonics is known

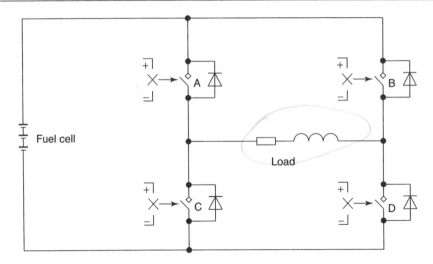

Figure 10.9   H-bridge inverter circuit for producing single-phase alternating current.

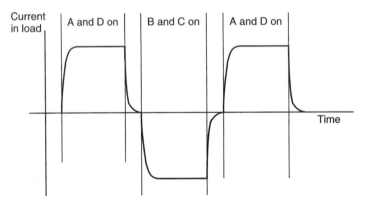

Figure 10.10   Current/time graph for a square-wave switched single-phase inverter.

as the Fourier analysis. For example, it can be shown that a square wave of frequency f can be expressed by the equation

$$v = \sin(\omega t) - \frac{1}{3}\sin(3\omega t) + \frac{1}{5}\sin(5\omega t) - \frac{1}{7}\sin(7\omega t)$$

$$+ \frac{1}{9}\sin(9\omega t) \ldots \text{etc.} \qquad\qquad\qquad\qquad [10.3]$$

where $\omega = 2\pi f$

So, the difference between a voltage or current waveform and a pure sine wave may be expressed in terms of higher-frequency harmonics imposed on the fundamental frequency.

The problem is that these higher-frequency harmonics can have harmful effects on other equipment connected to the grid and on cables and switchgear owing to high harmonic currents. Among the most serious effects are possible damage to protective equipment and disturbance of control systems. They can also cause inefficiencies in electric motors, unpleasant noise in all types of machines, damage to computers and other electronic equipment, picture interference in TV sets, and other undesirable effects. For this reason, there are now regulations concerning the 'purity' of the waveform of an AC current supplied to the grid. Unfortunately, these standards vary in different countries and circumstances. However, they are all expressed in terms of the amplitude of each harmonic relative to the amplitude of the fundamental frequency. One widely accepted standard is IEC 1000-2-2 (Heier, 1998). The maximum percentage of each harmonic, up to the 50th, for this standard is given in Table 10.2.

As can be seen from equation 10.3, a square wave exceeds the limits on the third harmonic by a factor of over 6, as well as others. So, how can a purer sinusoidal voltage and current waveform be generated? Two approaches are used, pulse-width modulation and the more modern tolerance-band pulse-inverter technique.

The principle of pulse-width modulation is shown in Figure 10.11. The same circuit as shown in Figure 10.9 is used. In the positive cycle, only switch D is on all the time, and switch A is on intermittently. When A is on, current builds up in the load. When A is off, the current continues to flow, because of the load inductance, through switch D and the 'free-wheeling' diode in parallel with switch C, around the bottom right loop of the circuit.

In the negative cycle a similar process occurs, except that switch B is on all the time, and switch C is 'pulsed'. When C is on, current builds in the load, and when off it continues to flow – though declining – through the upper loop in the circuit, and through the diode in parallel with switch A.

Table 10.2 Maximum permitted harmonic levels as a percentage of the fundamental in the low-voltage grid up to the harmonic number 50 according to IEC 1000-2-2. (Reproduced by kind permission from Heier S. (1998) *Grid Integration of Wind Energy Conversion Systems*, John Wiley & Sons, Chichester.)

v		2	3	4	5	6	7	8	9	10
%		2.0	5.0	1.0	6.0	5.0	5.0	0.5	1.5	0.5
v	11	12	13	14	15	16	17	18	19	20
%	3.5	0.2	3.0	0.2	0.3	0.2	2.0	0.2	1.5	0.2
v	21	22	23	24	25	26	27	28	29	30
%	0.2	0.2	1.5	0.2	1.5	0.2	0.2	0.2	0.63	0.2
v	31	32	33	34	35	36	37	38	39	40
%	0.6	0.2	0.2	0.2	0.55	0.2	0.53	0.2	0.2	0.2
v	41	42	43	44	45	46	47	48	49	50
%	0.5	0.2	0.49	0.2	0.2	0.2	0.46	0.2	0.45	0.2

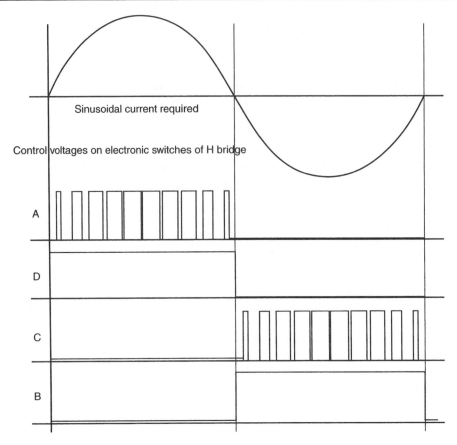

Figure 10.11 Pulse-width modulation switching sequence for producing an approximately sinusoidal alternating current from the circuit of Figure 10.9.

The precise shape of the waveform will depend on the nature (resistance, inductance, capacitance) of the load, but a typical half cycle is shown in Figure 10.12. The waveform is still not a sine wave, but is a lot closer than that of Figure 10.10. Clearly, the more pulses there are in each cycle, the closer will be the wave to a pure sine wave and the weaker will be the harmonics. A commonly used standard is 12 pulses per cycle, and generally this gives satisfactory results.

In modern systems, the switching pulses are generated by microprocessor circuits. This has led to the adoption of a more 'intelligent' approach to the switching of inverters called the *tolerance band pulse* method. This method is illustrated in Figure 10.13. The output voltage is continuously monitored, and compared with an internal 'upper limit' and 'lower limit', which are sinusoidal functions of time. In the positive cycle, switch D (in Figure 10.9) is on all the time. Switch A is turned on, and the current through the load rises. When it reaches the upper limit A is turned off, and the current flows on, though declining, through the diode in parallel with C, as before. When the lower limit is reached,

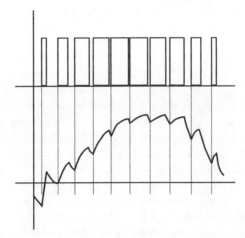

Figure 10.12   Typical voltage/time graph for a pulse-modulated inverter.

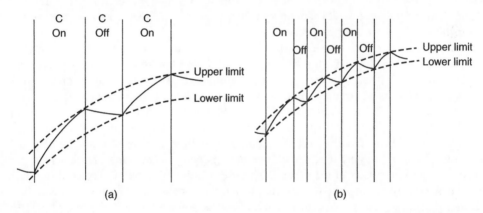

Figure 10.13   Typical voltage/time waveforms when using the tolerance band pulse inverter technique.

switch A is turned on again, and the current begins to build up again. This process is continuously repeated, with the voltage rising and falling between the tolerance bands.

In Figure 10.13, the on/off cycle is shown in (a) for a wide tolerance band and in (b) for a narrow tolerance band. It should also be appreciated that the resistance and inductance of the load will also affect the waveform, and hence the frequency at which switching occurs. This is thus an adaptive system that always keeps the deviation from a sine wave, and hence the unwanted harmonics, below fixed levels.

The only significant disadvantage of the 'tolerance band' regulation method is that it is possible for the pulsing frequency to become very high. Because most of the losses in the system occur at the time of switching, while the transistors move from off to on and on to off, this can lead to lower efficiency. The well-tried pulse-width modulation method is still widely used, but the tolerance band technique is becoming more widespread, especially for mains-connected inverters, where the nature of the load will not vary as much as in some other cases.

10.3.2 Three phase

In almost all parts of the world, electricity is generated and distributed using three parallel circuits, the voltage in each one being out of phase with the next by 120°. While most homes are supplied with just one phase, most industrial establishments have all three phases available. So, for industrial CHP systems, the DC from the fuel cell will need to be converted to three-phase AC.

This is only a little more complicated than the single phase. The basic circuit is shown in Figure 10.14. Six switches, with freewheeling diodes, are connected to the three-phase transformer on the right. The way in which these switches are used to generate three similar but out-of-phase voltages is shown in Figure 10.15. Each cycle can be divided into six steps. The graphs of Figure 10.16 show how the current in each of the three phases changes with time using this simple arrangement. These curves are obviously far from being sine waves. In practice, the very simple switching sequence of Figure 10.15 is modified, using pulse-width modulation or tolerance band methods, in the same way as for the single-phase inverters described above.

The modern three-phase inverter is built along similar lines whether it is for high or low power, and whether it is 'line-commutated' (i.e. the timing signals are derived from the grid to which it is connected) or 'self-commutated' (i.e. independent of the grid). Indeed, the same basic circuit is used whatever modulation method is used (pulse-width or tolerance band). The basic circuit is shown in Figure 10.17. The signals used to turn the switches on and off are taken from the microprocessor. Voltage and current sense signals may be taken from the three phases, the input, each switch, or other places. Digital signals from other sensors may be used. Also, instructions and information may be sent to and received from other parts of the system. The use of this information may be different in every case, but the *hardware* will essentially be the same – the circuit of Figure 10.17. Inverter units have thus become like many other electronic systems – a standard piece of

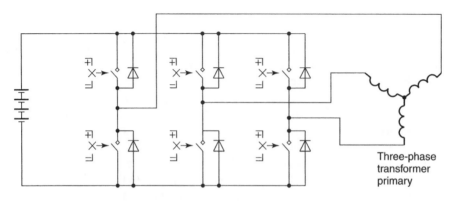

Figure 10.14   Three-phase inverter circuit.

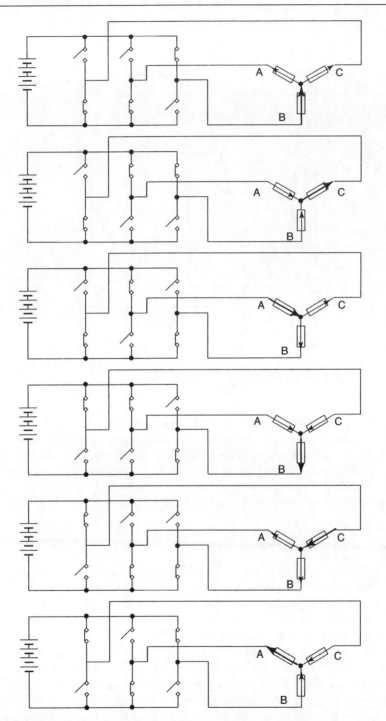

Figure 10.15   Switching pattern to generate three-phase alternating current.

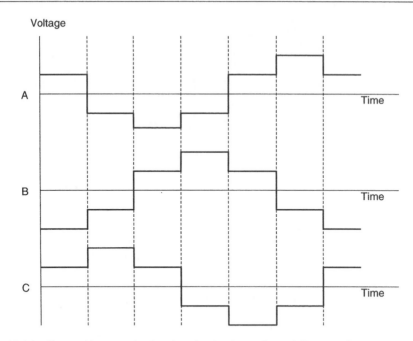

Figure 10.16 Current/time graphs for the simple three-phase AC generation system shown in Figure 10.15, assuming a resistive load. One complete cycle for each phase is shown. Current flowing *out* from the common point is taken as *positive*.

hardware that is programmed for each application. The switches would vary depending on the power requirements, but the same controller could be suitable for a large range of powers. The same inverter hardware, differently programmed, would thus be suitable for fuel cells, solar panels, wind-driven generators, Stirling cycle engines, and any other distributed-electricity generator. Inverters thus follow the trend seen throughout electronic engineering – lower prices, less electronics circuit design, and more programming.

10.3.3 Regulatory issues and tariffs

It is not sufficient simply to generate AC electricity and connect your generator to the grid. Quite properly, any private system has to conform to certain standards before it can be allowed to supply electricity. One of the most important of these is the level of the harmonics, an issue that we have already discussed. However, there must also be protection systems to protect the fuel cell, the inverter, and the grid from faults such as short circuits and lightning strikes. A particularly important feature is that the CHP system must disconnect from the grid in the event of the main power to the grid failing. This is to protect the grid-service personnel when carrying out fault repair or maintenance and to prevent the generator from feeding a short circuit with a damaging high current, the so-called 'fault current'. A generator is capable of feeding a local fault with considerably increased currents, when compared with that from a distant central power station. Switchgear and

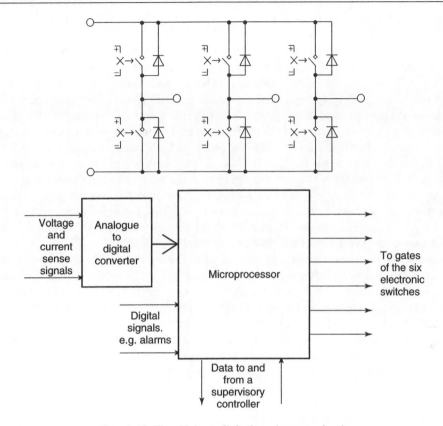

Figure 10.17 'Universal' 3-phase inverter circuit.

cabling may need expensive upgrading to cope with these increased currents that may flow under fault conditions. The local utility will have its own regulations and standards that equipment must conform to. In the United Kingdom, one such is the G59 standard. In the United States, equipment will nearly always need to be 'UL listed'. (UL refers to the Underwriters' Laboratory, a safety standards body.)

Where a fuel cell is generating energy for use in a mains electric grid, it will always be as part of a CHP system. The production of electricity will thus be linked with the production of heat. It is very likely that the demands for these two quantities will not be entirely synchronised, and so there will be times when the system may be producing less electricity than needed, or there may be times of high heat demand when there is an excess of electricity supply. Such problems can be solved by maintaining a connection to the local mains utility grid and drawing and supplying electricity as needed. The vast majority of CHP systems do this.

The precise regulations covering grid-connection (intertie) equipment will, unfortunately, vary between countries and between states within countries. In the United States, the Public Utilities Regulatory Policies Act (PURPA) of 1978 requires electric utilities to purchase electricity generated by small-power producers (SPPs) who run CHP systems.

Furthermore, the price of this electricity is fixed by a body independent of the electric utilities, a state utility regulatory agency, which operates under federal guidelines. The concept of 'avoided cost' is used as the basis for the utility power purchase rates. This is the rate that it costs the utility to generate the electricity itself. However, those rates will naturally be lower than the 'retail' rate at which they sell power.

For this reason, a CHP system connected to the grid will usually have two metres – one measuring the electricity consumed and the other the electricity provided. The charges paid will sometimes vary according to time. It is generally more expensive to generate and supply electricity during periods of high demand. Peaks of demand occur daily and vary according to the types of clients a utility has, on the climate, and on weather conditions. CHP systems are particularly suited to helping with the winter morning's peak that occurs in many areas, though they are not so well suited to reduce the summer afternoon peak that can occur in warmer climes. Such time of use (TOU) charges may apply to both electricity bought and sold, which could be advantageous to some CHP systems.

The two-metre system puts the SPP at a disadvantage.[1] However, in order to reduce investment in power generation equipment and to encourage the development of SPPs with their environmental advantages, some utilities and states in the United States have encouraged and instituted so-called 'net metering'. One metre is used, which winds back when electricity is supplied to the grid. The SPP is thus effectively paid the same for the electricity it supplies as it pays for what it consumes. Political pressure for the extension of these more favourable billing arrangements is likely in the years ahead, and is certainly well under way in the United States. In Europe too there are local suppliers who will offer net metering.

10.3.4 Power factor correction

A major advantage of distributed generators of electricity, especially those with higher power, is the ability to correct the power factor of a customer, or even a small district. The problem of power factors arises when the current consumed is not exactly in phase with the voltage. For example, the current might lag behind the voltage, as in Figure 10.18. The current can be split into two components, an in-phase component and a 90° out-of-phase component. This 90° out-of-phase component consumes no power, and is known as *reactive* power. Although it consumes no power locally, it does increase the power dissipated in the resistive electric cables distributing the electricity. This reactive power is, in a sense, 'free' to *generate* but not free to *distribute*. For this reason, it is particularly advantageous if it can be generated locally. Inverters fed from DC supplies are particularly suitable for this. The switching pulses in the pulse width modulation (PWM) or tolerance band inverter circuits are driven so that the current produced is deliberately out of phase with the line voltage.

Other small electricity generators, such as wind-driven systems, can generate this reactive power. However, it is a particular feature of fuel cell CHP systems, as they will always be very close to the point of usage of electricity.

---

[1] For example, in Oxford, in 2002, one local utility charges 7p and pays only 2p, per kilowatthour.

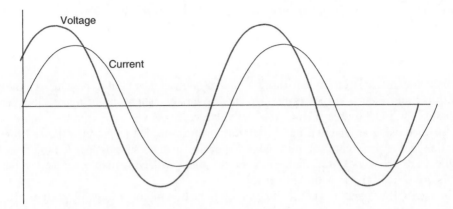

Figure 10.18 Voltage and current out of phase. The reactive power can be locally generated by distributed power systems such as fuel cells.

10.4 Electric Motors

10.4.1 General points

The importance of electric motors in fuel cell systems can hardly be overstated. For example, the 2-kW system described in Section 4.7, and shown in Figure 4.29, features three electric motors just to make it work. The final electrical power will often be delivered to an electric motor also. This is certainly the case for the 200-kW system also outlined in Section 4.7, which features at least four electric motors.

Electric motors are, of course, very widely used in engineering products. For example, there are about 20 electric motors in a typical modern motor car. Most of these motors are ordinary 'brushed' DC motors, using permanent magnets (PMs). It is assumed that the reader is familiar with this type of motor. They are very suitable for occasional use applications – which is the common mode of use in, for example, motor vehicles. To take a specific case, what is the lifetime use of the motors used to adjust the position of the external mirrors of a car? A few minutes at most. If motors of this type have a short life, or are inefficient, it is not of great importance. The motors used in fuel cells are *not* like this, they typically circulate reactant gases or cooling fluids, and are in use *all the time* the fuel cell is in use, and so they should be of the highest possible efficiency and have the longest possible life. In addition, the presence of volatile fuels like hydrogen means that the sparks that inevitably arise occasionally with brushes should be avoided. For these reasons we will only be considering 'brushless' motors in this section.

At least half the electricity generated in the power systems of developed countries is consumed by just one type of electric motor – the induction motor (Walters, 1999). This type of motor requires an AC supply, three phase for preference. Such a three-phase supply can easily be generated by an inverter as described in Section 10.3.2, and so they are sometimes used with fuel cells. A brief description of this type of motor is given in Section 10.4.2. Although hugely successful, and generally highly efficient, the induction motor is not the best in terms of power density and efficiency. Two other modern motors are also used in fuel cell systems; these are the brushless DC motor described in Section 10.4.3 and the switched reluctance motor (SRM) described in Section 10.4.4.

10.4.2 The induction motor

The induction motor is very widely used in industrial machines of all types. Its technology is very mature. Induction motors require an AC supply, which might make them seem unsuitable for a DC source such as fuel cells. However, as we have seen, AC can easily be generated using an inverter, and in fact the inverter needed to produce the AC for an induction motor is no more complicated or expensive than the circuits needed to drive the brushless DC or switched reluctance motors. So, these widely available and very reliable motors are suited for use with fuel cells.

The principle of operation of the three-phase induction motor is shown in Figures 10.19 and 10.20. Three coils are wound right around the outer part of the motor, known as the stator, as shown in the top of Figure 10.19. The rotor usually consists of copper or aluminium rods, all electrically linked (short-circuited) at the end, forming a kind of cage, as also shown in Figure 10.19. Although shown hollow, the interior of this cage rotor will usually be filled with laminated iron.

The three windings are arranged so that a positive current produces a magnetic field in the direction shown in Figure 10.20. If these three coils are fed with a three-phase alternating current, as in Figure 10.20, the resultant magnetic field rotates anti-clockwise, as shown at the bottom of Figure 10.20.

This rotating field passes through the conductors on the rotor, generating an electric current.

A force is produced on these conductors carrying an electric current, which turns the rotor. It tends to 'chase' the rotating magnetic field. If the rotor was to go at the same speed as the magnetic field, there would be no relative velocity between the rotating field and the conductors, and so no induced current, and no torque. The result is that the torque-speed graph for an induction motor has the characteristic shape shown in Figure 10.21. The torque rises as the angular speed 'slips' behind that of the magnetic field, up to an optimum slip, after which the torque declines somewhat.

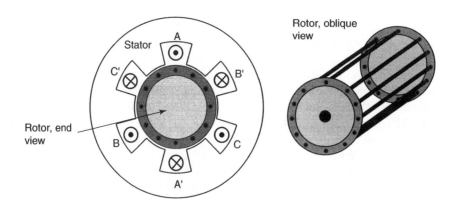

Figure 10.19   Diagram showing the stator and rotor of an induction motor.

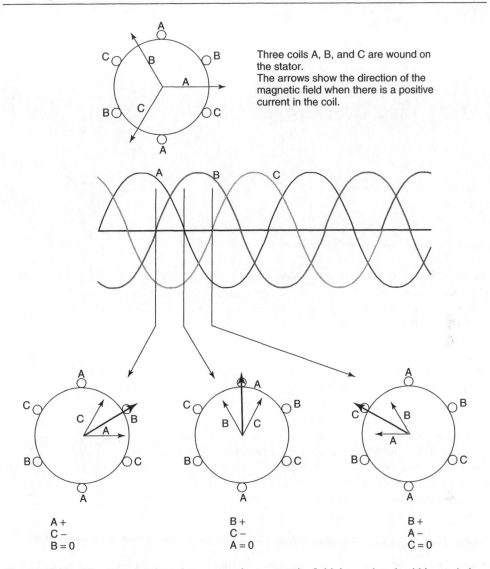

Three coils A, B, and C are wound on the stator.
The arrows show the direction of the magnetic field when there is a positive current in the coil.

A +
C −
B = 0

B +
C −
A = 0

B +
A −
C = 0

Figure 10.20 Diagrams to show how a rotating magnetic field is produced within an induction motor.

The winding arrangement of Figures 10.19 and 10.20 is known as *two-pole*. It is possible to wind the coils so that the magnetic field has four, six, eight, or any even number of poles. The speed of rotation of the magnetic field is the supply frequency divided by the number of pole pairs. So, a four-pole motor will turn at half the speed of a two-pole motor, given the same frequency AC supply, a six-pole motor a third the speed, and so on. This gives a rather inflexible way of controlling speed. A much better way is to control the frequency of the three-phase supply. By using a circuit such as that of Figure 10.17,

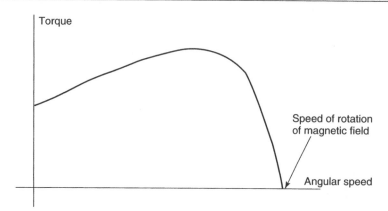

Figure 10.21   Typical torque/speed curve for an induction motor.

this is easily done. The frequency does not precisely control the speed, as there is a 'slip' depending on the torque. However, if the angular speed is measured, and incorporated into a feedback loop, the frequency can be adjusted to attain the desired speed.

The maximum torque depends on the strength of the magnetic field in the gap between the rotor and the coils on the stator. This depends on the current in the coils. A problem is that as the frequency increases the current reduces, if the voltage is constant, because of the inductance of the coils having an impedance that is proportional to the frequency. The result is that, if the inverter is fed from a fixed voltage, the maximum torque is inversely proportional to the speed. This is liable to be the case with a fuel cell system.

Induction motors are very widely used. Very high volume of production makes for a very reasonably priced product. Much research has gone into developing the best possible materials. Induction motors are as reliable and well developed as any technology. However, the fact that a current has to be induced in the motor adds to the losses, with the result that induction motors tend to be a little (1 or 2%) less efficient than the motors described below.

10.4.3 The brushless DC motor

The brushless DC motor (BLDC motor) is really an AC motor! The current through it alternates, as we shall see. It is called a *brushless DC motor* because the alternating current *must* be of variable frequency and so derived from a DC supply and because its speed/torque characteristics are very similar to the ordinary 'with brushes' DC motor. As a result of the 'brushless DC' not being an entirely satisfactory name, it is also, very confusingly, given different names by different manufacturers and users. The most common of these is 'self-synchronous AC motor', but others include 'variable frequency synchronous motor', 'permanent magnet synchronous motor', and 'electronically commutated motor' (ECM).

The basis of operation of the BLDC motor is shown in Figure 10.22. Switches direct the current from a DC source through a coil on the stator. The rotor consists of a permanent

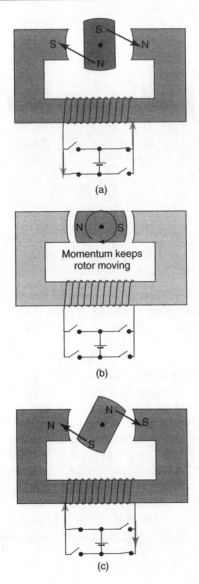

Figure 10.22 Diagram showing the basis of operation of the brushless DC motor.

magnet. In Figure 10.22a the current flows in the direction that magnetises the stator so that the rotor is turned clockwise, as shown. In Figure 10.22b the rotor passes between the poles of the stator, and the stator current is switched off. Momentum carries the rotor on, and as shown in Figure 10.22c, the stator coil is re-energised, but the current and hence the magnetic field are reversed. So the rotor is pulled on round in a clockwise direction. The process continues, with the current in the stator coil alternating.

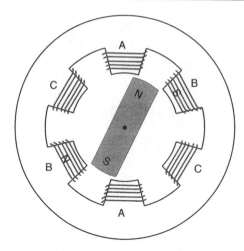

Figure 10.23 Diagram showing an arrangement of three coils on the stator of a BLDC motor.

Obviously, the switching of the current must be synchronised with the position of the rotor. This is done using sensors. These are often Hall-effect sensors that use the magnetism of the rotor to sense its position, but optical sensors are also used.

A problem with the simple single coil system of Figure 10.22 is that the torque is very unsteady. This is improved by having three (or more) coils, as in Figure 10.23. In this diagram, coil B is energised to turn the motor clockwise. Once the rotor is between the poles of coil B, coil C will be energised, and so on.

The electronic circuit used to drive and control the coil currents is usually called an *inverter* – and it will be the same as, or very similar to, our 'universal inverter' circuit of Figure 10.17. The main control inputs to the microprocessor will be the position sense signals.

A feature of these BLDC motors is that the torque will reduce as the speed increases. The rotating magnet will generate a back emf in the coil that it is approaching. This back emf will be proportional to the speed of rotation and will reduce the current flowing in the coil. The reduced current will reduce the magnetic field strength, and hence the torque. Eventually, the size of the induced back emf will equal the supply voltage, and at this point the maximum speed would have been reached. Notice also that this type of motor can very simply be used as a generator of electricity, and for regenerative or dynamic braking.

Although the current through the motor coils alternates, there must be a DC supply, which is why these motors are generally classified as 'DC'. They are very widely used in computer equipment to drive the moving parts of disc storage systems and fans. In these small motors the switching circuit is incorporated into the motor with the sensor switches. However, they are also used in higher power applications, with more sophisticated controllers (as in Figure 10.17), which can vary the coil current (and hence torque) and thus produce a very flexible drive system. Some of the most sophisticated electric vehicle drive motors are of this type, and one is shown in Figure 10.24. This is a 100-kW, oil-cooled motor, weighing just 21 kg.

Figure 10.24   100-kW, oil-cooled BLDC motor for automotive application. This unit weighs just 21 kg. (Photograph reproduced by kind permission of Zytek Ltd.)

These BLDC motors need a strong permanent magnet for the rotor. The advantage is that currents do not need to be induced in the motor (as with the induction motor), making them somewhat more efficient and giving a slightly greater specific power. The control electronics needed is essentially identical to those for the induction motor. However, the PM rotor does add significantly to the cost of these motors.

10.4.4 Switched reluctance motors

Although only recently coming into widespread use, the switched reluctance (SR) motor is, in principle, quite simple. The basic operation is shown in Figure 10.25 below. In Figure 10.25a, the iron stator and rotor are magnetised by a current through the coil on the stator. Because the rotor is out of line with the magnetic field, a torque will be produced to minimise the air gap and make the magnetic field symmetrical. We could lapse into rather 'medieval' science and say that the magnetic field is 'reluctant' to cross the air gap, and seeks to minimise it. Medieval or not, this is why this type of motor is called a *reluctance motor*.

At the point shown in Figure 10.25b, the rotor is aligned with the stator, and the current is switched off. Its momentum then carries the rotor on round over quarter of a turn, to the position as shown in Figure 10.25c. Here the magnetic field is reapplied, in the same direction as before. Again, the field exerts a torque to reduce the air gap and to make the field symmetrical, which pulls the rotor on round. When the rotor lines up with the stator again, the current would be switched off.

In the SRM, the rotor is simply a piece of magnetically soft iron. Also, the current in the coil does not need to alternate. Essentially then, this is a very simple and potentially low-cost motor. The speed can be controlled by altering the length of time that the current is on in each 'power pulse'. Also, since the rotor is not a permanent magnet, there is no back emf generated in the way it is with the BLDC motor, which means that higher

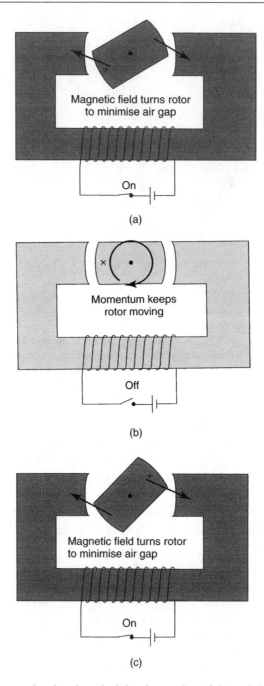

Figure 10.25 Diagram showing the principle of operation of the switched reluctance motor.

speeds are possible. In the fuel cell context, this makes the SR motor particularly suitable for radial compressors and blowers.

The main difficulty with the SR motor is that the timing of the turning on and off of the stator currents must be much more carefully controlled. For example, if the rotor is 90° out of line, as in Figure 10.22a, and the coil is magnetised, no torque will be produced, as the field would be symmetrical. So, the torque is much more variable, and as a result early SR motors had a reputation for being noisy.

The torque can be made much smoother by adding more coils to the stator. The rotor is again laminated iron, but has 'salient poles', that is, protruding lumps. The number of salient poles will often be 2 less than the number of coils. Figure 10.26 shows this principle. In Figure 10.26a, coil A is magnetised exerting a clockwise force on the rotor. When the salient poles come into line with coil A, the current in A is switched off. Two other salient poles are now nearly in line with coil C, which is energised, keeping the rotor smoothly turning. Correct turning on and off of the currents in each coil clearly needs good information about the position of the rotor. This is usually provided by sensors, but modern control systems can do without these. The position of the rotor is inferred from the voltage and current patterns in the coils. This clearly requires some very rapid and complex analysis of the voltage and current waveforms, and is achieved using a special type of microprocessor called a *digital signal processor*.[2]

An example of a rotor and stator from an SR motor is shown in Figure 10.27. In this example the rotor has eight salient poles.

The stator of an SR motor is similar to that in both the induction and the BLDC motor. The control electronics are also similar – a microprocessor and some electronic switches, along the lines of Figure 10.17. However, the rotor is significantly simpler, and so cheaper and more rugged. Also, when using a core of high magnetic permeability the torque that can be produced within a given volume exceeds that produced in induction motors (magnetic action on current) and BLDC motors (magnetic action on permanent magnets) (Kenjo, 1991, p. 161). Combining this with the possibilities of higher speed means that a higher power density is possible. The greater control precision needed for the currents in the coils makes these motors somewhat harder to apply on a 'few-of' basis, with the result that they are most widely used in cost-sensitive mass-produced goods such as washing machines and food processors. However, we can be sure that their use will become much more widespread.

Although the peak efficiency of the SR motor may be slightly below that of the BLDC motor, SR motors maintain their efficiency over a wider range of speed and torque than any other motor type.

10.4.5 Motors efficiency

It is clear that the motor chosen for any application should be as efficient as possible. How can we predict what the efficiency of a motor might be? It might be supposed that

---

[2] Although they were originally conceived as devices for processing audio and picture signals, the control of motors is now a major application of digital signal processors. BLDC motors can also operate without rotor position sensors in a similar way.

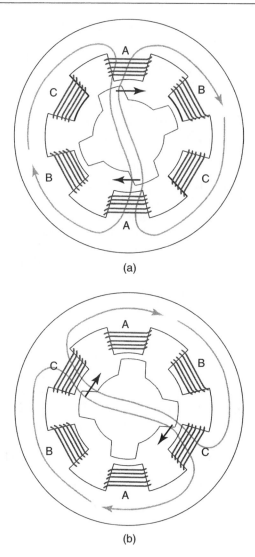

(a)

(b)

Figure 10.26 Diagram showing the operation of an SR motor with a four-salient pole rotor.

the *type* of motor chosen would be a major factor, but in fact it is not. Other factors are much more influential than whether the motor is BLDC, SR, or induction.

An electric motor is, in energy terms, fairly simple. Electrical power is the input and mechanical work is the desired output, with some of the energy being converted into heat. The input and output powers are straightforward to measure – the product of voltage and current for the input, and torque and angular speed at the output. However, the efficiency of an electric motor is not so simple to measure and describe as might be supposed. The problem is that it can change markedly with different conditions, and there is no

Figure 10.27 The rotor and stator from an SR motor. (Photograph reproduced by kind permission of SR Drives Ltd.)

single internationally agreed method of stating the efficiency of a motor (Auinger 1999).[3] Nevertheless, it is possible to state some general points about the efficiency of electric motors – the advantages and disadvantages of the different types, and the effect of motor size.

The first general point is that motors become more efficient as their *size* increases. Table 10.3 shows the efficiency of a range of three-phase, four-pole induction motors. The efficiencies given are the minimum to be attained before the motor can be classified 'Class 1' efficiency under European Union regulations. The figures clearly show the effect of size. While these figures are for induction motors, exactly the same effect can be seen with other motor types, including the BLDC and the SR types.

The second factor that has more control over efficiency than motor type is the *speed* of a motor. Higher speed motors are more efficient than lower speed motors. The reason is that one of the most important losses in a motor is proportional to torque, rather than power, and a lower speed motor will have a higher torque for the same power, and hence higher losses.

A third important factor is the cooling method. Motors that are liquid-cooled run at lower temperatures, which reduces the resistance of the windings, and hence improves efficiency, though this will only affect things by about 1%.

Another important consideration is that the efficiency of an electric motor might well be very different from any figure given in the specification, if it operates well away from optimum speeds and torque. In some cases an efficiency map, like that of Figure 10.28, may be provided. This is based on a real BLDC motor. The maximum efficiency is 94%,

---

[3] The nearest to such a standard is IEC 34-2.

Table 10.3 The minimum efficiency of
4-pole 3-phase induction motors to be
classified as Class 1 efficiency under EU
regulations. Efficiency measured according to IEC 34.2

Power kW	Minimum efficiency %
1.1	83.8
2.2	86.4
4	88.3
7.5	90.1
15	91.8
30	93.2
55	94.2
90	95.0

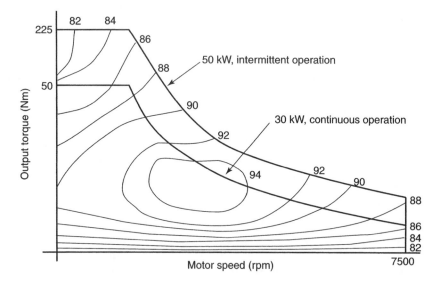

Figure 10.28   The efficiency map for a 30-kW BLDC motor.

but this efficiency is only obtained for a fairly narrow range of conditions. It is perfectly
possible for the motor to operate at well below 90% efficiency.

So, we can say that the efficiency of a good quality motor will be quite close to the
figures given in Table 10.3 for all motor types, even if they are not induction motors. The
efficiency of the BLDC and SR motors is likely to be 1 to 2% higher than that for an
induction motor, since there is far less loss in the rotor. SR motor manufacturers also claim
that their efficiency is maintained over a wider range of speed and torque conditions.

10.4.6 Motor mass

A motor should generally be as small and light as possible while delivering the required power. As with the case of motor efficiency, the type of motor chosen is much less important than other factors when it comes to the specific power and power density of an electric motor.

Figure 10.29 is a chart showing typical specific powers for different types of motors at different powers. Taking the example of the BLDC motor, it can be seen that the cooling method used is a very important factor. The difference between the air-cooled and liquid-cooled BLDC motor is most marked. The reason is that the size of the motor has to be large enough to dispose of the heat losses. If the motor is liquid-cooled, then the same heat losses can be removed from a smaller motor.

We would then expect that efficiency should be an important factor. A more efficient motor could be smaller, since less heat disposal would be needed. This is indeed the case, and as a result all the factors that produce higher efficiency, and which were discussed in the previous section, also lead to greater specific power. The most important of these are as follows:

- Higher *power* leads to higher efficiency, and hence higher specific power. This can be very clearly seen in Figure 10.29. (However, note that the logarithmic scale tends to make this effect appear less marked.)

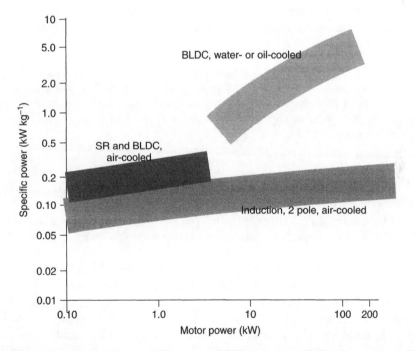

Figure 10.29 Chart to show the specific power of different types of electric motors at different powers. The power here is the continuous power. Peak specific powers will be about 50% higher. Note the logarithmic scales. (This chart was made using data from several motor manufacturers.)

Table 10.4 The mass of some 37-kW induction motors, from the same manufacturer, for different speeds. The speed is for a 50-Hz AC supply

Speed rpm	Mass kg
3000	270
1500	310
1000	415
750	570

- Higher *speed* leads to higher power density. The size of the motor is most strongly influenced by the motor *torque*, rather than the *power*. The consequence is that a higher-speed, lower-torque motor will be smaller. So if a low-speed rotation is needed, a high-speed motor with a gearbox will be lighter and smaller than a low-speed motor. A good example is an electric vehicle, where it would be possible to use a motor directly coupled to the axle. However, this is not done, and a higher-speed motor is connected by (typically) a 10:1 gearbox. Table 10.4 below shows this, by giving the mass of a sample of induction motors of the same power but different speeds.
- The more efficient *motor types*, SR and BLDC, have higher power density than the induction motor.

The curves of Figure 10.29 give a good idea of the likely power density that can be expected from a motor, and can be used to estimate the mass. The lines are necessarily broad, as the mass of a motor will depend on many factors other than those we have already discussed. The material the frame is made from is of course very important, as is the frame structure.

10.5 Fuel Cell/Battery or Capacitor Hybrid Systems

The use of batteries in association with a fuel cell can reduce the cost of a fuel cell–based power system. This is especially the case when powering certain types of electronic equipment.

The essence of a fuel cell hybrid is that the fuel cell works quite close to its maximum power at all times. When the total system power requirements are low, then the surplus electrical energy is stored in a rechargeable battery or capacitor.[4] When the power requirements exceed those that can be provided by the fuel cell, then energy is taken from the battery or capacitor. This presupposes that the power requirements are quite variable.

---

[4] The advantages of the capacitor are that the charge/discharge cycle is more efficient and much faster. The disadvantages are that they store much less energy for a given space and are more expensive per watthour stored.

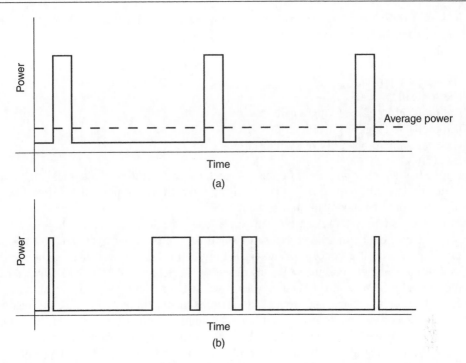

Figure 10.30 Power/time graphs for two different systems suited to a hybrid fuel cell/battery power supply.

Indeed, in cases in which the power requirement is quite constant the fuel cell–battery hybrid system offers no serious advantages.

The easiest hybrid systems to design are those in which the electrical power requirements are highly variable, yet also predictable. Such a situation is illustrated in Figure 10.30a, and can occur with certain data-logging equipment, with data transmitters, certain types of telecommunications equipment, and with land- or buoy-based navigation equipment. For fairly long periods the device is in 'standby' mode, and the fuel cell will be recharging the battery. During the 'transmit' periods, the battery supplies most of the power. The fuel cell operates more-or-less continuously at the average power, and is thus simple to specify. The battery requirements are also clear; they must provide sufficient power and hold enough energy for just one 'high-power pulse', and must be able to be recharged in the period between high-power pulses.

In many more cases the electrical power requirements are not only highly variable but also unpredictable. A good example is the mobile telephone. Such a situation is shown in Figure 10.30b. The higher power pulses can be more frequent and longer lasting or less frequent and shorter lasting. It would still be possible to have a fuel cell running at the long-term average power, but the battery would need to be considerably larger, since it might need to provide several high-power pulses without any time to recharge. It might also be advantageous to increase the power of the fuel cell above the long-term average power, so that the battery recharges more quickly.

In such hybrid systems the two main variables are thus the fuel cell power and the battery or capacitor capacity. In the case of a mobile telephone, these would be chosen on the basis of the following:

- The standby power
- The 'on-call' power
- The probability of receiving a call
- The probable length of a call
- The acceptable probability of system failure.

Notice that the only way to bring this last factor to zero is to have the fuel cell power equal to the 'on-call' power and to abandon the hybrid concept altogether. Most likely, the system would be modelled on a computer.

In essence, such systems can be said to be using the fuel cell as a battery charger. This is liable to be a common early use of small fuel cells, either hydrogen powered or methanol-powered. The concept is shown in Figure 10.31. A controller of some sort is needed to prevent overcharging. Also, as discussed in Section 10.2, a DC/DC converter would normally be necessary. Such a system would be particularly attractive as a range extender for mobile phones, and would suit the direct methanol fuel cell, since the average power, and hence fuel cell power, is so low. It could be very easy to refuel such a small fuel cell with methanol. The battery could be the standard mobile phone battery, with the fuel cell and its integral DC/DC converter in a separate charger unit (Hockaday, 1999). A portable computer is also another very likely candidate for such fuel cell/battery systems.

In some cases the models for designing such hybrid systems are quite well developed, because they have already been used for solar panel–battery hybrid systems. These are very widely used in telecommunications and navigation equipment. In comparison, the model needed for a fuel cell–based system is much simpler, since with the solar panel both the consumption *and generation* of power vary in a mixture of random and predictable ways. With fuel cell–based hybrid systems, the primary generation of electrical power can at least be relied upon and accurately modelled.

Such hybrid systems, in which the peak or battery power is much greater than the average power (and hence fuel cell power), are sometimes known as 'hard' hybrid systems.

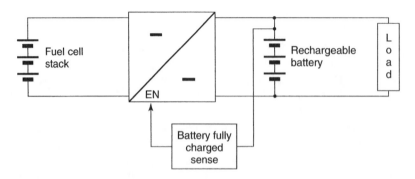

Figure 10.31   Simple fuel cell/battery hybrid system.

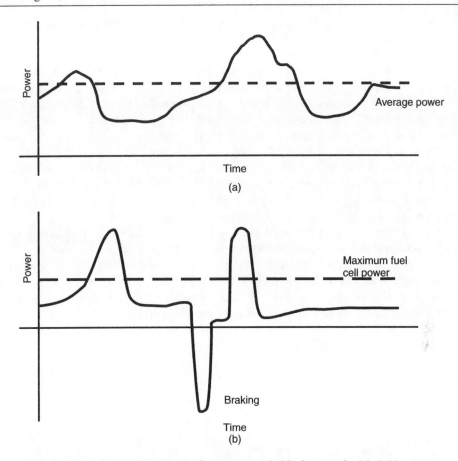

Figure 10.32   Power/time graphs for systems suitable for a 'softer' hybrid system.

By contrast, a 'soft' hybrid is one in which the battery power and energy storage are quite low compared to the fuel cell power. Such systems might be found in vehicles and boats, for example, where the peak power and the average power are not so different. An example power/time graph is shown in Figure 10.32a. The fuel cell power is sufficient most of the time, but the battery can 'lop-the-peaks' off the power requirement and can substantially reduce the required fuel cell capacity. During times when the fuel cell power exceeds the power demand, the battery is recharged, as before. This sort of power demand is characteristic of urban electric vehicles, where the peaks correspond to occasions such as accelerating from traffic lights, yet most of the time the vehicle will be proceeding slowly and steadily, or else it will be stationary.

A further possibility is illustrated in the power/time graph of Figure 10.32b. Here an electric vehicle is using the motor as a brake – all the electric motors described in the previous section can be used in this way. In *dynamic* braking, the motor is used as a generator, the motion energy is converted into electrical energy, but is then simply passed through a resistor and converted to heat. In *regenerative* braking, the electrical energy is passed to a rechargeable battery, to be used later to run the motor. Regenerative braking is clearly the better option from the point of view of system efficiency, but it does presuppose a hybrid system with a rechargeable battery. Such a hybrid system would also need a fairly sophisticated control system. For one thing, the flow of power from the fuel cell to the battery would need to be properly controlled, since the situation is much more variable, with sometimes only a small amount of surplus fuel cell power. Secondly, the battery's 'normal' state would have to be somewhat less than fully charged, so that it could absorb the electrical energy supplied by the motors during braking. The motor controller would also need to be a full 'four quadrant' type.[5] The controller would need to be more complex than the system of Figure 10.31. Figure 10.33 shows what would be required in system diagram form. In particular, a measurement of the battery state of charge would be needed – rather than just a 'fully charged' indication. The energy flows to and from the different parts of the system are much more complex.

In this last case, in which the battery is supplying fairly short-term power peaks, and also absorbing power from regenerative braking, the rate of transfer of charge in and out of the battery is liable to be a problem. This is where the use of special high capacity or 'super' capacitors are particularly attractive. These are being developed specially for such 'soft' hybrid applications, both for fuel cells and internal combustion engines. Their energy density is much less than rechargeable batteries, but their power density is much greater, typically $2.5\,kW\,kg^{-1}$, and they can also be used for at least 500,000 charge/discharge cycles (Harri et al., 1999). Büchi et al. (2002) describe a very impressive fuel cell–super

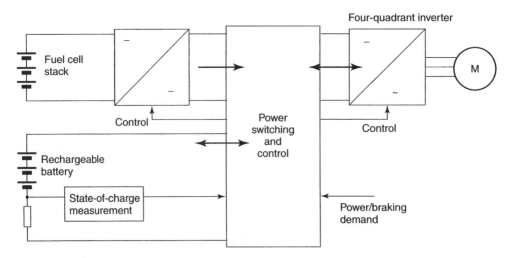

Figure 10.33 System diagram for a hybrid fuel cell/battery system. The bold arrows indicate energy flows. The resistor in series with the rechargeable battery is a current sense resistor used to measure the charge in and out.

---

[5] Four quadrants refers to the four possible modes of motoring – forwards accelerating, braking while going forwards, backwards accelerating, braking while going backwards.

capacitor hybrid, and show that the fuel cell power remains quite steady even when the power demand of the vehicle is highly variable.

The decision of whether to go for a hybrid system is influenced by two opposing characteristics of fuel cells. The main driving force in favour is that fuel cells are very expensive 'per watt', and so it makes sense to have them working at their full-rated power for as much time as possible, to get the maximum possible value from the investment. On the other hand, one of the advantages of fuel cells over most other fuel energy converters is that at part load their voltage, and thus their efficiency,[6] increases. Thus we should rate the fuel cell at the maximum required power, and if it operates at only part load for much of the time, so much the better: the efficiency will have improved. It is fair to say, that at the time of writing, the question of cost weighs more heavily in the balance.

In this discussion we have just considered fuel cells in partnership with an energy storage device. However, there are of course a huge number of other possibilities, for example, a solar panel, fuel cell, and battery. Such systems have been successfully deployed in roadside variable message systems (Gibbard, 1999), and could have many other applications in remote power supplies. The fuel cell could compensate for the somewhat unreliable nature of solar energy in many situations.

References

Auinger H. (1999) "Determination and designation of the efficiency of electrical machines", *Power Engineering Journal*, **13**(1), 15–23.

Büchi F., Tsukada A., Rodutz P., Garcia O., Ruge M., Kötz R., Bärtschi M., and Dietrich P. (2002) "Fuel cell supercap hybrid electric power train." *The Fuel Cell World 2002, Proceedings, European Fuel Cell Forum Conference*, Lucerne, pp. 218–231.

Gibbard H.K. (1999) "Highly reliable 50 watt fuel cell system for variable message signs." *Proceedings of the European Fuel Cell Forum Portable Fuel Cells Conference*, Lucerne, pp. 107–112.

Harri V., Erni P., and Egger S. (1999) "Super capacitors and batteries in applications with fuel cells." *Proceedings of the European Fuel Cell Forum Portable Fuel Cells Conference*, Lucerne, pp. 245–252.

Heier S. (1998) *Grid Integration of Wind Energy Conversion Systems*, John Wiley & Sons, Chichester.

Hockaday R. and Navas C. (1999) "Micro-fuel cells for portable electronics." *Proceedings of the European Fuel Cell Forum Portable Fuel Cells Conference*, Lucerne, pp. 45–54.

Kenjo T. (1991) *Electric Motors and Their Controls*, Oxford University Press, Oxford.

Spiegel R. J., Gilchrist T., and House D. E. (1999) "Fuel cell bus operation at high altitude", *Proceedings of the Institution of Mechanical Engineers*, Part A, **213**, 57–68.

Walters D. (1999) "Energy efficient motors, saving money or costing the earth?", *Power Engineering Journal*, Part 1, **13**(1), 25–30, Part 2, **13**(2), 44–48.

---

[6] See Section 2.4.

11

Fuel Cell Systems Analysed

11.1 Introduction

Throughout this book we have been concerned with the components of a fuel cell system. We began with a consideration of the fuel cells and how each fuel cell is engineered into a stack. We have considered the elements of fuel processing and the various balance of plant components, which together make up a complete fuel cell system. In trying to explain how fuel cell systems work, we should also consider some of the broader issues of system design and analysis. What follows in this final chapter is therefore a discussion of energy systems and how fuel cell systems can be viewed as an ideal way of converting one energy form (chemical energy) into another (electrical energy). We shall consider the issues of fuel choice and the application of system design and analysis. Finally, we will look at examples of real fuel cell systems and how these have been modelled, designed, and engineered.

Over and above the issues of system design and construction, the issue that the early developers and manufacturers will focus on is that of the cost of the materials and manufacturing of their systems. The reason for this is that despite all their advantages in terms of environmental impact compared with other technologies, most fuel cell systems have commercial competitors. Moreover, most of the competing technologies are in mature markets. In the transport arena we have the internal combustion engine, a highly engineered and developed technology that is of relatively low cost compared to current fuel cell systems. In the stationary power market, large generators based on gas or steam turbine technology can produce electricity at lower cost than most fuel cell systems. Such competing technologies set the cost targets that fuel cell systems will need to meet to become commercially viable. Several economic studies of fuel cell systems have been carried out in recent years. These are important to developers, but since much of the information on fuel cell component costs is proprietary at present, and the methods used are not specific to fuel cell systems, a detailed discussion of the economics of fuel cell systems is outside the scope of this book.

Fuel Cell Systems Explained, Second Edition James Larminie and Andrew Dicks
© 2003 John Wiley & Sons, Ltd ISBN: 0-470-84857-X

11.2 Energy Systems

Before we analyse fuel cell systems, we should try and understand the concept of energy systems. Any system has *inputs* and *outputs*, and Figure 11.1 illustrates a general view of the world's energy system, with its inputs – the various fuels and energy sources, and its outputs – the energy services and consumer needs that mankind requires, such as heating, lighting, transport and so on. The technologies that convert the raw fuel into energy services all sit within the energy system. Some are more efficient than others in converting one form of energy to another, and some are less intrusive on our world than others. Fuel cell systems as part of a much larger energy system are beneficial because they can convert fuel into electricity efficiently and with little intrusion into the atmosphere. Yet, any technology has to be manufactured and we also need to be aware of the environmental and societal effects that manufacturing processes have on the environment. Vehicle manufacturers today need to consider the whole life cycle cost of their products (Weiss et al., 2000), which includes evaluating the cost of recycling or disposing of components such as tyres, steel, aluminium, and plastics used in their construction. The cost of platinum exhaust catalysts has to be considered together with the cost of reclaiming the platinum from spent catalysts. So it is with fuel cell systems, and this is where fuel cells can have a lead over other technologies.

When we are designing and analysing fuel cell systems, we need to be aware of all the inputs and outputs that may be involved. These may be, for example,

Inputs	Outputs
During system manufacture:	During disposal and/or recycle:
Precious metals	Precious metals
Organic polymers	Organic polymers
Ceramics (e.g. zirconia)	Ceramics
Metals (Ni, Cu, steel, etc.)	Metals
Alkali carbonates (for molten carbonate (electrolyte) fuel cell (MCFC))	Alkali carbonates
Phosphoric acid (phosphoric acid (electrolyte) fuel cell (PAFC))	Phosphoric acid
During use:	
Fossil fuel (e.g. natural gas)	Electrical power
(Water) and purification filters	Water, spent filters
Air	Heat
Cost of site location	Local waste heat (losses)
Maintenance materials and costs	CO_2 and other gaseous emissions
Stack replacement	Noise

When we are describing fuel cell systems and comparing one with another, it is essential that we define what the inputs and outputs are. Perhaps the most important aspect of this concerns the fuel. A stack developer may describe his system as a system that runs on hydrogen. Another developer may include a natural gas reformer in his system. Reformer and stack do not necessarily have to be integrated in one package. For stationary applications this may be desirable, but for transportation systems there may be benefits

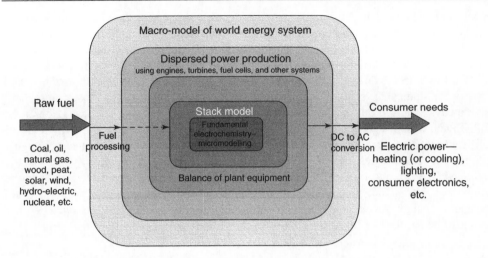

Figure 11.1 Representation of energy systems.

in carrying out reforming away from the vehicle and storing hydrogen on the vehicle, as is the case with the first fuel cell bus trials.

11.3 Well-To-Wheels Analysis

11.3.1 Importance of well-to-wheels analysis

In Chapter 8 we discussed how hydrogen could be made from a variety of hydrocarbon fuels. We also discussed electrolysis as a means of generating hydrogen from electricity. In the future, as more renewable energy sources are developed, such as wind, wave, and solar power, hydrogen could be made by electrolysis using such renewable energy sources. In the case of these three renewable sources, only oxygen is produced along with hydrogen when water is electrolysed – it is the only by-product. In the case of hydrocarbon fuels from fossil sources, carbon dioxide is also produced when the fossil fuel is converted into hydrogen. This cannot be avoided, whatever type of fuel processing is carried out, with the exception of pyrolysis in which the principal by-product is carbon. With the pressure to reduce global emissions of carbon dioxide into the atmosphere, the production of hydrogen from carbon-containing fuels needs careful consideration. Figure 8.1 showed various routes for converting fossil fuels into hydrogen, but we need to examine not only what conversion technology is used but also where the conversion takes place, since this will determine where the carbon dioxide is released into the atmosphere. For example, cars or buses running on hydrogen stored aboard the vehicle may only have a wisp of steam or water coming out of their exhausts and therefore be termed *zero-emission vehicles*. However, in the generation of hydrogen for these vehicles, carbon dioxide may have had to be released into the atmosphere, where natural gas perhaps has been reformed into hydrogen. In other words, the carbon dioxide has been discharged somewhere else in the chain that links the oil or gas well to the vehicle electric drive train. System developers, therefore, need to understand that you cannot just consider emissions from the vehicle in isolation; we need to look at the whole system from 'well-to-wheels.' In addition, each time that fuel is converted, or packaged, or transported, there is an associated loss of energy. The more conversion and transportation steps there are, the greater is the likely energy loss.

'Well-to-wheels' analysis, therefore, forms a key aspect of fuel cell systems, and fuel cycle studies have been published in recent years. Perhaps the most comprehensive analysis carried out recently was by General Motors, Argonne National Laboratory, BP, Exxon Mobil, and Shell with the support of the US Department of Energy's Office of Transportation Technologies (2001). The study was carried out to evaluate emerging propulsion technologies, that is, advanced internal combustion engines, hybrid electric vehicles (HEVs), and fuel cell vehicles (FCVs). Also evaluated were new fuels, and the study was carried out to provide aid for public policy development as well as to aid business strategy development for the supporting companies. Similar fuel cycle studies can also be carried out for testing the viability of stationary power systems.

11.3.2 Well-to-tank analysis

The big issue with vehicles running on hydrogen is the infrastructure required to deliver the hydrogen. In Chapter 8 we described a number of different routes for producing hydrogen. In addition, several hydrogen supply options exist for vehicles, for example,

- hydrogen generated by steam reforming of natural gas in a large centralised plant, which is then delivered as liquid hydrogen by trailer to filling stations,
- hydrogen generated by steam reforming of natural gas in a large centralised plant, which is then delivered by pipeline to filling stations,
- hydrogen generated as a by-product from oil refineries, industrial ammonia plant, and so on,
- hydrogen produced at the filling station by small-scale steam reformers running on pipeline natural gas,
- hydrogen produced by electrolysers located at the filling stations.

Several groups have studied the various options and carried out economic analyses and environmental assessments of the different pathways to hydrogen. Some results of studies in the United States have been summarised by Ogden (2002). The consensus view of these studies conducted in the United States is that in the near term hydrogen generated by steam reforming of natural gas in small, localised reformers is the best option. Where there is no natural gas supply, hydrogen generation by electrolysers might be the preferred alternative. As a hydrogen infrastructure emerges in the future, there may come a time when centralised production would be economic with the added bonus of being able to collect and sequester the carbon dioxide that is generated from the reformer. Many fuel cycle analyses or well-to-wheels studies have been published in the United States (Wang, 1999), in the United Kingdom (Hart, 1998), and elsewhere, and a conference on the subject was held in Nice in 2001.

In the well-to-wheels study carried out by Argonne National Laboratory, GM, and others (2001) for the North American market, 15 different vehicles were considered. These included conventional and HEVs with spark-ignition and compression-ignition engines, as well as hybridised and non-hybridised FCVs with and without on-board fuel processors. All 15 vehicles were configured to meet the same performance requirements. Thirteen fuels were considered in detail (selected from 75 different fuelling options or pathways). These include low-sulphur gasoline, low-sulphur diesel, crude-oil-based naphtha, Fischer–Tropsch (FT) naphtha,[1] liquid or compressed gaseous hydrogen, based on five different pathways, compressed natural gas, methanol, and neat and blended (E85) ethanol. The 13 fuels and 15 vehicles gave rise to 27 different pathways from 'well-to-wheels' analysed in the study (Figure 11.2). These were carried out in various stages as illustrated in Figure 11.3.

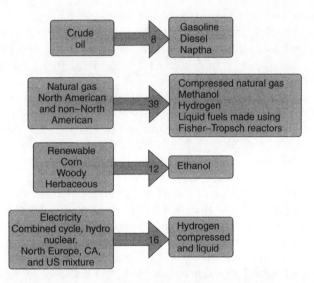

Figure 11.2   The 75 different pathways investigated in the GM (2001) well-to-wheels study.

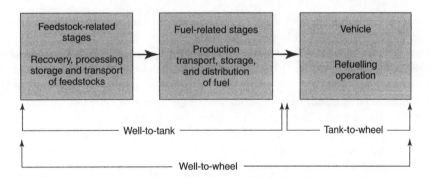

Figure 11.3   Different parts of the analysis carried out in the GM (2001) study.

---

[1] Fischer–Tropsch fuels are those made artificially using the FT process. Basically a fuel such as biomass, or even natural gas, is steam reformed using the methods described in Chapter 8. The product hydrogen and carbon dioxide are then reacted, over catalysts developed by Fischer and Tropsch, to produce liquid fuels such as octane, C_9H_{20}, $C_{10}H_{22}$, etc.

11.3.3 Main conclusions of the GM well-to-wheels study

The GM study applies to fuels and vehicles in North America, and it can be expected that differences will be found with studies carried out elsewhere. For example, in some areas of the world there will be relative abundance of certain fuel feedstocks over others. Vehicle requirements are also not the same, worldwide. Nevertheless, some key findings from the North American study are worth reporting, since they will influence the development of transportation systems in that part of the world, and also act as a benchmark for other studies:

Total energy use: For the same amount of energy delivered to the vehicle tank, of all the fuel types studied, petroleum-based fuels and compressed natural gas exhibit the lowest energy losses from the well-to-tank (WTT). Methanol, FT naphtha, FT Diesel, G.H2 from natural gas and corn-based ethanol are subject to moderate WTT energy losses. Liquid H_2 from, natural gas, electrolysis H_2 (gaseous and liquid), electricity generation and cellulosic bioethanol are subject to large WTT energy losses.

Greenhouse gas(GHG)emission: Liquid hydrogen (produced in both central plants and filling stations) and compressed gaseous hydrogen produced by electrolysis can be energy inefficient and can generate a large amount of GHG emissions. Ethanol (derived from renewable cellulose sources such as corn) offers a significant reduction in GHG emissions. Other fuel options had moderate energy efficiencies and GHG emissions.

Tank-to-wheels efficiency: Fuel cell systems use less energy than conventional power trains, because of the intrinsic high efficiency of the stacks. Hydrogen-based FCVs exhibit significantly higher fuel economy than those employing on-board fuel processors.

Overall well-to-wheels efficiency: Hybrid systems offer consistently higher fuel economy than conventional vehicles.

Fossil energy use: This follows similar patterns to the total energy use, that is, for petroleum, coal, and natural gas. Ethanol derived from cellulose (bioethanol) exhibits a high total energy use but ethanol production involves the burning of lignin, a non-fossil energy for heat.

Figure 11.4 summarises one aspect of the study, that is, the well-to-wheels energy consumption of the major pathways investigated.

A European study has now been undertaken by GM. This assessed 36 fuel pathways and 18 propulsion concepts for the 2010 time frame, from conventional engines to advanced concepts under European driving conditions. A principal finding of this study was that FCVs running on hydrogen produced from natural gas could be attractive in terms of well-to-wheel gas emissions, depending on the source of the natural gas. However, optimum results are realised when renewable energies such as biomass or wind power are used to produce the hydrogen. To a lesser extent, natural gas vehicles offered improvements relative to conventional gasoline and diesel vehicles. When natural gas was used to produce methanol for an on-board reformer FCV, no well-to-wheel benefits were seen relative to conventional gasoline or diesel internal combustion engine vehicles, or gasoline reformer FCVs. The European study was based on an Opel Zafira minivan, whereas in the study carried out by GM for the United States, a Chevrolet Silverado pickup was used as the reference vehicle. When interpreting the results of such studies we need to look at the

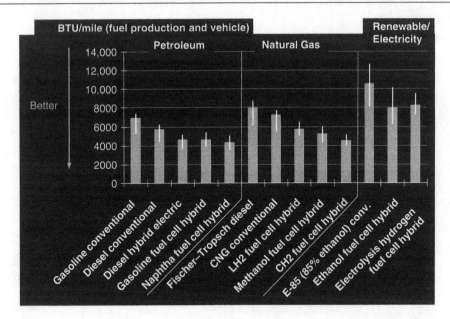

Figure 11.4 Well-to-wheel energy consumption for the GM North America study.

assumptions made, particularly those relating to the vehicle performance and the drive train used.

11.4 Power-Train or Drive-Train Analysis

FCVs are a special type of electric vehicle, and unlike internal combustion engine vehicles, which we are all familiar with, they are constructed in a different manner. In a conventional diesel or gasoline-fuelled vehicle, energy from the fuel is transmitted from the engine to the wheels by a mechanical power train. In an FCV the power train is electrical. In a hybrid vehicle it may be a combination of the two. A conventional vehicle power train is illustrated in Figure 11.5.

A simple FCV parallel hybrid power train is shown in Figure 11.6. In this case the fuel cell provides most of the power for the motor and can keep the battery charged when required. Provided the battery is fully charged, it can provide extra power for the motor when it is required for acceleration. The battery in such a power train is relatively small when compared with the fuel cell. Büchi et al. (2002) describe a vehicle built along these lines, with a supercapacitor electrical energy store instead of a battery.

An alternative is the series hybrid power train as illustrated in Figure 11.7. In this case, the battery provides all the power and therefore needs to be large. On the other hand, the fuel cell acts only as a charger for the battery and can therefore be relatively low power depending on the driving conditions. It only has to provide the average vehicle power. Such systems are particularly suitable for urban delivery vehicles, and have been used with some success with alkaline fuel cells.

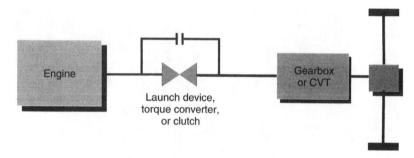

Figure 11.5 Conventional ICE vehicle power train.

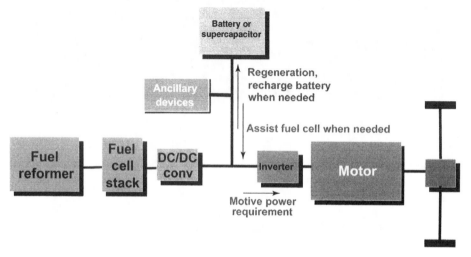

Figure 11.6 Fuel cell parallel hybrid power train configuration.

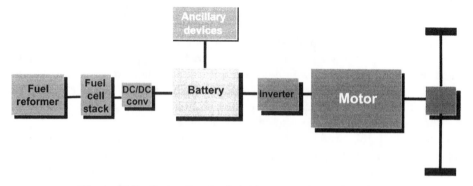

Figure 11.7 Fuel cell series hybrid power train configuration.

It will become apparent that there are many subtleties to the design of power trains. For example, with electric vehicles it is not essential that an electric motor transmit its energy to two driving wheels via the axle. It is possible to put motors on each wheel. Similarly, there are many different types of motors. Powerful computer software is now available that enables the engineer to model these variations without actually building anything. Nevertheless, the proof of the design is in constructing complete systems, and modelling can only go part way to providing an analysis of the vehicle.

11.5 Example System I – PEMFC Powered Bus

An example of an FCV system that has been very successfully demonstrated is the fuel cell engine for buses designed and manufactured by Ballard Power Systems. We have already mentioned this system in Chapter 4, and a more up-to-date version of the engine that we will consider here is shown in Figure 4.32. The system shown in Figure 11.8 is the year 2000 version of the one being used in buses now. We discuss that older system because much technical data has been provided, which (hardly surprisingly) cannot be said of the current systems. *Where the data being presented is 'hard' data provided by Ballard, it is printed in italics.* At other times the figures have to be deduced from the theory outlined in this book.

This section should have you looking back to diagrams and equations in Chapters 9 and 10.

The operating voltage of the stack (5) is from about *450 to 750 V. This is maintained at between 650 and 750 V using an up-chopper* (boost switch mode regulator) as in Figure 10.6. This unit is labelled (1) in Figure 11.8. We can therefore suppose that the system being used is as outlined at the end of Section 10.2, and illustrated in Figure 10.7, with the converter kicking in when the voltage reaches 650 V. There will also be several step-down regulators (as in Figure 10.3) to provide lower voltages for various sub-systems such as the controller (2) and the cooling system (3). There is also a *12-V battery used when starting the system*, which will be charged from the fuel cell. We might estimate an overall efficiency of about 95% for these electrical subsystems, making losses of about 13 kW.

The system is controlled using a programmable logic controller (PLC) (2). It is beyond the scope of this book to give an account of these very widely used control devices. They are used in all types of industrial control applications and are programmed in 'ladder logic' – an easy to learn symbolic programming method. One PLC can simultaneously control several control loops.

The fuel cell system is *water cooled, with a 'radiator' and electrically operated fan* (3). *The heat output is rated as 1.3 MBTU h⁻¹*, which is equivalent to 381 kW. This cooling system removes heat from the electric motor, and all the associated inverters and regulators, which are also water-cooled. As we shall see, the reactant air supply to the fuel cell also has to be cooled. *An ion exchange filter is used to keep the cooling water pure, preventing it from becoming a conductor of electricity*. Clearly then, no anti-freeze can be used, so the *system must be kept above zero, which is done using a heater, connected to the mains when the bus is garaged*.[2] We note that almost all the losses, including those

---

[2] This is one important area where current systems are different. They do not have this limitation.

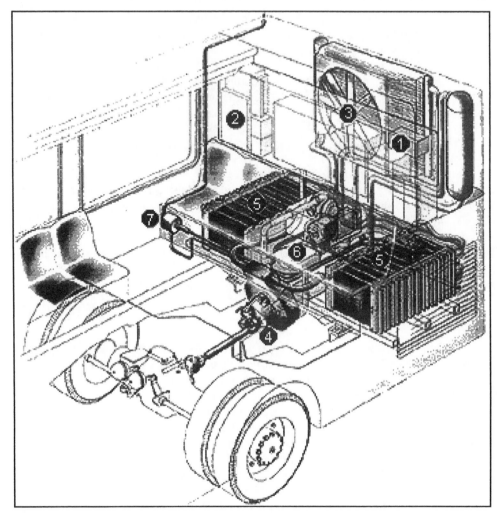

Figure 11.8 Fuel cell engine for buses based on 260-kW PEM fuel cell. (Diagram reproduced by kind permission of Ballard Power Systems.)

in the air intake system and the electrical systems are dealt with by this cooling system. Loss of 381 kW, therefore, seems a reasonable waste heat rate for a 260-kW fuel cell.

In Section 9.11, we introduced the idea of the 'effectiveness' of a cooling system. It was noted that a figure of between 20 and 30 should be obtainable. So, we can speculate that the cooling system, which disposes of up to 381 kW of heat might consume about 381/20, that is, about 20 kW of electrical power. This would be used to drive the fan and the various pumps moving the water around the different subsystems.

The electric motor (4) is a Brushless DC (BLDC) motor (Section 10.4.3), water-cooled, rated at 160 kW continuous. The inverter circuit uses IGBTs. (See Section 10.2.1.) The inverter is also water-cooled. The combined efficiency of the inverter and motor is 92%.

The continuous power rating is 160 kW, so for short periods it will be able to produce higher powers. Exactly what the peak power is we cannot tell, but it will be in the region of 200 kW. There is evidence (Spiegel et al., 1999) that some models of this fuel cell engine for buses use water-cooled induction motors, instead of BLDC. This illustrates very well what was said in the previous chapter – that there is not a great deal to choose between the three main brushless motor types. Induction motors are more rugged and lower cost; BLDC are slightly more efficient and compact. *Dynamic braking* (see Section 10.5) *is used to reduce wear on the vehicle brakes*, but this is not a hybrid system. *The electrical energy heats a resistor that is in the cooling system and effectively works as an immersion heater*. This same resistor, as has been mentioned above, is used in very cold conditions to prevent the fuel cell freezing by being plugged into an external supply. *The motor is coupled to the forward running drive shaft via a 2.43:1 gear*. This is connected to the bus back axle via a differential, which will have a gear ratio of about 5:1.

The fuel cell stacks (5) *are arranged as two units connected in parallel. Each half consists of 10 stacks, each of power 13 kW. The maximum electrical power is thus 260 kW. The cells are operated at 90°C, and both the fuel and the air are humidified*, as explained in Sections 4.4.5 and 4.4.6 of Chapter 4. *The operating pressure of the stacks is increased when higher power is needed, up to a maximum of 207 kPa above ambient pressure*. Each half unit probably consists of about 750 cells in series. They are constructed in the mainstream fashion for PEMFC, that is, with graphite, or graphite/polymer mixture, bipolar plates. They are *water-cooled*, as we would expect from Section 4.5.3.

It can be seen that the air delivery system (6) forms a substantial part of the system. At maximum power, we can speculate that the average cell voltage V_c will be about 0.62 V. (We go a little above 0.6 V as the system is pressurised, and fairly warm for a PEMFC.) The air stoichiometry λ will be about 2. Using equation A2.4 we can estimate the air flow rate to be

$$\text{Air flow rate} = 3.57 \times 10^{-7} \times 2 \times \frac{260{,}000}{0.62} = 0.3 \text{ kg s}^{-1}$$

This figure is also given in the product information. In Chapter 9 we have derived equations for the power needed to compress air, and also for the temperature rise. A reasonable estimate for the compressor efficiency η_c is 0.6. If we also assume that the entry temperature of the air is 300 K, and the pressure is 100 kPa, then, from equation 9.4 we have

$$\Delta T = \frac{300}{0.6}\left(\left(\frac{307}{100}\right)^{0.286} - 1\right) = 189 \text{ K}$$

And also, from equation 9.7,

$$\text{Compressor power} = 1004 \times \frac{300}{0.6}\left(\left(\frac{307}{100}\right)^{0.286} - 1\right) \times 0.3 = 57{,}000 \text{ W}$$

So we have a very substantial temperature rise, which will need to be compensated for by cooling the air before entry to the fuel cell. No doubt, some of this cooling is accomplished by the humidification process. However, we still need 57 kW of power for the compressor. *This is done in two stages. The first is a Lysholm or screw compressor, as in Section 9.2, and Figures 9.2b and 9.7, with the final pressure boosting being provided by a turbocharger, as in Sections 9.8 and 9.9.* To find the power provided by the turbine we make the following assumptions:

- The turbine efficiency is 0.6.
- The exit temperature of the oxygen depleted air is 90°C, 363 K, which is the operating temperature of the fuel cell stack.
- The exit pressure is somewhat less than the entry pressure, otherwise there would be no through flow, say 180 kPa above atmospheric pressure, 280 kPa total.

The exit air-flow rate will be greater than the entry air because of the addition of water, both because of humidification and by the addition of product water. The mass flow rate is calculated as follows:

The exit air flow rate (excluding water) is given by equation A2.5, and using the same values as above, this gives

$$\text{Exit air flow rate} = (3.57 \times 10^{-7} \times 2 - 8.29 \times 10^{-8}) \times \frac{260,000}{0.62} = 0.265 \text{ kg s}^{-1}$$

The exit mass flow is increased by the water content. If we assume that the exit air is at about 100% humidity, then $P_w = 70$ kPa, from Table 4.1.

Since the total pressure of the exit air is 280 kPa, then the pressure of the dry air is $280 - 70 = 210$ kPa. Thus, from equation 4.4,

$$\text{The humidity ratio } \omega = 0.622 \times \frac{70}{210} = 0.21$$

Hence the mass flow rate of water leaving the fuel cell is

$$\dot{m}_W = \omega\dot{m}_a = 0.21 \times 0.265 = 0.0557 \text{ kg s}^{-1}, \text{ from equation 4.1.}$$

As an aside, we see that we can use this result to estimate the rate at which water was added to the reactant gases at entry. The rate of water production from the electrochemical reaction is given by equation A2.10 in Appendix 2, which gives, in this case

$$\text{Rate of water production} = 9.34 \times 10^{-8} \times \frac{260,000}{0.62} = 0.0392 \text{ kg s}^{-1}$$

The difference between these two figures, that is, $0.0557 - 0.0392 = 0.0165$ kg s^{-1}, is an estimate of the rate at which water must enter the cell. Some of this water would have been in the air anyway, while the rest is added by the humidification process. If the humidity of the entry air is about 70%, then it can be shown that approximately two-thirds of the water is added via humidification.[3] This example illustrates the beneficial effect of

---

[3] It is left as an exercise for the reader to show this. All the necessary formulas are given in Sections 4.4.2 and 4.4.3.

pressure on the humidification process. The proportion of excess water that would have been added if the pressure was lower would be much greater.[4]

Returning to the exit air, the total exit flow rate is the dry air-flow rate plus the water flow rate, giving $0.265 + 0.056 = 0.32 \, \text{kg s}^{-1}$. This is only a little greater than the entry flow rate.

We can now use equation 9.10 to find the available turbine power. At the end of Section 9.8, we showed that $1100 \, \text{J kg}^{-1} \, \text{K}^{-1}$ was a good estimate for c_p and 1.33 for γ. If we assume that the turbine exit pressure is still somewhat above air pressure, say 150 kPa, then

$$\text{Turbine power} = 1100 \times 0.6 \times 363 \times \left(\left(\frac{150}{280} \right)^{0.275} - 1 \right) \times 0.32 = -12,100 \text{ watts}$$

The minus sign indicates that the power is given out. This $\sim 12 \, \text{kW}$ of power would make a useful contribution to the $57 \, \text{kW}$ needed. However, if we look at the temperature change, we see that it might not be possible to harness all this power. From equation 9.9, we see that the temperature change through the turbine would be

$$\Delta T = 0.6 \times 363 \times \left(\left(\frac{150}{280} \right)^{0.275} - 1 \right) = -34 \text{ K}$$

This would bring the exit gas temperature down to about 55°C. Bearing in mind that the air will have been more or less saturated as it left the fuel cell, we would anticipate a good deal of condensation in the turbine, which would inhibit its performance. We should therefore perhaps round down our estimated power from the turbine to the still by no means negligible $10 \, \text{kW}$. The power from the motor driving the screw compressor will therefore be about $47 \, \text{kW}$. This is a very substantial parasitic power loss, and largely explains why the traction motor mentioned above is rated at $160 \, \text{kW}$, whereas the fuel cell is $260 \, \text{kW}$. The other major losses are the cooling system (estimated at 20-kW parasitic losses) and the electrical sub-systems, estimated at $13 \, \text{kW}$.

The fuel feed system (7) is one of the simplest parts of this fuel cell system. *Hydrogen compressed to a pressure of 24,800 kPa is stored in cylinders on the roof of the bus. This is regulated down to the same pressure as the air feed, 207 kPa maximum, in two stages. The pressure regulation system incorporates an ejector, as described in Section 9.10, to circulate the hydrogen fuel through the stack.*

We might suppose that there would be a considerable temperature drop as the hydrogen gas pressure is reduced, which is what is normally observed with depressurising gases. If the hydrogen behaved as a perfect gas, then we could use the standard equation

$$\frac{T_2}{T_1} = \left(\frac{P_2}{P_1} \right)^{\frac{\gamma-1}{\gamma}}$$

---

[4] Another useful exercise for the reader would be to repeat this calculation substituting some lower value for the exit total pressure, such as 120 kPa, for the 280 kPa used above. It will be found that the entry water rate will have to be about $0.19 \, \text{kg s}^{-1}$, that is greater by a factor of over 10.

for adiabatic changes to find the exit temperature – and we would estimate it to be about 80 K – very cold! In fact the hydrogen behaviour is far from that of a perfect gas. The so-called 'Joule-Thompson' effect comes in, and there is actually a very modest temperature rise, of about 7°C, in the pressure regulation system.

11.6 Example System II – Stationary Natural Gas Fuelled System

11.6.1 Introduction

Fuel cell power plants for the stationary market can be broadly divided into two types, power-only and cogeneration systems. The power-only type include back-up or uninterruptible power systems fuelled by hydrogen, as well as various niche systems, for example, those designed for traffic lights (and as produced by H Power in the United States) and remote signalling. Cogeneration systems produce useful heat (or cooling) as well as electricity, and these systems are more complex. Portable systems for consumer electronic items present their own unique problems relating to miniaturising the fuel processing and closely integrating the stack and control systems. Many of the system issues for portable applications have already been covered in Chapters 4, 5, and 6.

In this section we focus on stationary power plants that are fuelled by natural gas and which provide energy in the form of heat as well as electricity. We know that for the PEMFC, the fuel requirements are more stringent than for the MCFC or solid oxide fuel cell (SOFC), and so that makes for a more interesting example to consider for this part of the book. Some systems designs for MCFC and SOFC have already been considered in Chapter 8. In designing a stationary plant at the kilowatt scale, the designer usually begins with drawing up a flow sheet, and example process flow sheets have been given in Figures 8.5 and 8.6 for PAFC and PEMFC systems. Having carried out an analysis of the steady state operation of such a flow sheet, the designer then needs to consider issues of how the cell is started up, how it responds to load changes, and what happens as the cell degrades. All of these features can be modelled using computer programs such as Hysis™, MATLAB® Simulink®, Cycle-Tempo (Technical University of Delft) and Aspentech™. The latter will model both steady state processes as well as dynamic changes. Another useful modelling platform is TRNSYS, which is used for dynamic modelling of thermal energy systems. Although this is perhaps more suitable for studying how fuel cell systems will integrate with other energy systems such as PV cells, wind turbines, electrolysers and so on, some fuel cell components are currently being evaluated within TRNSYS. Once the designer has modelled the system he will understand the basic requirements for the stack and each item of the balance of plant (BOP). From this a design specification is drawn up for the plant and a detailed design can begin. To illustrate the issues involved in the genesis of a system, we will consider in the next section the design of a stationary PEM fuel cell system for application at a scale of around 50 kW.

11.6.2 Flow sheet and conceptual systems designs

From our knowledge of a PEMFC stack (Chapter 4) and fuel reforming (Chapter 8), we can draw up a basic diagram showing the components of a natural gas fed fuel cell system. Figure 11.9 gives a simple example. In modelling this flow sheet, we need to make various assumptions about the process steps. In this example these are as follows:

1. Fuel input – this is natural gas fed at a flow rate of $1 \, kmol \, h^{-1}$, roughly equivalent to 100-kW supply. Initially the system will be modelled using pure methane, but we will make provision for using real gas compositions.
2. Steam reforming will be the means of conversion, with an initial steam:carbon ratio of 3. This is higher than the minimum and enables the reformer to run in a carbon-safe regime. The reformer will operate with an outlet temperature of 750°C to maximise hydrogen production.
3. Equilibrium reactors are used to model all the process steps.
4. Desulphurisation of the natural gas will be carried out at atmospheric pressure and temperature by an absorber and therefore will not be considered in this example system. Many natural gases will need more aggressive desulphurisation treatment at high temperatures and a more complex flow sheet will result (see e.g. Figure 8.6).
5. Pre-reforming will be carried out to reduce the concentration of high molecular weight hydrocarbons in the feed gas to the main reformer reactor. Reducing these hydrocarbons can also lead to a reduction in the steam usage, increasing plant efficiency (Amor, 1999). To do this, we shall use a low-temperature adiabatic reformer reactor containing a nickel catalyst and operating between 300 and 250°C.
6. The operating pressure of the system will be a little above atmospheric pressure, allowing for pressure drops through each of the fuel processor elements of some 10 mbar. The feed pressure for natural gas will therefore be 1.7 bar. Higher pressures (4 bar) are often used in PEM systems to reduce the costs of reactors, heat recovery, and compression (Arvindan et al., 1999). The issue of pressure in fuel cell systems has been analysed by Virji (1998).
7. There will be two stages of shift. The high-temperature reactor will contain a catalyst of iron oxide operating at an inlet temperature of 400°C and the low-temperature reactor containing a 30% CuO, 33% ZnO, 30% alumina catalyst (Arvindan et al., 1999) operating at 200°C.
8. A preferential oxidation (PROX) unit. This is modelled in the Aspen flow sheet code by two stoichiometric reactors, one to perform the PROX of carbon monoxide and the subsequent one to remove the remaining oxygen via reaction with hydrogen. These have been removed in the flow sheet shown below and shown as a single unit, as would be experienced in practice.
9. Heat for the reforming reaction is provided by combustion of exhaust gas from the anode of the fuel cell, supplemented by fresh natural gas, as required.

With these boundary conditions and set operating parameters we can draw up a conceptual process flow sheet as shown in Figure 11.9. This has been simulated using Aspentech, and the results are shown in the stream data beneath the flow diagram. From this we can see what the flow rates are in each section of the fuel processor and through the stack. Although not essential in this system, we have also calculated the composite heating and cooling curves for the hot and cold streams, as shown in Figure 11.10. The reader should refer back to Section 7.2.4 in which the concept of 'pinch' technology is described.

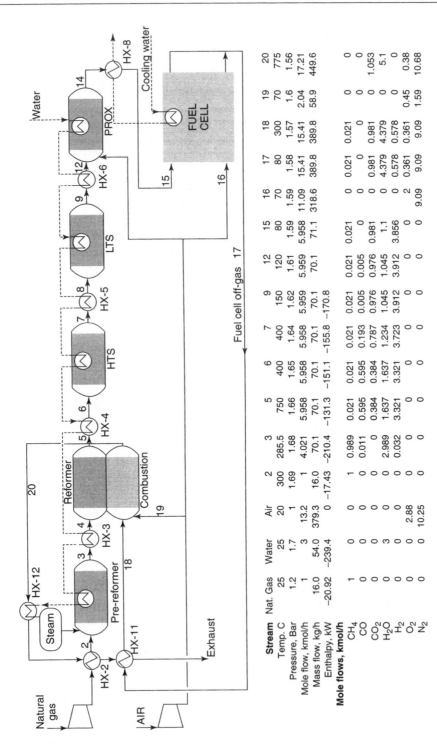

Figure 11.9 Flow sheet for an example PEM fuel cell system for a stationary application with about 100 kWe output. Note that for clarity some stream numbers have been omitted from the table.

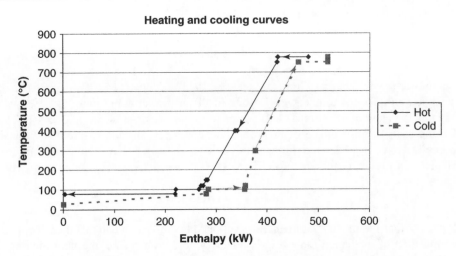

Figure 11.10 Composite heating and cooling curves for streams of the system shown in Figure 11.9.

By carrying out this analysis we have verified the optimum way of networking the various heat exchangers.

By running a computer simulation of the flow sheet, we now know how many heat exchangers are required and what duty each one has. We know how many reactors are needed, and we know what flow rates and compositions of gases are flowing through them. From this, and the knowledge of the catalysts and absorbers, we can calculate the sizes of the reactors and begin to understand how the system will fit together.

In our example, desulphurised natural gas supplied at 1.2 bar is compressed to 1.7 bar (Comp1) to overcome pressure drops within the reactors and stack. This is preheated to 300°C (HX-2) before being fed into a pre-reformer. The product at 286°C gas from the pre-reformer is heated to 750°C by HX-3 when it is fed into the steam reformer reactor. Heat for the reformer is provided by combustion of preheated anode off gas. The reformer product is cooled (HX-4) before being fed to a high-temperature shift reactor operating at 400°C and then further cooled (HX-5) before being fed to a low-temperature shift reactor at 150°C. The product is cooled further (HX-6) before being fed to a PROX unit at 120°C. The gas from the PROX unit is cooled to the stack temperature of 80°C by a final heat exchanger (HX-8). In practice some simplification of the process may be possible and some heat exchangers eliminated. The Aspen output shows us that the loads on the compressors are 311 W for the natural gas compressor and 5.5 kW for the air compressor. This tells us straight away that from a component supply point of view the natural gas compressor may be difficult to obtain whereas an air blower may be more readily available.

The fuel being fed to the stack contains hydrogen at a flow rate of 3.856 kmol h^{-1}. (See gas steam 15 in Figure 11.9.) This is equivalent to an energy flow of 259 kW, using the lower heating value (LHV) of hydrogen, being 241.83 kJ mol^{-1}. We assume that the fuel utilisation is 85% and that we want to run the stack at an operating point of 0.65 V. Under these conditions the efficiency of the stack can be found from the equation given in Section 2.4, remembering that we are using the LHV.

$$\text{Efficiency} = \mu = \mu_f \frac{V_c}{1.25} = 0.85 \frac{0.65}{1.25} = 0.442 = 44.2\% \text{ (ref. LHV)}$$

Therefore, the DC output from the stack is $0.442 * 259 = 114.5\,\text{kW}$. Let us take a reasonable figure for the efficiency of a state-of-the-art power conditioner of 95%, which means that the alternating current (AC) power from the stack is $114.5 * 0.95 = 109.8\,\text{kW}$.

On the basis of an estimate of 80% for their efficiency, the model predicted that the compressors have a power requirement of $5.48 + 0.31 = 5.89\,\text{kW}$. (Note, this figure is NOT given in Figure 11.9.) Therefore the net AC power delivered by the system will be

$$109.8 - 5.9 = 103.9\,\text{kW}.$$

We know that $1\,\text{kgmol}\,\text{h}^{-1}$ of methane is supplied to the system and that the net enthalpy of combustion of methane (Table 8.1) is $802.5\,\text{kJ}\,\text{mol}^{-1}$. Therefore, the net efficiency of the whole system is

$$\text{Efficiency} = \frac{103.9 \times 3600}{802.5 \times 1000} = 0.466 = 46.6\%$$

We can also estimate the size of stack required to deliver the $103.9\,\text{kW}$ of net AC power. Knowing that power is the product of voltage and current, we first determine the total current required by the stack. Since the stack produces $114.5\,\text{kW}$ of DC power, the total current of all the cells added together is

$$I = \frac{P}{V} = \frac{114.5 \times 1000}{0.65} = 176{,}153 \text{ A}$$

Referring to Figure 3.1, which gives the expected performance from a good modern PEMFC, we would expect the current density to be about $600\,\text{mA}\,\text{cm}^{-2}$ if the cell voltage is $0.65\,\text{V}$, or $0.6\,\text{A}\,\text{cm}^{-2}$. The total area of electrode in the fuel cell stack is thus

$$\text{Area} = \frac{\text{Current}}{\text{Current density}} = \frac{176{,}153}{0.6} = 293{,}590\,\text{cm}^2$$

If we make our cells with an active area of $15 \times 25 = 375\,\text{cm}^2$, this means that the stack will need to contain $293{,}890 \div 375 = 784$ cells. This sounds like a very large number, but at the end of Chapter 4 we noted that the Ballard stacks used in the fuel cell bus engines (Figure 4.32) had about 750 cells each. These have a similar power, so we should not be too surprised.

11.6.3 Detailed engineering designs

If we were to consider building our example system stationary power system we would now progress to the stage of detailed engineering design. This would address issues such as start-up and shutdown. For example, we may need to ask the following questions:

- Do we need to provide additional heat to get some reactors such as the reformer up to temperature? Do we need a supply of hydrogen to condition the reforming catalysts before start-up?
- How will the system perform at part load, and how will it respond to changes in load? To investigate this, some dynamic modelling may be required.
- Do we need to provide a purge gas for when the stack is shut down, or held hot (important for the MCFC)?
- What do we require in the way of control valves, thermocouples, and other sensors?
- What type of control system will we need?

After consideration of these questions, detailed mechanical design drawings will be made of all of the BOP items that need to be sourced or fabricated. A basic electrical line diagram will be drawn up that shows how the various electrical and control items link together, with safety features included. A hazard identification (hazid) and/or hazard and operability study (hazop)[5] or similar safety analysis will also be carried out. These considerations are outside the scope of this book but it is hoped that the reader would have gained enough understanding of the systems to be able to inquire further.

Many studies have been published describing analysis of stationary systems, covering all the major types of fuel cells. For PEM fuel cells, one of the most recent is that by Wallmark (2002), and for a discussion of PEMFC stack modelling the reader should consult Amphlett et al., 1995. Examples of PAFC, MCFC, and SOFC systems are given in Parsons (2000), and we have referred in earlier chapters to other examples of system designs. Many discussions of system analysis are also available, for example, a recent analysis of energy and exergy in simple SOFC systems has been carried out by Chan et al. (2002).

11.6.4 Further systems analysis

We have so far discussed the modelling and analysis of the fuelling options, and what might be termed the *macro-modelling* of the fuel cell system. If we refer back to Figure 11.1 there remain two other areas of importance concerning systems analysis. The first is the analysis of the outputs from the fuel cell system. In the case of transportation systems, this means looking at the drive duty of the vehicle, the miles or kilometres travelled per day, the terrain, the load, and so on. For stationary systems we need to consider the heating (or cooling) duties as well as the power requirements for the load envisaged. These are highly specific and will depend on the building and its use, for example, houses and flats have a very different electrical load profile to shops, offices, and schools.

Our final consideration of system analysis returns us to the beginning, that is, to look at the modelling of the fuel cell itself and at fuel cell stacks. This has been an important area of research since the advent of process simulation software in the 1980s. Application

---

[5] A hazard and operability study (Hazop), is a standard hazard analysis technique used worldwide in the process industries in the preliminary safety assessment of new systems or modifications to existing ones. The Hazop study is a detailed examination, by a group of specialists, of components within a system to determine what would happen if that component were to operate outside its normal design mode. Each component will have one or more parameters associated with its operation such as pressure, flow rate, or electrical power. The Hazop study looks at each parameter in turn and uses guide words to list the possible off-normal behaviour such as 'more', 'less', 'high', 'low', 'yes', or 'no'. The effect of such behaviour is then assessed.

of the Nernst equation and the theory given in Chapters 2 and 3 should enable the reader to write a simple code with an Excel spreadsheet that will predict quite accurately the open-circuit voltage of cells running on hydrogen, assuming that the cell is isothermal and one-dimensional. The whole subject becomes more complicated when we want to model the behaviour of a two-dimensional cell or indeed a three-dimensional stack, and the fact that voltage losses occur as we have described in Chapter 3. To account for the voltage losses, there are two approaches: one is to simply use data that has been obtained experimentally and to devise an empirical relationship between voltage and current. The second is to analyse the losses in detail using the Tafel equation and knowledge of the materials properties and electrochemical kinetics and estimate the total cell voltage loss by summing up all the individual losses. This latter fundamental approach has been used with some success for modelling PEM fuel cells (Amphlett, 1995), but a particular difficulty occurs when trying to account for the transport of water (Berning, Lu, and Djilali, 2002).

With high-temperature cells there is no water management issue, and cathodic polarisation can virtually be ignored. The voltage of an SOFC, for example, can be fairly accurately determined from equation 3.10 given in Chapter 3 assuming that the cell is running on hydrogen. However, if steam reforming is also occurring on the anode then we need to be able to account for this. A simple approach is to assume that the steam reforming reaction is fast and occurs first, and that it comes to equilibrium over the anode before any electrochemical reactions occur. We then use equation 3.10 to calculate what the cell voltage will be for the resulting equilibrium gas mixture. A little thought will tell you that this approach is naïve. In a real cell, as the hydrogen is consumed, the fuel gas composition will change – hydrogen concentration decreases on moving from inlet to outlet of the cell. At the outlet the concentration will be low because most of the hydrogen is consumed, whereas at the inlet of the cell there is no hydrogen because we are feeding steam and methane. So what determines the average concentration of hydrogen within the SOFC anode? Indeed what determines the voltage of an internally reforming cell? If we do assume that the steam reforming reaction occurs first, what will the temperature profile in the cell or stack look like? We know that large temperature gradients are best avoided if we want stacks with long lifetimes.

We can easily measure the composition of exhaust gases from cells such as the SOFC and from this we can evaluate our models, but from a practical point of view it is very difficult indeed to monitor what exactly is happening within high-temperature fuel cell stacks. The mechanism of internal reforming is subject to much debate and there are conflicting models in the literature. Certainly there is much to learn about the fundamental processes at the heart of the fuel cell.

11.7 Closing Remarks

It is hoped that the descriptions of stationary and mobile systems given in this chapter has had the reader looking back and forth through many different parts of this book. It takes us back to where we started, illustrating that fuel cell systems require a broad engineering knowledge and are highly interdisciplinary. We hope that we have introduced the many different technologies that underpin the operation of fuel cell systems. In all areas there is, of course, much more that could have been said, but having outlined the basic explanation of how and why things behave in the way they do, we hope that you have a good, broad understanding of all the main components of a fuel cell system. If you find yourself specialising in one aspect of fuel cell technology, we hope that this book has helped you understand the work of your colleagues and given you a good starting point for further studies in your particular area.

We are now witnessing the birth of a major new industry, one that has arisen through the dedicated effort of many people. Over the past few years there has been an unprecedented growth in interest, with major financial investments led by companies such as the vehicle makers Daimler-Chrysler, Ford, General Motors, and Toyota together with oil majors such as Shell, Atlantic Richfield (ARCO), and Texaco. As Geoffrey Ballard, founder of Ballard Power Systems has remarked, 'the fuel cell did not come into being because one person had an idea.' The dawn of the fuel cell industry has arisen 'because dozens of people chose to work together to make an idea come into existence (Koppel, 1999).'

References

Amphlett J.C., Mann B.A., Peppley P.R., and Roberger A. (1995) "A practical PEM fuel cell model for simulating vehicle power sources." *Proceedings of the Battery Conference on Applications and Advantages*, pp. 221–226.

Amor J.N. (1999) Review – the multiple roles for catalysis in the production of H_2, *Applied Catalysis A: General*, **176**, 159–176.

Arvindan N.S., Rajesh B., Madhivanan M., and Pattabiraman R. (1999) "Hydrogen generation from natural gas and methanol for use in electrochemical energy conversion systems (fuel cell)", *Indian Journal of Engineering and Materials Sciences*, **6**, 73–86.

Berning T., Lu D.M., and Djilali N. (2002) "Three-dimensional computational analysis of transport phenomena in a PEM fuel cell", *Journal of Power Sources*, **106**, 284–294.

Büchi F., Tsukada A., Rodutz P., Garcia O., Ruge M., Kötz R., Bärtschi M., and Dietrich P. (2002) "Fuel cell supercap hybrid electric power train." *The Fuel Cell World 2002, Proceedings, European Fuel Cell Forum Conference*, Lucerne, pp. 218–231.

Chan S.H., Low C.F., and Ding O.L. (2002) "Energy and exergy analysis of simple solid-oxide fuel cell power systems", *Journal of Power Sources*, **103**, 188–200.

Hart D. and Bauen (1998) *"Further Assessment of the Environmental Characteristics of Fuel Cells and Competing Technologies"*, Report # ETSU F/02/00153/REP, Department of Trade and Industry, Energy Technology Support Unit, New and Renewable Energy Programme, England.

Koppel T. (1999) *Powering the Future – The Ballard Fuel Cell and the Race to Change the World*, John Wiley & Sons, Toronto, Canada.

Ogden J.M. (2002) *"Review of Small Stationary Reformers for Hydrogen Production"*, IEA Report no. IEA/H2/TR-092/002.

Spiegel R.J., Gilchrist T., and House D.E. (1999) "Fuel cell bus operation at high altitude", *Proceedings of the Institution of Mechanical Engineers*, **213**, Part A, pp. 57–68.

Virji M.B.V., Adcock P.L., Mitchell P.J., and Cooley G. (1998) "Effect of operating pressure on the system efficiency of a methane-fuelled solid polymer fuel cell (SPFC) power source", *Journal of Power Sources*, **71**, 337–347.

Wallmark C. and Alvfors P. (2002) "Design of stationary PEFC system configurations to meet heat and power demands", *Journal of Power Sources*, **106**, 83–92.

Weiss M.A., Heywood J.B., Drake E.M., Schafer A., and AuYeung F.F. (2000) *"On the Road on 2020A Life-Cycle Analysis of New Automobile Technologies"*, Energy Laboratory Report # MIT EL 00–003, Energy Laboratory, Massachusetts Institute of Technology Cambridge, Massachusetts, 02139–4307.

Appendix 1

Change in Molar Gibbs Free Energy Calculations

A1.1 Hydrogen Fuel Cell

This section explains the calculations of $\Delta \overline{g}_f$ for the reaction

$$H_2 + \tfrac{1}{2}O_2 \rightarrow H_2O$$

that was given in Chapter 2.

The Gibbs function of a system is defined in terms of the entropy and the enthalpy:

$$G = H - TS$$

Similarly, the molar Gibbs energy of formation, the molar enthalpy of formation, and the molar entropy are connected by the equation

$$\overline{g}_f = \overline{h}_f - T\overline{s}$$

In this case, it is the *change* in energy that is important. Also, in a fuel cell, the temperature is constant. So we can say that

$$\Delta \overline{g}_f = \Delta \overline{h}_f - T\Delta \overline{s} \qquad [A1.1]$$

The value of $\Delta \overline{h}_f$ is the difference between $\overline{h}_f$ of the products and $\overline{h}_f$ for the reactants. Thus, for the reaction $H_2 + \tfrac{1}{2}O_2 \rightarrow H_2O$ we have

$$\Delta \overline{h}_f = (\overline{h}_f)_{H_2O} - (\overline{h}_f)_{H_2} - \tfrac{1}{2}(\overline{h}_f)_{O_2} \qquad [A1.2]$$

Fuel Cell Systems Explained, Second Edition James Larminie and Andrew Dicks
© 2003 John Wiley & Sons, Ltd ISBN: 0-470-84857-X

Similarly, $\Delta\bar{s}$ is the difference between $\bar{s}$ of the products and $\bar{s}$ of the reactants. So, in this case

$$\Delta\bar{s} = (\bar{s})_{H_2O} - (\bar{s})_{H_2} - \tfrac{1}{2}(\bar{s})_{O_2} \qquad [A1.3]$$

The values of $\bar{h}_f$ and $\bar{s}$ vary with temperature according to the equations given below. These standard equations are derived using thermodynamic theory, and their proof can be found in most books on engineering thermodynamics (Balmer, 1990). In these equations the subscript to $\bar{h}$ and $\bar{s}$ is the temperature, and $\bar{c}_p$ is the molar heat capacity at constant pressure. 298.15 K is standard temperature.

The molar enthalpy of formation at temperature T is given by

$$\bar{h}_T = \bar{h}_{298.15} + \int_{298.15}^{T} \bar{c}_P \, dT \qquad [A1.4]$$

The molar entropy is given by

$$\bar{s}_T = \bar{s}_{298.15} + \int_{298.15}^{T} \tfrac{1}{T}\bar{c}_P \, dT \qquad [A1.5]$$

The values for the molar entropy and enthalpy of formation at 298.15 K are obtainable from thermodynamics tables (Keenan and Kaye, 1948), and are given in Table A1.1. These values are at standard pressure.

To use equations A1.4 and A1.5, we need to know the values of the molar heat capacity at constant pressure $\bar{c}_p$. Over a range of temperatures, $\bar{c}_p$ is not constant. Nevertheless, empirical equations for $\bar{c}_p$ are obtainable and given in many thermodynamics texts (Van Wylen, 1986), and the equations given below are accurate to within 0.6% over the range 300 to 3500 K.

For steam,

$$\bar{c}_p = 143.05 - 58.040T^{0.25} + 8.2751T^{0.5} - 0.036,989T$$

For hydrogen, H_2

$$\bar{c}_p = 56.505 - 22,222.6T^{-0.75} + 116,500T^{-1} - 560,700T^{-1.5}$$

Table A1.1 Values of $\bar{h}_f$ in J mol^{-1} and $\bar{s}$ in J mol^{-1} K^{-1}, at 298.15 K, for the hydrogen fuel cell

	$\bar{h}_f$	$\bar{s}$
H_2O (liquid)	−285,838	70.05
H_2O (steam)	−241,827	188.83
H_2	0	130.59
O_2	0	205.14

Table A1.2 Sample values for $\Delta \bar{h}_f$, $\Delta \bar{s}$, and $\Delta \bar{g}_f$, for the reaction $H_2 + \frac{1}{2}O_2 \rightarrow H_2O$. Temperatures are in Celsius, other figures are in $kJ \, mol^{-1}$

Temperature	$\Delta \bar{h}_f$	$\Delta \bar{s}$	$\Delta \bar{g}_f$
100	−242.6	−0.046,6	−225.2
300	−244.5	−0.050,7	−215.4
500	−246.2	−0.053,3	−205.0
700	−247.6	−0.054,9	−194.2
900	−248.8	−0.056,1	−183.1

For oxygen, O_2

$$\bar{c}_p = 37.432 + 2.010,2 \times 10^{-5}T^{1.5} - 178,570T^{-1.5} + 2,368,800T^{-2}$$

All these equations for $\bar{c}_p$ are in $J g \, mole^{-1} K$. They can be substituted into equations A1.4 and A1.5, giving functions that can be readily integrated and thus evaluated at any temperature T. This is done to derive values for $\bar{h}_f$ and $\bar{s}$ for steam, hydrogen, and oxygen. These values are then substituted into equations A1.2 and A1.3. This gives values for $\Delta \bar{h}_f$ and $\Delta \bar{s}$, which are finally substituted into equation A1.1, giving us the change in molar Gibbs energy of formation $\Delta \bar{g}_f$. Sample values are shown in Table A1.2.

In the case of liquid water, standard values from Table A1.1 for $\bar{h}_f$ and $\bar{s}$ are used for 25°C. At 80°C, equations A1.4 and A1.5 are used again to find $\bar{h}_f$ and $\bar{s}$, but in this case we can assume that $\bar{c}_p$ is constant, since we are working over such a small temperature range.

A1.2 The Carbon Monoxide Fuel Cell

It is possible that in the higher-temperature fuel cells introduced in Chapter 6, the carbon monoxide gas generated from steam reforming of fuel such as methane is directly oxidised. The reaction is

$$CO + \frac{1}{2}O_2 \rightarrow CO_2$$

The method used, and the theory employed, for calculating the Gibbs free energy change is exactly the same as for the hydrogen fuel cell, except that the equations are altered to fit the new reaction. Oxygen features again as it did before, and the values of the molar specific heat capacity for carbon monoxide and carbon dioxide are given by

For CO: - $\bar{c}_p = 69.145 - 0.022,282T^{0.75} - 2,007.7T^{-0.5} + 5,589.64T^{-0.75}$

For CO_2: - $\bar{c}_p = -3.735,7 + 3.052,9T^{0.5} - 0.041,034T + 2.419,8 \times 10^{-6}T^2$

Table A1.3 Values of $\overline{h}_f$ in J mol^{-1} and $\overline{s}$ in J mol^{-1} K^{-1}, at 298.15 K, for the carbon monoxide fuel cell

	$\overline{h}_f$	$\overline{s}$
O_2	0	205.14
CO	−110,529	197.65
CO_2	−393,522	213.80

Table A1.4 Sample values for $\Delta\overline{h}_f$, $\Delta\overline{s}$, and $\Delta\overline{g}_f$, for the reaction $CO + \frac{1}{2}O_2 \rightarrow CO_2$. Temperatures are in Celsius, other figures are in kJ mol^{-1}

Temperature	$\Delta\overline{h}_f$	$\Delta\overline{s}$	$\Delta\overline{g}_f$
100	−283.4	−0.087,7	−250.7
300	−283.7	−0.088,8	−232.7
500	−283.4	−0.089,0	−214.6
700	−282.8	−0.088,7	−196.5
900	−282.0	−0.088,3	−178.5

Together with values from Table A1.3, these equations are used with equations A1.4 and A1.5 to find the molar enthalpies and entropies for the three gases in question.

The change in the molar enthalpy and molar entropy is then calculated using these two equations:

$$\Delta\overline{h}_f = (\overline{h}_f)_{CO_2} - (\overline{h}_f)_{CO} - \tfrac{1}{2}(\overline{h}_f)_{O_2}$$

$$\Delta\overline{s} = (\overline{s})_{CO_2} - (\overline{s})_{CO} - \tfrac{1}{2}(\overline{s})_{O_2}$$

The change in molar Gibbs free energy of formation is then calculated, as with the hydrogen fuel cell, using equation A1.1. Some example results are given below in Table A1.4.

References

Balmer R. (1990) Chapter 6, *Thermodynamics*, West, St Paul.
Keenan J.H. and Kaye J. (1948) *Gas Tables*, Wiley & Sons, New York.
Van Wylen G.J. and Sonntag R.E. (1986) *Fundamentals of Classical Thermodynamics*, 3rd ed., Wiley & Sons, New York, p. 688.

Appendix 2

Useful Fuel Cell Equations

A2.1 Introduction

In this appendix, many useful equations are derived. They relate to

- oxygen usage rate
- air inlet flow rate
- air exit flow rate
- hydrogen usage, and the energy content of hydrogen
- rate of water production
- heat production.

In many of the sections that follow, the term *stoichiometric* is used. Its meaning could be defined as 'just the right amount'. So, for example, in the simple fuel cell reaction

$$2H_2 + O_2 \rightarrow 2H_2O$$

exactly two moles of hydrogen would be provided for each mole of oxygen. This would produce exactly $4F$ of charge, since two electrons are transferred for each mole of hydrogen. Note that either or both the hydrogen and oxygen are often supplied at greater than the stoichiometric rate. This is especially so for oxygen if it is being supplied as air. If it was supplied at exactly the stoichiometric rate, then the air leaving the cell would be completely devoid of oxygen. Note also that reactants cannot be supplied at *less* than the stoichiometric rate.

This stoichiometry can be expressed as a variable, and the symbol λ is normally used. Its use can be put like this. If the rate of *use* of a chemical in a reaction is $\dot{n}$ moles per second, then the rate of *supply* is $\lambda \dot{n}$ moles per second.

To increase the usefulness of the formulas, they have been given in terms of the electrical power of the whole fuel cell stack P_e, and the average voltage of each cell in

Fuel Cell Systems Explained, Second Edition James Larminie and Andrew Dicks
© 2003 John Wiley & Sons, Ltd   ISBN: 0-470-84857-X

the stack V_c. The electrical power will nearly always be known, as it is the most basic and important information about a fuel cell system. If V_c is not given, it can be assumed to be between 0.6 and 0.7 V, as most fuel cells operate in this region (see Figures 3.1 and 3.2). If the efficiency is given, then V_c can be calculated using equation 2.5. If no figures are given, then using $V_c = 0.65$ V will give a good approximation. Estimate somewhat higher if the fuel cell is pressurised.

A2.2 Oxygen and Air Usage

From the basic operation of the fuel cell, we know that four electrons are transferred for each mole of oxygen. (Look back to equation 1.3.) So

$$\text{charge} = 4F \times \text{ amount of } O_2$$

Dividing by time, and rearranging

$$O_2 \text{ usage} = \frac{I}{4F} \text{ moles s}^{-1}$$

This is for a single cell. For a stack of n cells

$$O_2 \text{ usage} = \frac{In}{4F} \text{ moles s}^{-1} \qquad \text{[A2.1]}$$

However, it would be more useful to have the formula in kg s^{-1}, without needing to know the number of cells, and in terms of power, rather than current. If the voltage of each cell in the stack is V_c, then

$$\text{Power, } P_e = V_c \times I \times n$$

So,

$$I = \frac{P_e}{V_c \times n}$$

Substituting this into equation A2.1 gives

$$O_2 \text{ usage} = \frac{P_e}{4 \cdot V_c \cdot F} \text{ moles s}^{-1} \qquad \text{[A2.2]}$$

Changing from moles s^{-1} to kg s^{-1}

$$O_2 \text{ usage} = \frac{32 \times 10^{-3} \cdot P_e}{4 V_c F} \text{ kg s}^{-1}$$

$$= 8.29 \times 10^{-8} \times \frac{P_e}{V_c} \text{ kg s}^{-1} \qquad \text{[A2.3]}$$

This formula allows the oxygen usage of any fuel cell system of given power to be calculated. If V_c is not given, it can be calculated from the efficiency, and if that is not given, the figure of 0.65 V can be used for a good approximation.

However, the oxygen used will normally be derived from air, so we need to adapt equation A2.2 to air usage. The molar proportion of air that is oxygen is 0.21, and the molar mass of air is 28.97×10^{-3} kg mole^{-1}. So, equation A2.2 becomes

$$\text{Air usage} = \frac{28.97 \times 10^{-3} \times P_e}{0.21 \times 4 \times V_c \times F} \text{ kg s}^{-1}$$

$$= 3.57 \times 10^{-7} \times \frac{P_e}{V_c} \text{ kg s}^{-1}$$

However, if the air was used at this rate, then as it left the cell it would be completely devoid of any oxygen – it would all have been used. This is impractical, and in practice the airflow is well above stoichiometry, typically twice as much. If the stoichiometry is λ, then the equation for air usage becomes

$$\text{Air usage} = 3.57 \times 10^{-7} \times \lambda \times \frac{P_e}{V_c} \text{ kg s}^{-1} \qquad \text{[A2.4]}$$

The kilogram per second is not, in fact, a very commonly used unit of mass flow. The following conversions to 'volume at standard conditions related' mass flow units will be found useful. The mass flow rate from equation A2.4 should be multiplied by

- 3050 to give flow rate in standard m^3 h^{-1}
- 1795 to give flow rate in SCFM (or in standard ft^3 min^{-1})
- 5.1×10^4 to give flow rate in slm (standard L min^{-1})
- 847 to give flow rate in sls (standard L s^{-1})

A2.3 Air Exit Flow Rate

It is sometimes important to distinguish between the *inlet* flow rate of the air, which is given by equation A2.4 above, and the *outlet* flow rate. This is particularly important when calculating the humidity, which is an important issue in certain types of fuel cells, especially proton exchange membrane (PEM) fuel cells. The difference is caused by the consumption of oxygen. There will usually be more water vapour in the exit air, but we are considering 'dry air' at this stage. Water production is given in Section A2.5. Clearly

$$\text{Exit air flow rate} = \text{Air inlet flow rate} - \text{oxygen usage}$$

Using equations A2.3 and A2.4 this becomes

$$\text{Exit air flow rate} = 3.57 \times 10^{-7} \times \lambda \times \frac{P_e}{V_c} - 8.29 \times 10^{-8} \times \frac{P_e}{V_c} \text{ kg s}^{-1}$$

$$= (3.57 \times 10^{-7} \times \lambda - 8.29 \times 10^{-8}) \times \frac{P_e}{V_c} \text{ kg s}^{-1} \qquad \text{[A2.5]}$$

A2.4 Hydrogen Usage

The rate of usage of hydrogen is derived in a way similar to oxygen, except that there are two electrons from each mole of hydrogen. Equations A2.1 and A2.2 thus become

$$H_2 \text{ usage} = \frac{In}{2F} \text{ moles s}^{-1} \qquad [A2.6]$$

and

$$H_2 \text{ usage} = \frac{P_e}{2V_c F} \text{ moles s}^{-1} \qquad [A2.7]$$

The molar mass of hydrogen is 2.02×10^{-3} kg mole^{-1}, so this becomes

$$H_2 \text{ usage} = \frac{2.02 \times 10^{-3} \cdot P_e}{2V_c F}$$

$$= 1.05 \times 10^{-8} \times \frac{P_e}{V_c} \text{ kg s}^{-1} \qquad [A2.8]$$

at stoichiometric operation. Obviously, this formula only applies to a hydrogen-fed fuel cell. In the case of a hydrogen/carbon monoxide mixture derived from a reformed hydrocarbon, things will be different, depending on the proportion of carbon monoxide present. The result can be transformed to a volume rate using the density of hydrogen, which is 0.084 kg m^{-3} at normal temperature and pressure (NTP).

In addition to the rate of usage of hydrogen, it is often also useful to know the electrical energy that could be produced from a given mass or volume of hydrogen. The list provided in Table A2.1 gives the energy in kilowatthour, rather than joules, as this is the measure usually used for electrical power systems. In addition to the 'raw' energy per kilogram and standard litre, we have also given an 'effective' energy, taking into account the efficiency of the cell. This is given in terms of V_c, the mean voltage of each cell. If an equation with the efficiency is needed, then use the formula derived in Section 2.5,

$$\text{efficiency} = \frac{V_c}{1.48}$$

Table A2.1 'Raw' and effective energy content of hydrogen fuel. To obtain the lower heating value, multiply the HHV figures given by 0.846

Form	Energy content
Specific enthalpy (HHV)	1.43×10^8 J kg^{-1}
Specific enthalpy (HHV)	39.7 kWh kg^{-1}
Effective specific electrical energy	$26.8 \times V_c$ kWh kg^{-1}
Energy density at STP (HHV)	3.20 kWh m^{-3} = 3.20 Wh SL^{-1}
Energy density at NTP (HHV)	3.29 kWh m^{-3} = 3.29 Wh SL^{-1}

Note: higher heating value, HHV; standard temperature and pressure, STP.

A2.5 Water Production

In a hydrogen-fed fuel cell, water is produced at the rate of one mole for every two electrons. (Revisit Section 1.1 if you are not clear why.) So, we again adapt equation A2.2 to obtain

$$\text{Water production} = \frac{P_e}{2 \cdot V_c \cdot F} \text{ moles s}^{-1} \qquad [\text{A2.9}]$$

The molecular mass of water is 18.02×10^{-3} kg mole^{-1}, so this becomes

$$\text{Water production} = 9.34 \times 10^{-8} \times \frac{P_e}{V_c} \text{ kg s}^{-1} \qquad [\text{A2.10}]$$

In the hydrogen-fed fuel cell, the rate of water production more or less has to be stoichiometric. However, if the fuel is a mixture of carbon monoxide with hydrogen, then the water production would be less – in proportion to the amount of carbon monoxide present in the mixture. If the fuel was a hydrocarbon that was internally reformed, then some of the product water would be used in the reformation process. We saw in Chapter 7, for example, that if methane is internally reformed, then half the product water is used in the reformation process, thus halving the rate of production.

It is sometimes useful to give an example figure to clarify a formula such as this. Let us take as an example a 1-kW fuel cell operating for 1 h, at a cell voltage of 0.7 V. This corresponds to an efficiency of 47% (ref. HHV) (from equation 2.5). Substituting this into equation A2.9 gives

$$\text{The rate of water production} = 9.34 \times 10^{-8} \times \frac{1000}{0.7}$$

$$= 1.33 \times 10^{-4} \text{ kg s}^{-1}$$

So the mass of water produced in 1 h is

$$= 1.33 \times 10^{-4} \times 60 \times 60 = 0.48 \text{ kg}$$

Since the density of water is 1.0 g cm^{-3}, this corresponds to 480 cm^3, which is almost exactly 1 pint. So, as a rough guide, 1 kWh of fuel cell generated electricity produces about 1 pint or 0.5 L of water.

A2.6 Heat Produced

Heat is produced when a fuel cell operates. It was noted in Chapter 2, section 2.4, that if all the enthalpy of reaction of a hydrogen fuel cell was converted into electrical energy then the output voltage would be

> 1.48 V if the water product was in liquid form

or   1.25 V if the water product was in vapour form.

It clearly follows that the difference between the actual cell voltage and this voltage represents the energy that is not converted into electricity – that is, the energy that is converted into heat instead.

The cases in which water finally ends in liquid form are so few and far between that they are not worth considering. So we will restrict ourselves to the vapour case. However, please note that this means we have taken into account the cooling effect of water evaporation. It also means that energy is leaving the fuel cell in three forms: as electricity, as ordinary 'sensible' heat, and as the latent heat of water vapour.

For a stack of n cells at current I, the heat generated is thus

$$\text{Heating rate} = nI\,(1.25 - V_c) \text{ W}$$

In terms of electrical power, this becomes

$$\text{Heating rate} = P_e\left(\frac{1.25}{V_c} - 1\right) \text{ W} \qquad [\text{A2.11}]$$

Index

Fuel Cell Systems Explained, Second Edition James Larminie and Andrew Dicks
© 2003 John Wiley & Sons, Ltd   ISBN: 0-470-84857-X